EPISTLES OF THE BRETHREN OF PURITY

EPISTLES OF THE BRETHREN OF PURITY

On Arithmetic and Geometry

An Arabic Critical Edition and English Translation of EPISTLES 1 & 2

Edited and Translated by
Nader El-Bizri

With a Foreword

OXFORD
UNIVERSITY PRESS
in association with
The Institute of Ismaili Studies

OXFORD
UNIVERSITY PRESS

Great Clarendon Street, Oxford OX2 6DP

Oxford University Press is a department of the University of Oxford.
It furthers the University's objective of excellence in research, scholarship,
and education by publishing worldwide in

Oxford New York

Auckland Cape Town Dar es Salaam Hong Kong Karachi
Kuala Lumpur Madrid Melbourne Mexico City Nairobi
New Delhi Shanghai Taipei Toronto

With offices in

Argentina Austria Brazil Chile Czech Republic France Greece
Guatemala Hungary Italy Japan Poland Portugal Singapore
South Korea Switzerland Thailand Turkey Ukraine Vietnam

Published in the United States
by Oxford University Press Inc., New York

First published 2012

British Library Cataloguing in Publication Data
Data available

Library of Congress Cataloguing in Publication Data
Data available

ISBN 978-0-19-965560-1

1 3 5 7 9 10 8 6 4 2

Typeset and printed in Lebanon
on acid-free paper by
Saqi Books

The Institute of Ismaili Studies

The Institute of Ismaili Studies was established in 1977 with the object of promoting scholarship and learning on Islam, in historical as well as contemporary contexts, and a better understanding of its relationship with other societies and faiths.

The Institute's programmes encourage a perspective which is not confined to the theological and religious heritage of Islam, but seeks to explore the relationship of religious ideas to broader dimensions of society and culture. The programmes thus encourage an inter-disciplinary approach to the materials of Islamic history and thought. Particular attention is also given to issues of modernity that arise as Muslims seek to relate their heritage to the contemporary situation.

Within the Islamic tradition, the Institute's programmes seek to promote research on those areas which have, to date, received relatively little attention from scholars. These include the intellectual and literary expressions of Shiʿism in general, and Ismailism in particular.

In the context of Islamic societies, the Institute's programmes are informed by the full range and diversity of cultures in which Islam is practised today, from the Middle East, South and Central Asia, and Africa to the industrialised societies of the West, thus taking into consideration the variety of contexts which shape the ideals, beliefs and practices of the faith.

These objectives are realised through concrete programmes and activities organised and implemented by various departments of the Institute. The Institute also collaborates periodically, on a programme-specific basis, with other institutions of learning in the United Kingdom and abroad.

The Institute's academic publications fall into several distinct and interrelated categories:

1. Occasional papers or essays addressing broad themes of the relationship between religion and society, with special reference to Islam.
2. Monographs exploring specific aspects of Islamic faith and culture, or the contributions of individual Muslim figures or writers.
3. Editions or translations of significant primary or secondary texts.
4. Translations of poetic or literary texts which illustrate the rich heritage of spiritual, devotional, and symbolic expressions in Muslim history.
5. Works on Ismaili history and thought, and the relationship of the Ismailis to other traditions, communities, and schools of thought in Islam.
6. Proceedings of conferences and seminars sponsored by the Institute.
7. Bibliographical works and catalogues which document manuscripts, printed texts, and other source materials.

This book falls into category three listed above.

In facilitating these and other publications, the Institute's sole aim is to encourage original research and analysis of relevant issues. While every effort is made to ensure that the publications are of a high academic standard, there is naturally bound to be a diversity of views, ideas and interpretations. As such, the opinions expressed in these publications must be understood as belonging to their authors alone.

Epistles of the Brethren of Purity

The *Epistles of the Brethren of Purity* is published by Oxford University Press in association with the Institute of Ismaili Studies, London. This bilingual series consists of a multi-authored Arabic critical edition and annotated English translation of the *Rasāʾil Ikhwān al-Ṣafāʾ* (ca. tenth-century Iraq).

Previously published

Nader El-Bizri, ed. *The Ikhwān al-Ṣafāʾ and their 'Rasāʾil': An Introduction* (2008).

Ikhwān al-Ṣafāʾ. *The Case of the Animals versus Man Before the King of the Jinn: An Arabic Critical Edition and English Translation of Epistle 22*, ed. and tr. Lenn E. Goodman and Richard McGregor (2009).

Ikhwān al-Ṣafāʾ. *On Logic: An Arabic Critical Edition and English Translation of Epistles 10–14*, ed. and tr. Carmela Baffioni (2010).

Ikhwān al-Ṣafāʾ. *On Music: An Arabic Critical Edition and English Translation of Epistle 5*, ed. and tr. Owen Wright (2010).

Ikhwān al-Ṣafāʾ. *On Magic I: An Arabic Critical Edition and English Translation of Epistle 52A*, ed. and tr. Godefroid de Callataÿ and Bruno Halflants (2011).

In memoriam

my father, M. Mouhib El-Bizri (1927-2011)

Dr Nader El-Bizri, the General Editor of the *Epistles of the Brethren of Purity* series, is an Associate Professor in the Civilization Sequence Program at the American University of Beirut. Previously, he was a Principal Lecturer (Reader) at the University of Lincoln, and he has taught at the University of Cambridge, the University of Nottingham, the London Consortium, and Harvard University, in addition to having held research positions at the Institute of Ismaili Studies in London, and the Centre National de la Recherche Scientifique in Paris.

Dr El-Bizri's main areas of research are in Islamic intellectual history, phenomenology, and architectural humanities, all subjects on which he has both published and lectured widely and internationally. As well as contributing to various BBC radio and TV programs, he has also acted as a consultant to the Science Museum in London, the Aga Khan Trust for Culture in Geneva, and the Solomon Guggenheim Museum in New York and Berlin. He acts as an elected council member of the Société Internationale d'Histoire des Sciences et des Philosophies Arabes et Islamiques (CNRS, Paris), whilst maintaining active memberships in various international societies. Furthermore, he serves on the editorial boards of publications by Oxford University Press, Cambridge University Press, Springer, I. B. Tauris, and E. J. Brill. Besides his academic profile, he used to practise as a professional architect in offices in London, Cambridge, New York, and Beirut.

Contents

Foreword

The Brethren of Purity (Ikhwān al-Ṣafāʾ) were the anonymous members of a fourth-/tenth-century[1] esoteric fraternity of lettered urbanites that was principally based in the southern Iraqi city of Basra, while also having a significant active branch in the capital of the ʿAbbāsid caliphate, Baghdad. This secretive coterie occupied a prominent station in the history of scientific and philosophical ideas in Islam owing to the wide intellectual reception and dissemination of diverse manuscripts of their famed philosophically oriented compendium, the *Epistles of the Brethren of Purity* (*Rasāʾil Ikhwān al-Ṣafāʾ*). The exact dating of this corpus, the identity of its authors, and their doctrinal affiliation remain unsettled questions that are hitherto shrouded with mystery. Some situate the historic activities of this brotherhood at the eve of the Fāṭimid conquest of Egypt (ca. 358/969), while others identify the organization with an earlier period that is set chronologically around the founding of the Fāṭimid dynasty in North Africa (ca. 297/909).

The most common account regarding the presumed identity of the Ikhwān is usually related on the authority of the famed litterateur Abū Ḥayyān al-Tawḥīdī (ca. 320–414/930–1023), who noted in his *Book of Pleasure and Conviviality* (*Kitāb al-Imtāʿ waʾl-muʾānasa*) that these adepts were obscure 'men of letters': Abū Sulaymān Muḥammad b. Maʿshar al-Bustī (nicknamed al-Maqdisī); the *qāḍī* Abū al-Ḥasan ʿAlī b. Hārūn al-Zanjānī; Abū Aḥmad al-Mihrajānī (also known as Aḥmad al-Nahrajūrī); and Abū al-Ḥasan al-ʿAwfī. Abū Ḥayyān also claimed that they were the senior companions of a secretarial officer at the

1 All dates are Common Era, unless otherwise indicated; where two dates appear (separated by a slash), the first date is hijri (AH), followed by CE.

Būyid regional chancellery of Basra, known as Zayd b. Rifāʿa, who was reportedly an affiliate of the Brethren's fraternity and a servant of its ministry. Even though this story was reaffirmed by several classical historiographers in Islamic civilization, it is not fully accepted by scholars in terms of its authenticity. Furthermore, some Ismaili missionaries (*duʿāt*) historically attributed the compiling of the *Epistles* to the early Ismaili Imams Aḥmad b. ʿAbd Allāh (al-Taqī [al-Mastūr]) or his father, ʿAbd Allāh (Wafī Aḥmad), while also suggesting that the *Rasāʾil* compendium was secretly disseminated in mosques during the reign of the ʿAbbāsid caliph al-Maʾmūn (r. 198–218/813–833).

Encountering 'veracity in every religion', and grasping knowledge as 'pure nourishment for the soul', the Ikhwān associated soteriological hope and the attainment of happiness with the scrupulous development of rational pursuits and intellectual quests. Besides the filial observance of the teachings of the Qur'an and hadith, the Brethren also reverently appealed to the Torah of Judaism and to the Gospels of Christianity. Moreover, they heeded the legacies of the Stoics and of Pythagoras, Hermes Trismegistus, Socrates, Plato, Aristotle, Plotinus, Nicomachus of Gerasa, Euclid, Ptolemy, Galen, Proclus, Porphyry, and Iamblichus.

The Brethren promoted a convivial and earnest 'companionship of virtue'. Their eschatological outlook was articulated by way of an intricate cyclical view of 'sacred' history that is replete with symbolisms and oriented by an uncanny hermeneutic interpretation of the microcosm and macrocosm analogy: believing that the human being is a microcosmos, and that the universe is a 'macroanthropos'. The multiplicity of the voices that were expressed in their *Epistles* reflects a genuine quest for wisdom driven by an impetus that is not reducible to mere eclecticism; indeed, their syncretism grounded their aspiration to establish a spiritual refuge that would transcend the sectarian divisions troubling their era.

In general, fifty-two epistles are enumerated as belonging to the *Rasāʾil Ikhwān al-Ṣafāʾ*, and these are divided into the following four parts: Mathematics, Natural Philosophy, Sciences of the Soul and Intellect, and Theology. The first part consists of fourteen epistles, and it deals with 'the mathematical sciences', treating a variety of topics in arithmetic, geometry, astronomy, geography, and music.

It also includes five epistles on elementary logic, which consist of the following: the *Isagoge*, the *Categories*, the *On Interpretation*, the *Prior Analytics*, and the *Posterior Analytics*. The second part of the corpus groups together seventeen epistles on 'the physical or natural sciences'. It thus treats themes on matter and form, generation and corruption, metallurgy, meteorology, a study of the essence of nature, the classes of plants and animals (the latter being also set as a fable), the composition of the human body and its embryological constitution, a cosmic grasp of the human being as microcosm, and also the investigation of the phonetic and structural properties of languages and their differences. The third part of the compendium comprises ten tracts on 'the psychical and intellective sciences', setting forth the 'opinions of the Pythagoreans and of the Brethren of Purity', and accounting also for the world as a 'macroanthropos'. In this part, the Brethren also examined the distinction between the intellect and the intelligible, and they offered explications of the symbolic significance of temporal dimensions, epochal cycles, and the mystical expression of the essence of love, together with an investigation of resurrection, causes and effects, definitions and descriptions, and the various types of motion. The fourth and last part of the *Rasāʾil* deals with 'the *nomic* or legal and theological sciences' in eleven epistles. These address the differences between the varieties of religious opinions and sects, as well as delineating the 'pathway to God', the virtues of the Ikhwān's fellowship, the characteristics of genuine believers, the nature of the divine *nomos*, the call to God, the actions of spiritualists, of jinn, angels, and recalcitrant demons, the species of politics, the cosmic hierarchy, and, finally, the essence of magic and talismanic incantations. Besides the fifty-two tracts that constitute the *Rasāʾil Ikhwān al-Ṣafāʾ*, this compendium was accompanied by a treatise entitled *al-Risāla al-jāmiʿa* (The Comprehensive Epistle), which acted as the *summa summarum* for the whole corpus and was itself supplemented by a further abridged appendage known as the *Risālat jāmiʿat al-jāmiʿa* (The Condensed Comprehensive Epistle).

In spite of their erudition and resourcefulness, it is doubtful whether the Brethren of Purity can be impartially ranked amongst the authorities of their age in the realms of science and philosophy. Their inquiries into mathematics, logic, and the natural sciences were recorded in the *Epistles* in a synoptic and diluted fashion, sporadically infused with gnostic, symbolic, and occult directives. Nonetheless, their accounts of religiosity, as well as their syncretic approach, together with their praiseworthy efforts to collate the sciences, and to compose a pioneering 'encyclopaedia', all bear signs of commendable originality.

In terms of the epistemic significance of the *Epistles* and the intellectual calibre of their authors, it must be stated that, despite being supplemented by oral teachings in seminars (*majālis al-ʿilm*), the heuristics embodied in the *Rasāʾil* were not representative of the most decisive achievements of their epoch in the domains of mathematics, natural sciences, or philosophical reasoning. Moreover, the sciences were not treated with the same level of expertise across the *Rasāʾil*. Consequently, this opus ought to be judged by differential criteria as regards the relative merits of each of its epistles. In fairness, there are signs of conceptual inventiveness, primarily regarding doctrinal positions in theology and reflections on their ethical-political import, along with signs of an intellectual sophistication in the meditations on spirituality and revelation.

The *Rasāʾil* corpus is brimming with a wealth of ideas and constitutes a masterpiece of mediaeval literature that presents a populist yet comprehensive adaptation of scientific knowledge. It is perhaps most informative in terms of investigating the transmission of knowledge in Islam, the 'adaptive assimilation' of antique sciences, and the historical evolution of the elements of the *sociology* of learning through the mediaeval forms of the popularization of the sciences and the systemic attempts to canonize them. By influencing a variety of Islamic schools and doctrines, the Brethren's heritage acted as a significant intellectual prompt and catalyst in the development of the history of ideas in Islam. As such, their work rightfully holds the station assigned to it amongst the distinguished Arabic classics and the high literature of Islamic civilization.

The composition of this text displays impressive lexical versatility, which encompasses the technical idioms of mathematics and logic, the heuristics of natural philosophy, and the diction of religious pronouncements and occult invocations, in addition to poetic verses, didactic parables, and satirical and inspirational fables. Despite the sometimes disproportionate treatment of topics, the occasional hiatus in proofs, irrelevant digressions, or instances of verbosity, the apparent stylistic weaknesses disappear, becoming inconsequential, when a complete impression is formed of the architectonic unity of the text as a whole and of the convergence of its constituent elements as a remarkable *oeuvre des belles lettres*.

Modern academic literature on the *Rasāʾil Ikhwān al-Ṣafāʾ* is reasonably extensive within the field of Islamic studies, and it continues to grow, covering works dating from the nineteenth century up to the present, with numerous scholars attempting to solve the riddles surrounding this compendium. The academic rediscovery of the *Rasāʾil* in modern times emerged through the monumental editorial and translation efforts of the German scholar Friedrich Dieterici between the years 1861 and 1872. Several printed editions aiming to reconstruct the original Arabic have also been established, starting with the *editio princeps* in Calcutta in 1812, which was reprinted in 1846, then a complete edition in Bombay between 1887 and 1889, followed by the Cairo edition of 1928, and the Beirut editions of 1957, 1983, 1995, and their reprints.[2] Although the scholarly contribution of these Arabic editions of the *Rasāʾil* is laudable, as they valuably sustained research on the topic, they are uncritical in character, and they do not reveal their manuscript sources. Consequently, the current printed editions do not provide definitive primary-source documentation for this classical text. Given this state of affairs, the Institute of Ismaili Studies (IIS) in London has undertaken the publication (in association with Oxford University Press) of a multi-authored, multi-volume Arabic critical edition and annotated English translation of the fifty-two epistles.

2 The principal complete editions of this compendium that are available in print consist of the following: *Kitāb Ikhwān al-Ṣafāʾ wa-Khullān al-Wafāʾ*, ed. Wilāyat Ḥusayn, 4 vols. (Bombay: Maṭbaʿat Nukhbat al-Akhbār, 1305–1306/ca. 1888); *Rasāʾil Ikhwān al-Ṣafāʾ*, ed. Khayr al-Dīn al-Ziriklī, with two separate introductions by Ṭaha Ḥusayn and Aḥmad Zakī Pasha, 4 vols. (Cairo: al-Maṭbaʿa al-ʿArabiyya bi-Miṣr, 1928); *Rasāʾil Ikhwān al-Ṣafāʾ*, ed. with introduction by Buṭrus Bustānī, 4 vols. (Beirut: Dār Ṣādir, 1957); and an additional version, *Rasāʾil Ikhwān al-Ṣafāʾ*, ed. ʿĀrif Tāmir, 5 vols. (Beirut: Manshūrāt ʿUwaydāt, 1995).

In preparation for the critical edition, reproductions of nineteen manuscripts were acquired by the IIS, and their particulars can be summarized as follows, with the corresponding Arabic sigla:

Bibliothèque nationale de France, Paris:
MS 2303 (1611 CE): [ر]
MS 2304 (1654 CE): [ز]
MS 6.647–6.648 (AH 695; Yazd): [د]

Bodleian Library, Oxford:
MS Hunt 296 (n.d.): [ج]
MS Laud Or. 255 (n.d.): [ح]
MS Laud Or. 260 (1560 CE): [خ]
MS Marsh 189 (n.d.): [غ]

El Escorial, Madrid:
MS Casiri 895/Derenbourg 900 (1535–1536 CE): [س]
MS Casiri 923/Derenbourg 928 (1458 CE): [ش]

Istanbul collections (mainly the Süleymaniye and associated libraries):
MS Atif Efendi 1681 (1182 CE): [ع]
MS Esad Efendi 3637 (ca. thirteenth century CE): [ن]
MS Esad Efendi 3638 (ca. 1287 CE): [أ]
MS Feyzullah 2130 (AH 704): [ف]
MS Feyzullah 2131 (AH 704): [ق]
MS Köprülü 870 (ca. fifteenth century CE): [ك]
MS Köprülü 871 (1417 CE): [ل]
MS Köprülü 981 (n.d.): [و]

Königliche Bibliothek zu Berlin:
MS 5038 (AH 600/1203 CE): [ب]

The Mahdavī Collection, Tehran:[3]
MS 7437 (AH 640): [ط]

The reconstruction of the contents of the *Rasāʾil* by way of a critical edition is undertaken through manuscript reproductions that are significantly distanced in time from the original, and these have proved to be traceable to a variety of transmission traditions that cannot be articulated with confidence in terms of a definitive *stemma codicum*.[4] The dexterity of the copyists, their deliberate tampering, or commendable exercise of restraint and relative impartiality, along with their scribal idioms, would have conditioned the drafting of the manuscripts. Such endeavours would also have been influenced by the intellectual impress of the prevalent geopolitical circumstances in which

3 It is worth noting that these acquisitions by the IIS, which consist of the oldest complete manuscripts, along with significant supplementary fragments of an early dating, were each carefully selected from over one hundred extant manuscripts, which are preserved in thirty-nine libraries and collections, noted in alphabetical order by country, as follows: *Egypt*: Dār al-Kutub, Arab League Library (possibly also in the Arab League offices in Tunis); *France*: Bibliothèque nationale de France; *Germany*: Königliche Bibliothek zu Berlin, Herzogliche Bibliothek zu Gotha, Eberhard-Karlis-Universität (Tübingen), Leipzig (Bibliotheca Orientalis), München Staatsbibliothek; *Iran*: Muṭahharī Library, Tehran University Central Library, Mahdavī Collection (private); *Ireland*: Chester Beatty Library; *Italy*: Biblioteca Ambrosiana, Biblioteca Vaticana; *Netherlands*: Bibliotheca Universitatis Leidensis; *Russia*: Institut des Langues Orientales (St Petersburg); *Spain*: Biblioteca del Monasterio San Lorenzo de El Escorial; *Turkey*: Süleymaniye, Aya Sofia, Amia Huseyn, Atif Efendi, Esad Efendi, Millet Library, Garullah, Köprülü, Kütüphane-i ʿUmūmī Defterī, Manisa (Maghnisa), Rashid Efendi (Qaysari), Topkapi Saray, Yeni Çami, Revan Kishk; *United Kingdom*: Bodleian Library, British Library, British Museum, Cambridge University (Oriental Studies Faculty Library), Institute of Ismaili Studies (including copies from the Hamdani, Zāhid ʿAlī, and Fyzee collections), Mingana Collection (Selly Oak Colleges Library, Birmingham), School of Oriental and African Studies (SOAS); *United States*: New York Public Library, Princeton University Library.

4 Within both the English translation and the Arabic edition here, the beginning of each folio of the MS Atif Efendi 1681 [ع] manuscript is indicated, starting at verso folio 5: for example, (fol. 5b) in the English translation, and |ع ٥ ظ| in the Arabic edition; or from recto folio 6, in the form: (fol. 6a) in the English translation, and |ع ٦ و| in the Arabic edition. The pagination of the Beirut (Dār Ṣādir) printed edition is also noted using square brackets, for example [p. 48] or [٤٨].

this text was transcribed, in addition to its channels of transmission. By widening the selection of the oldest manuscripts and fragments, based on the period of the copying, the levels of completeness and clarity, and the recommendations of past and present scholars who have consulted these collections, a suitably grounded critical edition will be produced, and a more reliable textual reconstruction will offer us improved access to the contents of the *Rasāʾil* beyond what is presently available through the printed editions (i.e., those from Bombay, Cairo, and Beirut). It is ultimately hoped that the collective authorial effort, in establishing the Arabic critical edition of the *Rasāʾil* and the first complete annotated English translation, will eventually render service to the academic community and lay a scholarly foundation for further studies dedicated to the Brethren's corpus and its impact on the history of ideas in Islam and beyond.

The present volume comprises the Arabic critical edition and annotated English translation with commentaries of Epistle 1: 'On Arithmetic' and Epistle 2: 'On Geometry', from the first part of the *Rasāʾil*, which focuses on the propaedeutical and mathematical sciences. This volume offers technical and epistemic analyses of mathematical concepts, and of their metaphysical underpinnings and bearings, as interpreted by the Ikhwān, in terms of investigating the properties of numbers and of geometric magnitudes, and based for the most part on the traditions derived from Nicomachus of Gerasa and Euclid. With a distinctly Pythagorean interpretation of mathematics within the symbolic order of mysticism, and also inspired by Islamic faith, the Ikhwān view arithmetic and geometry through a monotheistic spiritual lens. This outlook informed their treatment of mathematical knowledge as part of the quadrivium, which aimed at preparing the novices of their coterie to study logic and the sciences of nature and the soul, by way of preparation for the lofty topics of theology and metaphysics.

The content of this volume further offers enlightening perspectives from the standpoint of the sociology of knowledge, as regards the canonization and popularization of the mathematical sciences within the Islamic intellectual milieu, and particularly within the tenth-century urban culture of Mesopotamian and Syrian locales.

The *Epistles of the Brethren of Purity* series, and this present volume, would not have been realizable without the scholarly contributions of numerous esteemed colleagues who are participating in this challenging textual endeavour to establish the Arabic critical edition and annotated English translation with commentaries of the *Rasāʾil*. I am deeply thankful to all of them for their furtherance of this venture. I also express my gratitude to the distinguished members of the Editorial and Advisory Boards of the series for their continual academic support, especially to Professors Hermann Landolt, Wilferd Madelung, Ismail K. Poonawala, and Roshdi Rashed. Profound thanks must go as well to my dear colleagues at the Institute of Ismaili Studies, London, and to its Directors and Governors, past and present, for their generous sponsorship of this scholarly project. I would like to record here my indebtedness to Dr Farhad Daftary for his constant support of this institutional initiative. Most special thanks are due to my friend and colleague Ms Tara Woolnough for her meticulous copy-editing work and dedicated editorial care, and for her solicitous help in compiling the selected bibliography and establishing the English index. I also thank Mr Saleh al-Achmar for proofreading the Arabic text.

This volume would not have been completed without the motivating support of my wife, Dr Dina Kiwan, and the joy and blessings that our sons, Mouhib and Magdi, inspire in our life. As ever, I also relied on the encouragements of my parents, my sister, my two brothers, and on the thoughtfulness of my friends. I am truly indebted to all of them.

Nader El-Bizri
(General Editor, *Epistles of the Brethren of Purity*)

Beirut / London, June 2012

Introduction

Exordium

The Ikhwān al-Ṣafā' wa-Khillān al-Wafā' (the Brethren of Purity and the Friends of Loyalty) not only recognized the pedagogic, didactic, and epistemic merits of the mathematical sciences, but actually celebrated the symbolic order of their mystical significance and ontological bearings, reflecting the influence of Neopythagorean leitmotifs in their teachings, as mediated by Neoplatonist interpretations. Following the broader practices of their age, and in line also with the antique quadrivium tradition, the *Rasā'il Ikhwān al-Ṣafā'* (*Epistles of the Brethren of Purity*) proto-encyclopaedia begins with two treatises on pure mathematics; the first deals with arithmetic and the second focuses on geometry. These were grouped under the first part of the *Rasā'il*, which contained the propaedeutical and mathematical sciences (*al-ʿulūm al-riyāḍiyya al-taʿlīmiyya*). Even if the exact dating of the *Rasā'il* corpus is not settled in definitive terms, and the chronology of its compilation and composition is still debated by scholars in this specific subfield of academic research, a consensus in historiography emerges in terms of stating that the ninth and tenth centuries (third and fourth centuries of the hijrī calendar) constituted a foundational epoch for the exact sciences and mathematics in Islamic civilization. Whether the Ikhwān fraternity flourished towards the end of the ninth century, or continued to be operative across more than one generation until the last quarter of the tenth century, its adepts certainly would have been active in an intellectual milieu that witnessed remarkable developments in mathematics, and at highly specialized levels of inquiry.[1]

---

1 In addition to the writings of numerous learned scholars on the topic, the issue

In their epistles on arithmetic and geometry, the Ikhwān present technical and epistemic analyses of mathematical concepts and of their metaphysical underpinnings. They investigate in these tracts the properties of numbers and of geometric magnitudes, based primarily on the traditions of Nicomachus of Gerasa and of Euclid. With a Pythagorean construal of the ultimate principles of mathematics within the symbolic order of mysticism,[2] and inspired by their own ontological outlook on the articles of faith in Islam, the Ikhwān viewed arithmetic and geometry through a monotheistic spiritual lens. This informed their treatment of mathematical knowledge, in pedagogic and heuristic terms, as part of the essential disciplines of the classical quadrivium curriculum (conceptually attributed to Pythagoras of Samos [d. ca. 480 BCE], and 'institutionally' elaborated by Anicius Manlius Severinus Boethius [d. ca. 524 CE]),[3] which aimed at preparing novices to study logic, then to examine the sciences of nature and the soul, before graduating to the lofty topics of theology and metaphysics. Prior to engaging directly with the Ikhwān's epistles on arithmetic and geometry, it would help to contextualize their views within the intellectual milieu of their era and the more recent past by moving into the *spatium historicum* of their epoch's perspectives on the foundational developments in mathematics.

Mathematics in the ninth and tenth centuries

The legacies in mathematics that were established in the ninth and tenth centuries had a great impact on the emergence and unfolding

of dating has been discussed in some detail in *The Ikhwān al-Ṣafāʾ and their Rasāʾil: An Introduction*, ed. Nader El-Bizri (Oxford: OUP–IIS, 2008), Chapter 3, pp. 58–82; see also my 'Prologue' in that same edited volume on pp. 1–32, especially p. 3.

2 This aspect is analysed in detail by Yves Marquet in his last book, *Les 'Frères de la pureté', pythagoriciens de l'Islam* (Paris: EDIDIT, 2006), pp. 161–185; especially pp. 168–180 in connection with Epistles 1 and 2.

3 Supposedly coined by Boethius, the quadrivium grouped together four topics of initiatory study, namely, arithmetic, geometry, astronomy, and music, and was integrated as part of the curriculum of mediaeval monasteries. In reference to Boethius' reflections on arithmetic, geometry, and music, including his adaptation of the mathematical tradition of Nicomachus of Gerasa, see Boethius, *De Institutione Arithmetica, De Institutione Musica, Geometria*, ed. G. Friedlein (Leipzig: Teubner, 1867).

of later scientific and mathematical disciplines, not only in the lands of Islam, and through the medium of the Arabic language, but also in Europe across the late Middle Ages and the Renaissance, and up to the early modern period in the seventeenth century. The mathematicians of this classical epoch in Islamic civilization were inventive in their reception, transmission, adaptive assimilation, and expansion of Hellenic mathematics, at levels that remained unsurpassed until perhaps the seventeenth century, with the works of polymaths of the calibre of René Descartes, Gottfried Wilhelm Leibniz, Pierre de Fermat, and Bonaventura Francesco Cavalieri. The most notable contributions in the mathematical sciences in ninth- and tenth-century Islamic civilization arose from the research of what may be referred to as 'the school of Baghdad', which embodied the Archimedean-Apollonian legacy in mathematics — namely, as attributed to the investigation and re-interpretation of the methods and treatises of Archimedes of Syracuse (d. ca. 212 BCE) and of Apollonius of Perga (d. ca. 190 BCE). This cross-generational school was represented in the ninth century (third of the hijra) by mathematicians such as the famed Banū Mūsā (the illustrious three sons of Mūsā ibn Shākir: Muḥammad, Aḥmad, and al-Ḥasan), by the celebrated Ṣābi'an scholar Thābit ibn Qurra, and his reputed grandson Ibrāhīm ibn Sinān. In the tenth century (fourth of the hijra) this tradition continued with mathematicians like Abū Ja'far al-Khāzin, Abū Sahl Wayjan ibn Rustām al-Qūhī, and Abū al-'Alā' ibn Sahl.[4] This legacy ultimately culminated in the eleventh century with the geometrical and optical research of al-Ḥasan ibn al-Ḥasan ibn al-Haytham (known in Latinate renderings as 'Alhazen' or 'Alhacen'; d. ca. 1041 C.E.), and in his systematization of the newly emergent field of infinitesimal mathematics (*al-riyāḍiyyāt al-taḥlīliyya*), which determined the pre-history of the early modern infinitesimal calculus.[5] These polymaths worked in research teams

4 One should also note the great contributions in mathematics that were made in the tenth and eleventh centuries by the likes of al-Uqlīdisī, al-Būzjānī, Ibn Yūnus, Ibn Sīnā, al-Bīrūnī.

5 Regarding the developments from the ninth to the eleventh century in infinitesimal mathematics, I refer the reader to Roshdi Rashed, *Les mathématiques infinitésimales du IXe au XIe siècle, Volume 1: Fondateurs et commentateurs* (London: al-Furqān Islamic Heritage Foundation, 1996); Roshdi Rashed,

across several generations instead of being active as isolated individuals. Their treatises and findings were handed down within continual and progressive mathematical traditions for more than three centuries (from the times of the Banū Mūsā to those of Ibn al-Haytham). One should also not forget to mention the associated traditions of the algebraists who built their work on the early ninth-century legacy of Muḥammad ibn Mūsā al-Khwārizmī.

While it is worth emphasizing the impact that algebra had on the development of mathematics in the ninth and tenth centuries CE, it must be noted that there are no explicit indications in the *Rasā'il* that the Ikhwān were fully aware of the detailed aspects of the work of the algebraists, or that they were themselves seriously interested in elaborating on the science of algebra. This is the case, even though the discipline was systematized in the first quarter of the ninth century in al-Khwārizmī's compendious treatise on restoration and reduction, *Kitāb al-Jabr wa'l-muqābala* (The Book of Algebra and *Muqābala*; *Liber Muameti filii Moysi Alchoarismi de Algebra et Almuchabala*).[6] This discipline had its own method that was algorithmic (which itself carries the name of its founder, al-Khwārizmī). The Ikhwān mention the algebraists (*al-jabriyyūn*) in passing in their Epistle 1, 'On Arithmetic', under a section that discusses multiplication (*al-ḍarb*), roots (*al-judhūr*), and cubes (*al-muka''abāt*).[7] Furthermore, there are

Founding Figures and Commentators in Arabic Mathematics: A History of Arabic Sciences and Mathematics, ed. Nader El-Bizri, tr. Roger Wareham, with Chris Allen and Michael Barany (London: Routledge, 2011). See also Roshdi Rashed, *Geometry and Dioptrics in Classical Islam* (London: al-Furqān Islamic Heritage Foundation, 2005).

6 This is the title that was given in Latin by Gerard of Cremona; see Barnabas B. Hughes, 'Gerard of Cremona's Translation of al-Khwārizmī's *al-Jabr*: A Critical Edition', *Mediaeval Studies* 48 (1986), pp. 211–263. It was also rendered as *Liber Algebrae et Almuchabala* by Robert of Chester. Refer also to Barnabas B. Hughes, *Robert of Chester's Latin Translation of al-Khwārizmī's* al-Jabr. *A New Critical Edition*, Collection Boethius XIV (Stuttgart: F. Steiner Verlag Wiesbaden, 1989); Roshdi Rashed, *Al-Khwārizmī: The Beginnings of Algebra*, tr. Judith Field, revised translation Nader El-Bizri (London: Saqi Books, 2009), p. 11 — originally published with a French translation as *Al-Khwārizmī, Le commencement de l'algèbre* (Paris: Editions Albert Blanchard, 2007).

7 *Rasā'il*, vol. 1, p. 69. All *Rasā'il* references are to the Beirut Ṣadir edition of 1957, unless otherwise noted.

no indications of the Ikhwān's familiarity with the *Arithmetica* of Diophantus of Alexandria (fl. third century CE),[8] which was translated in the ninth century by Qusṭā ibn Lūqā, who in his turn relied for his rendition of the Greek text into Arabic on the technical lexicon that was developed in al-Khwārizmī's *Algebra*. The development of algebra was not restricted to the circle of the Bayt al-Ḥikma (the House of Wisdom) academy in Baghdad; a whole new generation of algebraists conducted research in the various branches of this discipline across the ninth and tenth century, including figures like Abū Bakr al-Karajī, who was also inspired by the Diophantine tradition but attempted to grant more autonomy to algebra from geometry. The sphere of algebra had applications in the calculation of inheritance and wills and of related obligations (*al-farā'iḍ*).[9] It was also used in the apportioning of irrigation channels, surveying and division of land, and various legal and juridical calculations. Indeed, al-Khwārizmī was educated in the Ḥanafī law, at some point towards the end of the eighth century, by Imam Abū Ḥanīfa himself. A new *mathesis* was established through the systematization of algebra, which required new classification systems of mathematical knowledge and disciplines that could no longer be sustained by the antique quadrivium model.[10] We can see the consequences of this in the theoretical reflections of al-Farābī and Ibn Sīnā on the classificatory systems of knowledge, not only in response to the Greek corpus, but also in connection with the various branches of mathematics and the associated novel themes in ontology. This aspect is not articulated with clarity by the Ikhwān, in the sense that their compendium is still based on a quadrivium curriculum, even though their epistle on geography is inserted as a separate theme alongside astronomy.

---

8 For the Latin translation of Diophantus' *Arithmetica*, refer to Paul Tannery, *Diophanti Alexandrini Opera Omnia*, 2 vols. (Leipzig: Teubner, 1893–1895; repr., 1974). Regarding the Arabic version of the text, which is accompanied by an annotated English translation, see Jacques Sesiano, *Books IV to VII of Diophantus'* Arithmetica *in the Arabic translation attributed to Qusṭā ibn Lūqā* (New York: Springer, 1982). For the analytic commentary on the mathematical contents, I refer the reader to Thomas L. Heath, *Diophantus of Alexandria: A Study in the History of Greek Algebra*, 2nd edition (New York: Dover, 1964).

9 Rashed, *Al-Khwārizmī: The Beginnings of Algebra*, pp. 7–8, 25–27.

10 Rashed, *Al-Khwārizmī: The Beginnings of Algebra*, p. vii.

During this era, the fields of geometry, arithmetic, and algebra were applied to one another, and this resulted in the development of novel branches in mathematical research. The investigations were also centred on conic sections, in view of measuring their volumetric magnitudes and surface areas, and in establishing detailed studies of parabolas and paraboloid solids, in addition to related analyses of the properties of ellipses, circles, curved surfaces, and portions of curved solids, including the examination of multifarious characteristics of the sphere and the cylinder. These domains of research were enmeshed with select mathematical problems that emerged in relation to unprecedented developments in the study and use of geometrical transformations (namely, similitude, translation, homothety, and affinity). They were also affected by the introduction of motion into geometry, in the context of investigating the conditions for drawing continuous curves, and the related studies in stereography, spherics,[11] and the perfect compass — as was the case with al-Qūhī's research, for example; this is in addition to the analysis of the anaclastic properties in the geometric modelling of lenses through the agency of conic, cylindrical, and spherical geometrical sections, in the sciences of dioptrics and catoptrics (respectively, the sciences of the refraction and reflection of light) — as it was undertaken, for instance, in the investigation of burning instruments (*al-ḥarrāqāt*) and indices of refraction by the tenth-century geometer Abū al-ʿAlāʾ ibn Sahl (predating the seventeenth-century so-called 'Snell's Law of Refraction').[12]

Ibn al-Haytham further advanced the findings of his predecessors. In response to the Archimedean problem of calculating areas and volumes,

11 This field relates to the intermediary sciences (*al-mutawassiṭāt*) that are situated between Euclid's *Elements* and Ptolemy's *Almagest*, like Theodosius' *Spherica* (*Kitāb al-ʾUkar*, ca. 100 BCE) and Menelaus' *Spherica* (*Kitāb al-Ashkāl al-kuriyya*, ca. 100 CE).

12 This refers to a principle that allows the determination of the angles of incidence and refraction when light rays pass through the boundary-limit between two transparent isotropic media that differ in their refractive indices, such as air and water, or between either of these and glass. This phenomenon was studied since the beginnings of the 1600s by many scientists, such as Thomas Harriot, Willebrord Snellius (Snell), René Descartes, Pierre de Fermat, and Christiaan Huygens. This principle is sometimes referred to as the 'Snell-Descartes Law' instead of the more common appellation, 'Snell's Law of Refraction'.

including the exhaustion technique and the reliance on numerical computation instead of merely following a theory of proportion, he reinvented the classical methods of proof. This encompassed research on the quadrature of the circle and of lunes (crescent-like shapes; *ashkāl hilāliyya*), the calculation of the volumes of paraboloids and of the sphere by way of the method of exhaustion, along with the extraction of square and cubic roots, and the investigation of the solid angle. In addition, he looked at determining the characteristics of surfaces, with the maximal volumetric magnitudes they envelop, and studying their isepiphanic aspects (i.e., of equal surface areas), while also investigating isoperimetric plane figures (i.e., those with equal perimeters) in terms of the surface areas that get delimited within their given perimeters.

Such endeavours reflected an emergent epistemic tendency to offer mathematical solutions to problems in theoretical philosophy and in Aristotelian physics, as exemplified in the work of al-Qūhī in the tenth century. This also finds a culminating expression in the eleventh century with Ibn al-Haytham's revolutionizing methods of scientific research, and in terms of his isomorphic combination of mathematics with physics in controlled experimentation (*al-iʿtibār*) with natural phenomena, and in his geometrization of the fundamental notions of natural philosophy, such as the conception of place (*al-makān*; *topos*) as an imagined/postulated geometric void *qua* spatial extension.[13]

---

13 I have examined related mathematical developments elsewhere, particularly with reference to optics and as exemplified by the œuvre of Ibn al-Haytham, including in the following studies: Nader El-Bizri, 'In Defence of the Sovereignty of Philosophy: al-Baghdādī's Critique of Ibn al-Haytham's Geometrisation of Place', *Arabic Sciences and Philosophy* 17, 1 (2007), pp. 57–80; 'A Philosophical Perspective on Alhazen's Optics', *Arabic Sciences and Philosophy* 15, 2 (2005), pp. 189–218; 'La perception de la profondeur: Alhazen, Berkeley et Merleau-Ponty', *Oriens–Occidens: Sciences, mathématiques et philosophie de l'antiquité à l'âge classique* (Cahiers du Centre d'Histoire des Sciences et des Philosophies Arabes et Médiévales, CNRS) 5 (2004), pp. 171–184; 'Ibn al-Haytham', in *Medieval Science, Technology, and Medicine: An Encyclopedia*, ed. Thomas F. Glick, Steven J. Livesey, and Faith Wallis (New York–London: Routledge, 2005), pp. 237–240.

Mathematics in the Epistles of the Brethren of Purity

The initiatory mathematical tracts, on arithmetic (*Risāla fī al-ʿadad; al-arithmāṭīqī*) and on geometry (*Risāla fī al-jūmaṭrīyā*), that commence the *Rasāʾil Ikhwān al-Ṣafāʾ* (*Epistles of the Brethren of Purity*)[14] were composed as epitomes in this field of knowledge. The pedagogic rationale behind these two epistles aimed at supporting further training in natural philosophy (*ṭabīʿiyyāt*; *phusikê*) and psychical sciences (*nafsāniyyāt*; *peri psukhês*; and the Latin being *de anima* — broadly conceivable as psychology), with a view to properly priming the novice to explore topics in theology (*ilāhiyyāt*; *theologikê*).[15] This approach was not restricted to the Ikhwān but reflected the classification of knowledge in their age according to ancient practices in canonizing the sciences and disciplines with their subfields of inquiry (as was the case most notably with figures such as al-Kindī, al-Khwārizmī, and al-Fārābī). In the first instance, the topics of arithmetic and geometry would develop the demonstrative reasoning skills of the apprentices within the Ikhwān's fraternity, including the training in demonstration that starts with the geometrical forms and is then further enhanced through logic. As they noted in Epistle 14,[16] 'He who wants to study logical demonstrations should first practise geometrical demonstrations' — for

14 As noted above, unless otherwise indicated all references are to the Ṣādir edition of the *Rasāʾil Ikhwān al-Ṣafāʾ* (hereafter cited as *Rasāʾil*). Besides the annotated English translation which is included in this present volume, I also refer the reader to an annotated English translation with commentaries of the epistle on arithmetic in Bernard R. Goldstein, 'A Treatise on Number Theory from a Tenth-Century Arabic Source', *Centaurus* 10 (1964), pp. 129–160, with the translated text on pp. 135–159. See also, Sonja Brentjes, 'Die erste Risāla der *Rasāʾil Ikhwān al-Ṣafāʾ* über elementare Zahlentheorie', *Janus* 71 (1984), pp. 181–274.

15 See *Rasāʾil*, vol. 1, p. 49. Moreover, as Yves Marquet has noted, the Pythagorean teachings not only shaped the Ikhwān's mystical and symbolic extrapolations of the results of the mathematical disciplines, but, moreover, they had an impact on the composition of many of their other epistles, including those grouped under the heading of theology. See Marquet, *Les 'Frères de la Pureté', pythagoriciens de l'Islam*, pp. 168–180.

16 See *Epistles of the Brethren of Purity. On Logic: An Arabic Critical Edition and Annotated English Translation of Epistles 10–14*, ed. and tr. Carmela Baffioni (New York–London: OUP–IIS, 2010), Epistle 14, Chapter 22, pp. 146–147 (English translation).

the reason that the latter is less complex,[17] given that its examples are perceivable by the sense of sight, as they illustrated by evoking Euclid's Proposition 32 and Proposition 47,[18] in Book I of his *Elements*.

The *Rasā'il Ikhwān al-Ṣafā'* is customarily judged as being akin to an encyclopaedia in its synoptic content, although the investigations in arithmetic and geometry within it are less advanced in their epistemic consequences than the contemporary accomplishments of mathematicians in mediaeval Islamic civilization across the ninth and tenth centuries, as highlighted above. However, one ought to add, in fairness, that the Ikhwān explicitly affirmed that their compendium represented 'the most concise of summaries' (*awradnāhā bi-awjaz mā yumkin min al-ikhtiṣār*),[19] and one should concede that its epitomes

17 The Ikhwān's view in this regard is debatable even on the grounds of the epistemic criteria of their contemporaries. Many mathematicians would have doubted the primacy of logic and its complexity over that of geometry; one could mention in this regard al-Qūhī's case as an example from the tenth-century School of Baghdad. The debate over the pre-eminence of logic over mathematics, or vice versa, continued till our age in the twentieth century. This is, for instance, exemplified by the view that mathematics transcends logical reasoning, as pointed out by Kurt Gödel in his 'incompleteness theorem' and in his endeavour to critically re-assess the *Principia Mathematica* of Alfred North Whitehead and Bertrand Russell, who attempted to derive all mathematical truths from a set of axioms and inferential rules in symbolic logic. Gödel argued his case in his article: 'Über formal unentscheidbare Sätze der "Principia Mathematica" und verwandter Systeme', *Monatshefte für Mathematik und Physik* 38 (1931), pp. 173–198. See also Alfred North Whitehead and Bertrand Russell, *Principia Mathematica*, 3 vols., 2nd ed. (Cambridge: CUP, 1925–1927).

18 Proposition 32 in Book I of Euclid's *Elements* reads as follows: 'In any triangle, if one of the sides be produced, the interior angle is equal to the two interior and opposite angles, and the three interior angles of the triangle are equal to two right angles'; Euclid, *The Thirteen Books of the Elements*, tr. Thomas L. Heath, 3 vols. (Cambridge: CUP, 1925; repr., New York: Dover, 1956), vol. 1, p. 316.

Proposition 47 in Book I of Euclid's *Elements* reads as follows: 'In right-angled triangles, the square on the side subtending the right angle is equal to the squares on the sides containing the right angle'. This proposition is commonly associated with what is known as 'Pythagoras' Theorem' regarding the square on the hypotenuse. This theorem is known in populist expressions as 'the theorem of the nymph [or bride]' (*theorem tês numphês*) or 'the two-horned' (*dhū al-qarnayn*) on account of its geometric shape in the course of demonstration. See Euclid, *The Thirteen Books of Euclid's Elements*, tr. Heath, vol. 1, pp. 417–418.

19 *Rasā'il*, vol. 1, p. 77.

were supplemented by oral teachings in seminars (*majālis al-ʿilm*). Nonetheless, the Ikhwān's epistles on arithmetic and geometry gave only rudimentary summaries (*awjaz ikhtiṣār*) of the classical branches in mathematics of their era, excluding a direct treatment of algebra.

In the remaining parts of this introduction, my analysis will be principally explanatory. I shall attempt to elucidate the technical aspects of Epistles 1 and 2, while also endeavouring to situate the contents of both tracts within the context of the development of the mathematical sciences in the ninth and tenth centuries. In particular, I will focus on a detailed exposition of the various arithmetical and geometrical propositions of the Ikhwān, as well as indicating their textual sources, which originated from the ancient Greek heritage and were mediated through the flourishing of mathematics after the founding of the Bayt al-Ḥikma (House of Wisdom) in Baghdad. I will also present some of the more significant aspects regarding the ontological and epistemological entailments of the Ikhwān's philosophical reflections on the intellective and symbolic bearings of the study of mathematics. This is coupled with an assessment of the pedagogic possibilities and heuristics that the Ikhwān's treatment of the mathematical sciences offers to those seeking to study logic or natural philosophy, and to those who ultimately inquire about the nature of the soul with a view to reflecting on the metaphysical fundamentals of theology and gnosis.

The analysis begins with Epistle 1: 'On Arithmetic' and goes on to examine Epistle 2: 'On Geometry', in both cases following the sequence that we find in the thematic subdivisions of the contents of these tracts. It must be noted here that the Ikhwān's views on mathematics reverberate throughout their *Rasāʾil*, even if the bulk of their dedicated analysis of arithmetic and geometry is concentrated in the first two epistles. Mathematical notions are encountered across all the epistles that constituted the Ikhwān's textual embodiment of their version of the quadrivium; these also figure in their investigation of proportions and ratios in Epistle 6. We are then shown how numbers are significant in grasping the bodily constitutions of animals in zoological analyses, as presented in the overture of Epistle 22, which constituted an intrinsic *scholium* preceding their celebrated ecological fable, which casts the oppressed and exploited animals in a court case against humanity before the King of the Jinn.[20] The properties of numbers are also evoked in that fabulist setting by a Byzantine speaker from Greece, who enumerates the merits of arithmetic. In Epistle 33, the Ikhwān again remind their readers about the significance of the Pythagorean ontology, in the sense that all classes of existents can be grasped analytically and in abstract form through arithmetical means. And in Epistle 41 they offer a synoptic set of definitions in arithmetic and geometry from the division on the psychical sciences (or psychology). Many aspects of the treatment of arithmetic and geometry in meta-mathematical contexts are intertwined with propositions in ontology and epistemology, and with perspectives on natural and psychical phenomena through the prism of the classical microcosm-macrocosm analogies — namely, that the human being is a 'micro-cosmos' (*al-insān ʿālam ṣaghīr*) and that the world is a 'macro-anthropos' (*al-ʿālam insān kabīr*). Mathematics and natural philosophy were subjected to overarching *logoi* that unified the compositional architectonics of the *Rasāʾil* as a text, and that embodied furthermore a symbolic worldview that served the Ikhwān's soteriology as well as its gnostic underpinnings, inspired mainly by Pythagorean arcana in combination with Neoplatonist dogmata.[21]

20 Refer to *Epistles of the Brethren of Purity. The Case of the Animals Before the King of the Jinn: An Arabic Critical Edition and Annotated English Translation of Epistle 22*, ed. and tr. Lenn E. Goodman and Richard McGregor (New York–London: OUP–IIS, 2009).

21 The theological mysticism in their treatment of arithmetic may also have been inspired by fragments and conceptual leitmotifs from the œuvre of the Syrian Neoplatonist Iamblichus Chalcidensis (i.e., of Chalcis), mainly in connection with the anonymous treatise *Theologoumena arithmêtikês* (Theology of Arithmetic), which is commonly ascribed to Iamblichus and possibly paraphrased in part from Nicomachus' teachings. See *Theologumena arithmaticae*, ed. Fridericus Astius (Leipzig: Libraria Weidmannia, 1817); edited by Vittorio de Faco with a subheading *Theological Principles of Arithmetic* (Leipzig: Teubner, 1922); and translated by Robin Waterfield as *The Theology of Arithmetic: On the Mystical, Mathematical, and Cosmological Symbolism of the First Ten Numbers* (Grand Rapids, MI: Phanes Press, 1988). The Ikhwān's focus on the centrality of mathematics is not only Pythagorean but also Platonist. This is best embodied

As they affirmed in Epistle 5 'On Music',[22] the Ikhwān endeavoured to demonstrate analogically how numbers are combined from the unit (i.e., the number 1) in arithmetic, and that the point in geometry is like the arithmetical unit. In addition, they attempted to show how, in astronomy, the Sun is amongst the heavenly bodies like the unit 1 is amongst numbers, and how the point is also situated amongst lines, planar figures, and volumetric solids in geometry. In Epistle 6, they sought to show how an equation constitutes a fundamental rule like the unit 1 in arithmetic and the point in geometry. In music, they indicated that the notes of rhythmic cycles are compounded from units, just as all numbers are compounded from the unit 1. Similarly, in logic, substance (*al-jawhar*) is like the unit 1, and the categories are like nine units, with substance, quality, quantity, and relation given primacy over place, time, position, state, action, and affection (passivity). The body itself was conceived by them as being constituted from substance, length, breadth, and depth, which are compounded in ways analogous to the structure of the absolute cosmic body.

Arithmetic

The epistle on arithmetic is divided into twenty-five chapters (*fuṣūl*), which treat topics such as: 'Properties of Numbers'; 'Whole [Numbers]

---

in ancient philosophical thought in the beginnings of dialogical philosophy by Plato. The *Timaeus* and the *Politeia* make it clear that mathematics occupied an elevated epistemic station and a noble ontological realm within the Platonist worldview. This was principally articulated in the theory of forms, and in response to pre-Socratic wisdoms. It is even claimed that, at the level of praxis and of didactic-pedagogic applications in the exclusive, dialectic Academy, an epigraphic pronouncement (ascribed to Plato and publicized at the entrance of his Academy) read, *Ageômetrêtos mêdeis eisitô* ('Let none but geometers enter'). It is believed that the *Scholarchês* of the Academy, in its old, middle, and new settings, was supposed to uphold the emphasis on the principality of mathematics and the knowledge of geometry in particular. See also Myles F. Burnyeat, 'Plato on Why Mathematics is Good for the Soul', in *Mathematics and Necessity: Essays in the History of Philosophy*, ed. Timothy Smiley, Proceedings of the British Academy, vol. 103 (Oxford: OUP, 2000), pp. 1–81.

22 Refer to *Epistles of the Brethren of Purity. On Music: An Arabic Critical Edition and Annotated English Translation of Epistle 5*, ed. and tr. Owen Wright (New York–London: OUP–IIS, 2010), pp. 104–105 (English translation), and pp. 53–55 (Arabic edition).

and Fractions'; 'Perfect, Deficient, and Abundant Numbers'; 'Amicable [Numbers]'; 'Multiplication [i.e., duplation; *taḍʿīf*] of Numbers'; 'Subdivisions' 'Whole Numbers'; 'Multiplication, Roots, and Cubes'; 'Square Numbers'; 'Root Numbers'; 'Propositions from Euclid's *Elements*, Book II'; 'Arithmetic and the Soul'; and 'The Purpose of the Sciences'. In analysing this epistle on arithmetic, I shall follow an approximately parallel thematic subdivision that is suitable to the scope of this study, to be noted under the sub-title 'Arithmetic', and to be followed by a roman numeral.

Arithmetic I[23]

The early development of arithmetic in mediaeval Islamic civilization originated in part from the systematization of *ḥurūf al-hind* (namely, the Hindi numerals associated with the *ghubār* dust-board methods of numeration). Following the ancient Greek mathematical traditions, the science of number (*arithmêtikê tekhnê*; *ʿilm al-ʿadad*) was eventually distinguished from the science of reckoning (*logistikê tekhnê*; *ʿilm al-ḥisāb*) as a result of the pragmatic and quotidian uses of calculation. For instance, Abū al-Wafāʾ al-Būzjānī (d. ca. 998 CE) composed his treatise *Kitāb Fī mā yaḥtāj ilayh al-kuttāb wa'l-ʿummāl min ʿilm al-ḥisāb* (What Scribes and Labourers Require of the Science of Calculation) to serve mercantile needs. The computational art was further elucidated by Abū al-Ḥasan Aḥmad ibn Ibrāhīm al-Uqlīdisī (d. ca. 980 CE) in his *Fī al-ḥisāb al-hindī* (On Indian Calculation).[24]

23 The analysis that is contained in the remaining parts of this introduction is based closely on my earlier investigation of Epistles 1 and 2 as it figured in my chapter, 'Epistolary Prolegomena: On Arithmetic and Geometry', in El-Bizri, ed., *The Ikhwān al-Ṣafāʾ and their 'Rasāʾil'*, Chapter 7, pp. 180–213.

24 The science of calculation (*ʿilm al-ḥisāb*) consisted of three main types, namely, hand reckoning (*ḥisāb al-yad*), finger reckoning (*ḥisāb al-ʿuqūd*), and mental reckoning (*al-ḥisāb al-hawāʾī*). For references, see Ahmad S. Saidan, 'The Earliest Extant Arabic Arithmetic: *Kitāb al-Fuṣūl fī al-Ḥisāb al-Hindī* of Abū al-Ḥasan Aḥmad ibn Ibrāhīm al-Uqlīdisī', *Isis* 57 (1966), pp. 475–490; *The Arithmetic of al-Uqlīdisī* (Dordrecht: D. Reidel, 1978); 'Numeration and Arithmetic', in *Encyclopedia of the History of Arabic Science*, ed. Roshdi Rashed with Régis Morélon, vol. 2 (London: Routledge, 1996), pp. 331–348. I also refer the reader to Rida A. K. Irani, 'Arabic Numeral Forms', *Centaurus* 4 (1955), pp. 1–12; and the synoptic survey of A. I. Sabra, ''Ilm al-ḥisāb', *EI2*, vol. 3, pp. 1138–1141.

The Ikhwān's number theory was inspired by the received Pythagorean tradition, and it was principally derived from the *Arithmêtikê eisagôgê* (*Introductio Arithmetica* in Latin)[25] of Nicomachus of Gerasa (ca. 60–120, Roman Syria),[26] who emphasized the anteriority of arithmetic with respect to geometry, given that geometry presupposes arithmetic and already implies it within its own disciplinary bounds. It is reported, moreover, that Nicomachus alluded to the composition of an 'Introduction to Geometry' that followed his *Arithmetic*, though that tract is lost.[27]

In emulation of Nicomachus' tradition, the Ikhwān interpreted the entailments of arithmetic through mystical symbolisms. They also relied on induction rather than proceeding by way of proofs (*barāhīn*) in this particular mathematical discipline.[28] In addition, they believed

25 Nicomachus' *Introduction to Arithmetic* was translated into Arabic by Thābit ibn Qurra, under the title *Kitāb al-Madkhal ilā ʿilm al-ʿadad*, with a manuscript preserved at the British Museum, as per the particulars: MS 426.15, (add. 74731), ff. 198, dated 1242; see *Arabische Übersetzung der Arithmêtikê Eisagôgê des Nikomachus von Gerasa*, ed. Wilhelm Kutsch (Beirut: Imprimerie Catholique, St. Joseph, 1959). Nicomachus' tract had a markedly Pythagorean mystical penchant, a tendency that was more emphatically articulated in his *Theologumena arithmeticae*. Nicomachus' *Introduction to Arithmetic* itself is published under the title *Nicomachi Geraseni Pythagorei Introductionis arithmeticae*, ed. Richard Hoche (Leipzig: Teubner, 1886). See also Nicomachus of Gerasa, *Introduction to Arithmetic*, trans. M. L. D'Ooge, with studies by F. E. Robbins and L. C. Karpinski (New York, 1926); Nicomachus of Gerasa, *Introduction arithmétique*, tr. Janine Bertier (Paris: J. Vrin, 1978); Leonardo Tarán, 'Nicomachus of Gerasa', in *Dictionary of Scientific Biography*, ed. Charles Coulston Gillispie et al., vol. 10 (New York: Charles Scribner's Sons, 1974), pp. 112–114.

26 *Rasāʾil*, vol. 1, p. 49. See also Carmela Baffioni, 'Citazioni di autori antichi nelle *Rasāʾil degli Ikhwān al-Ṣafāʾ*: il caso di Nicomaco di Gerasa', in *The Ancient Tradition in Christian and Islamic Hellenism: Studies on the Transmission of Greek Philosophy and Sciences Dedicated to H. J. Drossaart Lulofs on his Ninetieth Birthday*, ed. Gerhard Endress and Remke Kruk (Leiden: Leiden Research School, 1997), pp. 3–27; Muḥammed Jalūb Farhān, 'Philosophy of Mathematics of Ikhwān al-Ṣafāʾ', *Journal of Islamic Science* 15 (1999), pp. 25–53.

27 See *Introduction arithmétique*, tr. Bertier, pp. 8, 58.

28 In this context, one would evoke the structuring function assigned by the Ikhwān to the microcosm/macrocosm analogy and to magic: Epistle 26 'On the Microcosm' (*al-Insān ʿālam ṣaghīr*; the human being is a microcosm); Epistle 34 'On the Macrocosm' (*al-ʿĀlam insān kabīr*; the world is a 'macroanthropos'); and Epistle 52 'On Magic' (*Fī al-siḥr*). See Geo Widengren, 'Macrocosmos–Microcosmos Speculation in the *Rasāʾil Ikhwān al-Ṣafāʾ* and Some Ḥurūfī Texts',

that numbers corresponded foundationally with natural phenomena and worldly configurations in reference to signifiers associated with time, place, the elements, classes of existents (*al-mawjūdāt*), and the differential derivations of human language.[29] Their analogism, which was saturated with picture-based language, resulted in mistaking resemblances for explanations; consequently, their meta-mathematical speculation around arithmetic borders on 'empty verbalism', which produced pseudo-explanations that are hardly translatable into meaningful epistemic terms.[30] This was not their own shortcoming per se, given that it was inherited from the symbolic orders of Pythagorean and Hermetic doctrines, which had an impress on their conceptions of gnosis and magic. It is not surprising that this aspect of their thought attracted criticism, such as what was reported by the tenth-century litterateur Abū Ḥayyān al-Tawḥīdī, that the scholar Abū Sulaymān al-Manṭiqī al-Sijistānī ibn Bahrām asserted that the Ikhwān exercised great efforts without enriching our knowledge.[31]

The Ikhwān were explicit about their indebtedness to Pythagoras and Nicomachus in arithmetic,[32] to Euclid's *Elements* (*Stoikheia*; *Kitāb Uqlīdis fī al-uṣūl*) in geometry, and to the *Almagest* of Claudius Ptolemy (d. 165) in astronomy. And yet, although the influence of Nicomachus is evident throughout the Ikhwān's epistle on arithmetic, the subdivisions and their contents do not entirely overlap with those of Nicomachus' tract.[33] One of the determining distinctions between these texts may be

---

Archivio di Filosofia 1 (1980), pp. 297–312; Paul Casanova, 'Alphabets magiques arabes', *Journal Asiatique* 18 (1921), pp. 37–55, 19 (1922), pp. 250–262.

29 For instance, the fourfold schema (i.e., the four *murabbaʿāt*) described the correspondence of the elements, humours, temperaments, and seasons, following in this a tradition associated with Galenism. See *Rasāʾil*, vol. 1, pp. 51–52.

30 For an illuminating explanation of Pythagorean mathematics, I refer the reader to Jules Vuillemin, *Mathématiques pythagoriciennes et platoniciennes* (Paris: Albert Blanchard, 2001). See also Dominic J. O'Meara, *Pythagoras Revived: Mathematics and Philosophy in Late Antiquity*, 2nd ed. (Oxford: Clarendon Press, 1990).

31 Refer to Butrus al-Bustani's preface to the Beirut Ṣādir edition of 1957, p. 20.

32 The Ikhwān also acknowledged their indebtedness to Pythagoras in the mathematical application of harmonics in music.

33 As I have indicated earlier in this introduction, there are no explicit indications that the Ikhwān directly benefited from the *Arithmetica* of Diophantus of Alexandria (fl. third century), which was translated into Arabic by Qusṭā ibn

attributed to the stress on monism in the Ikhwān's *risāla*, versus a more pronounced emphasis by Nicomachus on dualism and on the associated interplay between identity *qua* sameness (the unit 1 *qua* the same) and difference *qua* otherness (the number 2 *qua* the other). Furthermore, the Ikhwān did not even mention Nicomachus' focal distinction between what he designated as a *noêtos arithmos* ('intelligible number'; what in Arabic could be coined as "*adad 'aqlī*') and the *epistêmonikos arithmos* (namely, the 'mathematical/scientific number'; what could be coined in Arabic as "*adad 'ilmī*'). The *noêtos arithmos* relates to numbers as they pertain to the workings of the Divine Artisan (the Greek Demiurge) and are ontologically distinguished from the realm of sensible beings, as, for instance, is highlighted in reference to the intelligible in Plato's *Timaeus* (27d–28a) and Aristotle's *Topics* (100b19). As for the *epistêmonikos arithmos*, it designates numbers in their abstract form (*monadikos arithmos*; 'abstract number') as mathematical objects or entities that regulate all beings and sustain the acquisition of rational knowledge as opposed to mere opinion.

Despite the existential-cum-metaphysical applications of numbers in modulating the hierarchical structuration of all beings as claimed by the Ikhwān, we do not find in their arithmetic any indication of Nichomachus' distinction between *noêtos arithmos* and *epistêmonikos arithmos*. This textual state of affairs may have resulted from the Ikhwān's theological inclination for a prudent eschewal of references to the Creator of monotheism that may be suggestive of similarities with the pagan Artisan Deity who generated the ordering of the universe and its beings from pre-existing and co-eternal worldly constituents. Based on theological (meta-mathematical) assessments of the merits of arithmetic, the Ikhwān asserted that the unit 1 (*al-wāḥid*, which precedes the first number, i.e., 2) was changeless and indivisible, and that it was the

---

Lūqā. It is, moreover, unclear whether the Ikhwān fully integrated investigations on numbers as they figured in Plato's *Timaeus*, or as in the exaggerated version of realism that was presented in the *Phaedo*. There is also no evidence that they were influenced by the objections raised by Aristotle against the Platonic thesis on the self-subsistence of numbers. See Thomas L. Heath, *Diophantus of Alexandria: A Study in the History of Greek Algebra* (reprinted in New York, 1964); Aristotle, *Metaphysics*, ed. David Ross (Oxford: Clarendon Press, 1997), 1083b23–29, 1085b34.

beginning and the end of numbers, similar to the way the Immutable Divine is related to all beings. Moreover, the Ikhwān believed that the propositions of arithmetic supported the unfolding of metaphysical interrogations regarding the question 'How does an infinite multitude begin with the One?' In response to this, they explicitly affirmed that the generation of numbers from the unit 1 ad infinitum offers lucid evidence (*dalīl*) of God's Unity (*waḥdāniyya*). This notion of unity can be linked to the symbolic order that underpinned the teachings of Neopythagoreans in *theourgia* (theurgy) and to the associated Neoplatonist conception of *henosis* as a union with the ultimate source of reality (as oneness with the One; *to en*) from which originates the notion of monad (*monas*).[34] This view is also confirmed in the context of Epistle 32, wherein the Ikhwān argue that the relationship of the Creator to all existents is like that of the unit 1 to all numbers, whilst the Active Intellect (*al-ʿaql al-faʿʿāl*) is like the number 2, the cosmic universal soul (*al-nafs al-kulliyya al-falakiyya*) like the number 3, and prime matter (*al-hayūlā al-ʾūlā*) like the number 4. Then all creatures (*al-khalāʾiq*) were fashioned in the same way that the rest of the numbers followed on from the number 4.[35] This is also affirmed from the onset in Epistle 5: 'On Music'.[36]

Arithmetic II

The Ikhwān linked their mystical number theory with an ontic-ontological investigation of the nature of existent beings, and followed in this a pattern in thinking akin to the Platonic tradition in emphasizing the importance of mathematics as a foundation for the study of philosophy and the teleological pursuit of happiness.

In commenting on the foundational unit 1, the Ikhwān described this arithmetical entity in general terms as a *shayʾ* (namely, a 'thing').[37] Moreover, they noted in the introduction to their first epistle that arithmetic involves a study of the attributes of numbers (*khawāṣṣ*

34 *Rasāʾil*, vol. 1, p. 54.
35 *Rasāʾil*, vol. 1, p. 54.
36 See *On Music*, p. 106 (English translation), and p. 55 (the Arabic edition); refer to what is also indicated below on the Ikhwān's reflections on the significance of the *tetraktys*, pp. 41, 121.
37 *Rasāʾil*, vol. 1, p. 49.

al-ʿadad) in view of elucidating the properties of (correlative) existents (*maʿānī al-mawjūdāt*). Ultimately, as the first of the abstract sciences, arithmetic involves the investigation of numbers for the purpose of grasping the reality of worldly beings (*al-mawjūdāt al-latī fī al-ʿālam*). Moreover, the general appellation the 'thing' (*shay'*; or in Greek, *ôn*) refers to the utterance 'one' (*wāḥid*), used either in concrete linguistic applications (*bi'l-ḥaqīqa*) or by way of metaphor (*bi'l-majāz*), and points to a monotheistic conception of divinity. In its proper application (or quotidian use), the one designates a thing that cannot be partitioned or divided, whilst metaphorically, the one is a multitude (*jumla*; *plêthos*) that acts as an undivided unity. Numerically, the one encompasses in its oneness the idea of a unit (*waḥda*; *monas*/*monad*), which grounds the applications of the art of *ḥisāb* ('calculation' or 'reckoning') in the addition and subtraction of numbers (*jamʿ al-ʿadad wa-tafrīquh*).

In this context, the Ikhwān's reflections on arithmetic are reminiscent of what is stated in Book VII (On the Fundamentals of Number Theory) of Euclid's *Elements*, wherein, Definition 1 notes that 'a unit [*monad*] is that by virtue of which each of the things that exist is called one'; and Definition 2, adds that 'a number is a multitude of compound units'. Thus, the idea of number (*ʿadad*; *arithmos*) was associated with the notion of a unit, and it was furthermore connected to the concept of a thing with the gathering of units into a multitude, namely, a whole that is made up of units, which refers to the quantitative 'iteration-number' that is needed in order to produce a given number *A*.[38] The unit remains therefore unaltered if it is multiplied by itself any number of times, or if it is separated from the multitude of numbers; such a unit is most solitary (*monôtatê*).

38 The quantitative 'iteration-number' ($a \geq 2$), designates the cumulation of unit-monads (E) 'as often as' (*hosakis*) it is needed to result in the number-arithmos (A). This may be notated as follows: $A \Rightarrow (aE)$, $a \geq 2$; which means 'as many units as there are in it' (*hosai eisin en autô monades*). For further particulars, see Ioannis M. Vandoulakis, 'Was Euclid's Approach to Arithmetic Axiomatic?', *Oriens–Occidens* 2 (1998), pp. 143–144. See also, in the same volume of the journal, Jean-Louis Gardies, 'Sur l'axiomatique de l'arithmétique Euclidienne', pp. 125–140.

Arithmetic III

Following Nicomachus' precedent, the Ikhwān divided numbers (*al-aʿdād*) into two principal kinds: (positive) integers (whole number; *ʿadad ṣaḥīḥ*) and fractions (*kusūr*). They also reaffirmed that the one is 'the origin of numbers' (*mabdaʾ al-ʿadad*), and that integers start with the smallest quantity (that is, 2) and increase ad infinitum by way of addition (*yadhhab bi'l-tazāyud bilā nihāya*).[39] An integer is thus generated out of the one and is led back by way of analysis to that unit, following a 'natural progression' (*yanshaʾ al-ʿadad al-ṣaḥīḥ min al-wāḥid wa-yanḥall ilayh ʿalā al-naẓm al-ṭabīʿī*).

Integers themselves are also classed according to four ranks (*marātib*), namely: the scale of ones (units), of tens, of hundreds, and the scale of thousands (respectively: *aḥād, ʿasharāt, miʾāt, ulūf*). These were designated by twelve simple utterances (*lafẓa basīṭa*) from which all numbers are named (i.e., the ten from *wāḥid* to *ʿashara*, plus *miʾa*, plus *alf*).[40] The Ikhwān also observed that each integer has the property of being equal to half the quantity that results from the sum of its two immediately adjacent numbers (*ḥāshiyatāh*), e.g., $5 = \frac{1}{2}(4 + 6)$; $6 = \frac{1}{2}(5 + 7)$; $7 = \frac{1}{2}(6 + 8)$; $8 = \frac{1}{2}(7 + 9)$; $9 = \frac{1}{2}(8 + 10)$; $10 = \frac{1}{2}(9 + 11)$. This can be expressed generally in the algebraic form: $n = \frac{1}{2}[(n + m) + (n - m)]$.

The Ikhwān's system of correspondence between numbers and letters of the alphabet was based on the common mathematical practices of their age, which emulated in part the classical Phoenician and Greek models, and in part the system, more familiar to us, of Roman numerals.

---

39 *Rasāʾil*, vol. 1, pp. 50, 68–69. The Ikhwān echoed a Greek conception of number, wherein 2 (*duas*) is taken to be the smallest number without qualification, as was noted in, for instance, Aristotle, *Physics*, ed. W. David Ross (Oxford: Clarendon Press, 1998), Delta 220a27–28.

40 In addition to these ranks, we also highlight the tradition inherited from the sexagesimal system that is believed to be of Babylonian provenance (in other words, the scale of 60, as opposed to our decimal system). This numerical order denoted numbers by letters, and was used in finger reckoning in reference to the divisors of 60: 2, 3, 4, 5, 6, 10, 12, 15, 20, and 30. This system was of importance for astronomy, principally in reckoning angles and subdivisions of circles: 30°, 45°, 60°, 90°, 120°, 180°, 240°, 300°, and 360°. It is also still used in measuring time: 60 seconds, 60 minutes, or 12 hours (as $\frac{1}{5}$ of 60), or 24 hours (as $\frac{2}{5}$ of 60).

The Ikhwān represented numbers through the following *abjad* sequence (*ḥurūf al-jumal*):[41]

م	ل	ك	ي	ط	ح	ز	و	ه	د	ج	ب	ا
40	30	20	10	9	8	7	6	5	4	3	2	1

Based on this system, the correspondence between numerals and alphabet letters is as follows:

1–10	ا ب ج د ه و ز ح ط ي
20–90	ك ل م ن س ع ف ص
100–900	ق ر ش ت ث خ ذ ض ظ
1,000–10,000	غ بغ جغ دغ هغ وغ زغ حغ طغ يغ

Of the sixteen numerical ranks (*marātib*) that the Ikhwān utilized, after the Pythagoreans, the largest set they noted in relation to usable quantities was that of quadrillions: 1,000,000,000,000,000 = 10^{15} (named *mārū ulūf ulūf ulūf ulūf ulūf*).[42] The sixteen ranks themselves started with units of 1 (*aḥād*) and ended with quadrillions, as follows: 1, 10, 10^2, 10^3, 10^4, 10^5, 10^6, 10^7, 10^8, 10^9, 10^{10}, 10^{11}, 10^{12}, 10^{13}, 10^{14}, 10^{15}.

Regarding fractions (*al-kusūr*), the Ikhwān noted that these started with the largest quantity and proceeded by way of division; hence, fractions point to the one as a part within an integer (w) that is greater than one ($w > 1$): e.g., 1 in 2 is a half (½; *niṣf*); 1 in 3 is a third (⅓; *thulth*); 1 in 4 is a quarter (¼; *rubʿ*). Moreover, fractions were divided into fractional portions, like, for instance, $^1/_{15}$, which is referred to as *thulth al-khums* (a third of a fifth; which can be notated as: ⅓ of ⅕ = $^1/_{15}$). The *kusūr* were also represented by way of Arabic letters (the *abjad* sequence; *ḥurūf al-jumal*) as follows:

41 The first use of '0' (*ṣifr*) as a placeholder within a positional base notation system is attributed by some to al-Khwārizmī, as noted in his *al-Ḥisāb al-hindī* (*Algoritmi de numero Indorum*). Throughout the tenth century, finger-reckoning arithmetic was the calculation system deployed by the business community, based on *ḥurūf al-jumal* (i.e., the *abjad* numeric system), with varying calligraphic forms.

42 *Rasāʾil*, vol. 1, pp. 51, 55–56.

ي	ط	ح	ز	و	ه	د	ج	ب
عشر	تسع	ثمن	سبع	سدس	خمس	ربع	ثلث	نصف
$\frac{1}{10}$	$\frac{1}{9}$	$\frac{1}{8}$	$\frac{1}{7}$	$\frac{1}{6}$	$\frac{1}{5}$	$\frac{1}{4}$	$\frac{1}{3}$	$\frac{1}{2}$

يه	يد	يج	يب	يا
ثلث الخمس	نصف السبع	جزء من	نصف السدس	جزء من
(a third of one-fifth)	(half of one-seventh)	(a part of thirteen)	(half of one-sixth)	(a part of eleven)
$\frac{1}{15}$	$\frac{1}{14}$	$\frac{1}{13}$	$\frac{1}{12}$	$\frac{1}{11}$

For instance, the *alif* represents 1; the *bāʾ*, $\frac{1}{2}$; the *jīm*, $\frac{1}{3}$; *dāl*, $\frac{1}{4}$; *hāʾ*, $\frac{1}{5}$; *wāw*, $\frac{1}{6}$; *zayn*, $\frac{1}{7}$; *ḥāʾ*, $\frac{1}{8}$; *ṭāʾ*, $\frac{1}{9}$; *yāʾ*, $\frac{1}{10}$ … *yab*, $\frac{1}{12}$ (half of one sixth; *niṣf al-suds*).[43]

Arithmetic IV

As regards the attributes of integers (*khawāṣṣ al-ʿadad al-ṣaḥīḥ*; properties of whole numbers), the Ikhwān followed Nicomachus in stating that 2 was the first of all numbers, as well as being the original numerical quantity initiating the series of even numbers (*azwāj*; *arithmos artios*), and the basis of their generation (hence, producing half of all numbers). After all, the number 2 was construed as being 'the first of the numbers', given that 1 is a unit used in counting numbers but is not itself considered a number (*ʿadad*) as such. As for 3, it is the first odd number (*fard*; *arithmos perissos*), and supposedly acts as the basis for the generation of one third of all numbers; while 4 is the first perfect square number (*awwal ʿadad majdhūr*), since $4 = 2 \times 2$ (in other words,

43 See *Rasāʾil*, vol. 1, pp. 51, 56. In a fraction, the monad is the numerator (top number in the fraction) with the integer ($w > 1$) as the denominator (lower numbers in the fraction); as for the factor, it is a number that divides into another number without resulting fractions (so, for example, the factors of 10 are: 1, 2, and 5; or those of 12 are: 1, 2, 3, 4, 6). Decimals are contrasted here with duodecimals, which are based on sets of 12, and have more divisors than the sets of 10.

4 being the first number that results from the multiplication of one number by itself). Moreover, 5 is the first 'automorphic' number (*ʿadad dāʾir*; it is also referred to as a 'round', 'circular', 'cyclical', or 'spherical' number); since whenever 5 is exponentiated, it reappears at the end of the resulting quantity, as in 25, 125, 625, 3,125, 18,625, 93,125, etc.; or such as 5, 25, 625, 390,625. Similarly, 6, which is the first perfect number (*tāmm*; *teleios*), is also an automorphic number, recurring in multiples of 6 as 36, 216, 1,296, 7,776, etc.; or such as 6, 36, 1,296.[44] The Ikhwān also held that 7 is the first 'complete number' (*ʿadad kāmil*), since, arguably, it carries the properties of all the numbers preceding it, and is equal to the first odd number added to the second even number (i.e., 3 + 4), and is also equal to the first even number added to the second odd number (i.e., 2 + 5), and is furthermore equal to the unit 1 added to the first perfect number (i.e., 1 + 6). The number 8, which is the first cubic number (*mukaʿʿab*; 'cube', or, 'perfect cube'), since it results from the multiplication of a *majdhūr* (i.e., a 'perfect square', like 4) with its *jadhr* (i.e., its 'square root', like 2), is also considered by the Ikhwān as being the first figurate solid number (*ʿadad mujassam*).[45] The number 8 is a product of the base 2 raised to an exponent 3 (i.e., $8 = 2^3$). This describes what came to be known later as a *logarithm*, in the sense that it refers to the power (exponent) to which a fixed number (the base) must be raised in order to equal a given number. The noble aspect of the number 8 (*faḍīlat al-thamāniya*) was given special attention by the Ikhwān in the music epistle in order to offer a critique of doctrines that emphasize the mystical mysteries of other numbers:[46] for example, the group they identified as the so-called Seveners (*al-musabbiʿa*), who advocated sevenfold structures in emphasizing the symbolism of the number 7, or the Trinitarian (*tathlīth*) Christian focus on the number 3, or the fourfold orders (*murabbaʿāt*; literally 'squares') of the naturalists in highlighting the significance of the number 4, or the pentagonal structures (*mukhammasāt*) of the Khuramiyya who pointed to the holiness of the number 5, or the fascination of the Indians with sets of the number 9 (*muttasiʿāt*), which is the first odd perfect square

44 *Rasāʾil*, vol. 1, p. 58.
45 *Rasāʾil*, vol. 1, p. 59.
46 *On Music*, pp. 132–136 (English translation), and pp. 103–111 (Arabic edition).

(*fard majdhūr*). As for the number 10, the Ikhwān noted that it is the first in *al-ʿasharāt* (the series 10–90). The number 11 is considered by the Ikhwān as being the first 'deaf' number (*ʿadad aṣamm*; *arhêtos/alogos*; or, in Latin, *surdus*) because it does not have a fractional part that carries a name of its own; they then listed other 'deaf' numbers: 13, 17, 19, 23, 29, 31, 37, 41, 43, 47, 53, 59, 61, 67, 71, 73, 79, 83, 89, 91, etc. However, they did not indicate how a 'deaf' number differs from a 'prime' number (i.e., one that has divisors of only the unit 1 and itself). They also omitted the numbers 3, 5, and 7 from the list of what they designate as 'deaf numbers'; perhaps because they preferred to group these according to other specific properties that they gave them (namely, that 3 is the first odd number, 5 the first automorphic number, and 7 the first complete number).

Moreover, the Ikhwān took the number 12 to be the first abundant/excessive number (*ʿadad zāʾid*; *hupertelês arithmos*), given that the sum of its divisors (*qua* factors; *ajzāʾuh*) results in a quantity larger than the number itself: i.e., $6 + 4 + 3 + 2 + 1 = 16 > 12$ (*kul ʿadad idhā jumiʿat ajzāʾuh wa kānat akthar mih summiya ʿadadan zāʾidan*). In contrast, with any number called deficient/defective (*ʿadad nāqiṣ*; *ellipês arithmos*), the sum of its divisors (*ajzāʾuh*) produces a quantity that is less than it (*al-ʿadad al-nāqiṣ idhā jumiʿāt ajzāʾuh kānat aqall minh*), such as 4, 8, or 10 (i.e., $2 + 1 = 3 < 4$; $4 + 2 + 1 = 7 < 8$; $5 + 2 + 1 = 8 < 10$).[47]

Arithmetic V

The Ikhwān divided even numbers into three categories: (i) powers of two (*zawj al-zawj*; *artiakis artios*); (ii) pairs of odd numbers (*zawj al-fard*; *artios perissos*); (iii) couples of pairs of odd numbers (*zawj zawj al-fard*; *perissartios*).[48] Following this classification, a *zawj al-zawj* takes the form $2n$, and it is always an even number that is divisible into two even integers, which can, in turn, be split into two equal halves that

47 *Rasāʾil*, vol. 1, p. 60. Refer also to the discussion of amicable numbers (*al-aʿdād al-mutaḥābbah*) as noted below, under the heading Arithmetic VII, and in *Rasāʾil*, vol. 1, pp. 65–66. Therein, 220 is an abundant number and 284 is a deficient number.

48 *Rasāʾil*, vol. 1, pp. 60–64.

are integers, e.g., $64 = 2^1 \times 32 = 2^2 \times 16 = 2^3 \times 8 = 2^4 \times 4 = 2^5 \times 2 = 2^6$. Moreover, in the series: 1, 2, 4, 8, 16, 32, 64, multiplying the numbers from the outer margins with each other results in equal quantities as we move towards the middle of the series: $1 \times 64 = 2 \times 32 = 4 \times 16$. As for a *zawj al-fard*, it takes the form $2(2p + 1)$, meaning that each can be expressed as a pair of odd numbers. For instance, each of these numbers: 6, 10, 14, 18, 22, 26, is divisible into two equal-half quantities only once, resulting in the respective values: 3, 5, 7, 9, 11, 13, which are, in their turn, indivisible into equal halves as integers (since the divisions of: 3, 5, 7, 9, 11, 13 result in fractions). As regards a *zawj zawj al-fard*, it takes the form: $2n(2p + 1)$, where $n \geq 2$ and $p \neq 0$; and it refers to numbers like: 12, 20, 24, 28, 36, 44, 52, 60, 68, whereby each is divisible into equal quantities twice, like: $20 = 2 \times 10 = 2\ (2 \times 5)$.

Furthermore, the Ikhwān considered the odd number (*ʿadad fard*) as either a prime number (*ʿadad awwal*; *prôtos arithmos*) or a composite (*ʿadad murakkab*; *sunthetos arithmos*). The former type refers to numbers like 3, 5, 7, 11, 13, 17, 19, 23, that are only divisible (without fractions) each by itself and by the unit 1; this definition was inspired by Book VII of Euclid's *Elements*, wherein it was noted that an in-composite (*asunthetos*) prime number (*prôtos arithmos*) is that which is measured by a unit alone (Def. 11),[49] and that numbers are *relatively prime* (*mutabāyina*) if they get measured by a unit alone as a common scale (Def. 12). The Ikhwān also referred to associated numbers (*ʿadad mushtarak*), like 9, 15, 21; whereby each is counted by the unit 1 and the common denominator 3. Ultimately, a 'prime number' is a 'natural number' (a positive integer $w > 1$) that is divisible only by itself (w) and by 1.

Regarding composite (*murakkab*) odd numbers, the Ikhwān highlighted that these have factors of the unit 1 as well as numbers, like 3, 5, 7, 9 in reference to the corresponding numbers: 9, 25, 49, 81 — whereby 9 is factored by 1 and by 3; 25 is factored by 1 and 5; 49 by 1 and 7; 81 by 1 and 9. This statement echoes Book VII of Euclid's *Elements* where it was noted that a *composite* is a number

49 This is also noted in Aristotle's *Posterior Analytics* (II.13, 96a36) and *Metaphysics* (1088a6).

that is measured by some other number (Def. 13),[50] and that numbers are *relatively composite* to one another if they are measured by some number as a common measure (Def. 14); such as 6 and 8 with 2 as common measure, or 6 and 9, with 3 as common measure.

Arithmetic VI

In their attempt to define the 'perfect number' (*al-ʿadad tāmm*), the Ikhwān were inspired by Proposition 36 in Book IX of Euclid's *Elements*, whereby it is noted that:

> If as many numbers as we please beginning from a unit be set out continuously in double proportion, until the sum of all becomes prime, and if the sum multiplied into the last make some number, the product will be perfect.

A perfect number is equal to the sum of its proper divisors, and can be expressed in the form: $2p^{-1} (2p - 1)$ where $p > 1$ is a prime number and $(2p - 1)$ is a prime.[51] For example, 6, 28, 496, 8,128 are perfect numbers (given in the same order as Nicomachus), with their corresponding primes for $p > 1$ being respectively: 2, 3, 5, 7; these result in the correlative primes for $2p - 1$, respectively: 3, 7, 31, 127.[52] The Ikhwān also noted that only one of the perfect numbers is to be found within a rank of numbers (*martaba*); for example, 6 in the series 1–9, 28 in the range 10–90, 496 in 100–900, and 8,128 in 1,000–9,000.[53]

As for the 'perfect square' (*al-ʿadad al-majdhūr*),[54] this was defined

50 See also Aristotle's *Metaphysics* 1020b3.

51 It was debated whether the quantity: $S = 2p - 1$ is always a prime number if p were a prime number. This was raised by Pierre de Fermat in a critical letter to Marin Mersenne. Refer to the *Oeuvres de Fermat*, ed. Paul Tannery and Charles Henry (Paris: Gauthier–Villars et fils, 1894), vol. 2, pp. 197–199. See also *The Thirteen Books of Euclid's Elements*, tr. Heath, vol. 2, p. 425.

52 See also Roshdi Rashed, 'Analyse combinatoire, analyse numérique, analyse diophantienne et théorie des nombres', in *Histoire des sciences arabes*, ed. Roshdi Rashed with Régis Morélon (Paris: Editions du Seuil, 1997), vol. 2, p. 87.

53 *Rasāʾil*, vol. 1, pp. 64–65.

54 A 'perfect square' is by mathematical convention also any 'square' (though this usage is rarer). However, in the case of offering an English rendering of *majdhūr* in this context, the use of 'perfect square' would allow for a distinction to be made with a *murabbaʿ* (*qua* 'square'). The algebraists used an additional word to designate a 'square', namely the technical term: *māl*. These variations reflect

by the Ikhwān as being a number having the quantity y^2, which, when added to twice its square root, i.e., $2y$ (*jadhrayh*), then added to the unit 1, results in a perfect square, n. The expression can be simplified to: $n = y^2 + 2y + 1$. They also noticed that if a perfect square, y^2, is diminished by twice its square root, $2y$, then added to 1, the remainder is a perfect square, m: $y^2 - 2y + 1 = (y - 1)^2 = m$.

Arithmetic VII

Following the Graeco–Arabic mathematical tradition, the Ikhwān considered two distinct integers as 'amicable' numbers (*mutaḥābān*) if the sum of the aliquot parts (i.e., proper divisors) of one is equal to the other number, and vice versa.[55] The mathematician Thābit ibn Qurra (whose work predates that of the Ikhwān, even allowing for chronologies situating their *Rasāʾil* at the earliest possible date) studied amicable numbers in the lemmas of his *Maqāla fī istikhrāj al-aʿdād al-mutaḥābba* (Treatise on the Determination of Amicable Numbers), which he composed in response to the legacies of Pythagoras and Euclid. The Ikhwān state that 220 and 284 constitute the smallest pair of amicable numbers.[56] They add that these are respectively abundant and deficient (*aḥadahumā zāʾid waʾl-ākhar nāqiṣ*), and that such pairs are rare (*nādirat al-wujūd*).[57] If we take Sn to be the sum of the aliquot parts of the dividend number n (in other words, the sum of the positive integer divisors of n, excluding n itself) then, $\Sigma_{220} = 1 + 2 + 4 + 5 + 10 + 11 + 20 + 22 + 44 + 55 + 110 = 284$ (and 220 is also an abundant number), and $\Sigma_{284} = 1 + 2 + 4 + 71 + 142 = 220$ (and 284 is also a deficient number).[58]

---

the distinctions between arithmetic, geometric and algebraic definitions, which in that period already had cross-disciplinary applications within mathematics. The use of these terms is more closely examined in the Arithmetic IX section below. See also *Rasāʾil*, vol. 1, p. 72.

55 *Rasāʾil*, vol. 1, pp. 65–66. As for equivalent numbers (*mutaʿādilān*), these consist of two distinct integers where the respective sums of the proper divisors of each is the same, e.g., 39 and 55, since $\Sigma 39\ (1 + 3 + 13) = \Sigma 55\ (1 + 5 + 11)$.

56 The discovery of this pair was ascribed by Iamblichus to Pythagoras.

57 *Rasāʾil*, vol. 1, pp. 65–66.

58 Rashed, 'Analyse combinatoire', p. 87. We would note that if $p < (2n - 1)$, where

Arithmetic VIII

The Ikhwān asserted that the multiplication of numbers or what may be referred to as 'duplication'/'duplation' (*taḍʿīf al-ʿadad*) increases ad infinitum (*bilā nihāya*). This process is evoked by them in terms of 'natural progression' (*al-naẓm ṭabīʿī*), as in the series: 1, 2, 3, 4, 5, 6, 7, 8, 9, 10, 11, 12, 13, 14, 15, ..., *n*.[59] It is, moreover, reflected in the progression of even numbers (*naẓm al-azwāj*), as in: 2, 4, 6, 8, 10, 12, 14, etc.; or through the progression of odd numbers (*naẓm al-afrād*), in the form: 1, 3, 5, 7, 9, 11, 13, 15, etc. Furthermore, this process involved a more direct form of 'multiplication' (*ḍarb*), that is of the type: $(a \times b)$, wherein the increase of the quantity a is produced through b iterations.

To elaborate on 'natural numbers' (positive integers), we note that these are arranged in an ascending series, where each term is greater than the one preceding it by a unit of 1, and the sum of such a set of numbers with an nth term would be:

$\Sigma = n/2 \times (n + 1) = 1 + 2 + 3 + \ldots + (n - 1) + n.$

Moreover, progressions can be arithmetical or geometrical;[60] whereby the former refers to a series of numbers each differing from its predecessor by a constant quantity of addition or subtraction d, while the latter designates a series of numbers, which stand with respect to each other in a constant ratio of multiplication or division r. An arithmetical progression takes the form:

$\Sigma = n/2 \times [2a + d\,(n - 1)],$

where Σ is the sum of the series, a its first term, and d the constant difference of addition or subtraction. The geometrical progression takes, in its turn, the following form:

$\Sigma = a + a(r) + a(r^2) + \ldots + a(r\,n^{-2}) + a(rn^{-1}) \ldots = a(rn - 1) / (r - 1),$

where Σ is the sum of the series, a its first term, and r the constant ratio of multiplication or division.

$(2n - 1)$ is prime, then $p\,(2n - 1)$ is an abundant number; we would also add that if $p > (2n - 1)$, where $(2n - 1)$ is prime, then $p\,(2n - 1)$ is a deficient number.

59 *Rasāʾil*, vol. 1, p. 66.

60 One should note that the Ikhwān did not explicitly mention any mathematical distinction between arithmetical and geometrical progressions.

Arithmetic IX

Concerning their reflections on square numbers (*al-aʿdād al-murabbaʿa; tetragônos arithmos*), the Ikhwān's definitions of the properties of 'multiplication', 'roots', and 'cubes' (*ḍarb*, *jadhr*, and *mukaʿʿabāt*, respectively) rested on Def. 16–19 of Book VII of Euclid's *Elements*. Therein, it was noted that two numbers multiplied by one another produce a plane number (*ʿadad musaṭṭaḥ*; *epipedos arithmos*), *c*. As in: $a \times b$, where $a \neq b$ (Def. 16).[61] Whereas, three numbers multiplied by each other result in a solid number (*ʿadad mujassam*; *stereou arithmou*): $x \times y \times z$ (Def. 17).[62] Moreover, a square number (*ʿadad murabbaʿ*; *arithmos tetragônos*) is taken to be the product of a number multiplied by itself: x^2 (Def. 18).[63] As for a cube number (*mukaʿʿab*), this is the result of a number multiplied by itself and then multiplied by itself again: x^3 (Def. 19).[64]

The Ikhwān also observed that when a square number (*al-ʿadad al-murabbaʿ*), be it a perfect square (*murabbaʿ majdhūr*) or not, is multiplied by any number whatsoever, the product is a solid number (*ʿadad mujassam*). If a perfect square (*murabbaʿ majdhūr*) is multiplied by a number less than its square root (*aqall min jadhrih*), the product is called a diminished solid number (*ʿadad mujassam libnī*); while, if a perfect square (*murabbaʿ majdhūr*) is multiplied by a number greater than its square root (*akthar min jadhrih*), its product is an augmented solid number (*ʿadad mujassam bīrī*). In addition, they indicated that if a rectangular number (*ʿadad murabbaʿ ghayr majdhūr*; imperfect

61 Definition 16 in Book VII of Euclid's *Elements* reads as follows: 'where two numbers having multiplied one another make some number, the number so produced is called plane [*epipedos*] and its sides are the numbers which have multiplied one another.'

62 Definition 17 of Book VII of Euclid's *Elements* reads as follows: 'where three numbers having multiplied one another make some number, the number so produced is solid [*stereos*] and its sides are the numbers which have multiplied one another'.

63 Definition 18 in Book VII of Euclid's *Elements* reads as follows: 'a square [*tetragônos*] number is equal multiplied by equal, or a number which is contained by two equal numbers'.

64 Definition 19 in Book VII of Euclid's *Elements* reads as follows: 'a cube [*kubos*] is equal multiplied by equal and again by equal, or a number which is contained by three equal numbers'.

square) was multiplied by its shorter side (*ḍilʿih al-aṣghar*) the product is a diminished solid (*mujassam libnī*), and if multiplied by its longer side (*ḍilʿih al-aṭwal*) the product is an augmented solid (*mujassam bīrī*); while if it is multiplied by a number smaller or greater than both of them, the product will be called a free solid (*mujassam lawḥī*).[65]

Arithmetic X

Although the Ikhwān affirmed that their 'number theory' was principally based on the mathematical traditions of Pythagoras and Nicomachus, it is also a truism that their arithmetic was Euclidean, as is explicitly manifest in their reflections on the first ten Propositions of Book II of the *Elements* (*al-Maqāla al-thāniya min Kitāb Uqlīdis fī al-uṣūl*), which may be summarized in the more modern algebraic form by the following quadratic equations (namely, equations containing a coefficient of x^2):[66]

II.1: $a(b + c + d) = ab + ac + ad$ (i.e., illustrating 'distributivity');[67]

II.2: $(a + b)(a + b) = (a + b)a + (a + b)b$;

II.3: $(a + b)b = ab + b^2$;

II.4: $(a + b)^2 = a^2 + b^2 + 2ab$;

II.5: $[(a + b)/2]^2 = ab + [(a - b)/2]^2$;

II.6: $(x + a)a + (x/2)^2 = [(x/2) + a]^2$;

II.7: $(a + b)^2 + b^2 = 2(a + b)b + a^2$;

II.8: $(2a + b)^2 = 4a(a + b) + b^2$;

II.9: $a^2 + b^2 = 2[(a + b)/2]^2 + 2[(a - b)/2]^2$;

II.10: $(a + x)^2 + x^2 = 2[(a/2 + x)^2 + (a/2)^2]$.

---

65 *Rasāʾil*, vol. 1, pp. 70–71.

66 For further particulars, see Euclid's *Elements*, Book II (1–10); see also Bernard R. Goldstein, 'A Treatise on Number Theory from a Tenth-Century Arabic Source', pp. 154–157; *Rasāʾil*, vol. 1, pp. 72–75.

67 To give a geometric flavour of how this proposition (II.1) reads in Euclid's *Elements*: 'If there are two straight lines, and one of them is cut into any number of segments whatever, then the rectangle contained by the two straight lines equals the sum of the rectangles contained by the uncut straight lines and each of the segments.' An algebraic illustration of this proposition can be also expressed as follows: $x(y_1 + y_2 + \ldots + y_n) = xy_1 + xy_2 + \ldots + xy_n$.

Arithmetic XI

Following the Greek traditions of the Pythagoreans and the Platonists, the Ikhwān pointed to the entanglement of arithmetic with the science of the soul *qua* psychology (*ʿilm al-nafs; de anima*).[68] As noted earlier, they held that the mathematical disciplines (*riyāḍiyyāt*) ought to be acquired as a foundation for studies in natural philosophy (i.e., physics; *ṭabīʿiyyāt*) as well as psychical sciences (*nafsāniyyāt*), which would ensure a learned approach to the treatment of theological topics (*ilāhīyyāt* and *nāmūsiyyāt*). Consequently, enquiring about the substance/essence of the soul (*jawhar al-nafs*) is a necessary stage on the path to studying theology.

Moreover, any inquiry into arithmetic points to the positing of a computing person, whereby numbers are grasped as being the quantity of the forms of things in the soul of the one who performs the counting or numbering (*kammiyyat ṣuwar al-ashyāʾ fī nafs al-ʿādd*).[69] The Ikhwān's observations in this regard were also reminiscent of what Aristotle noted regarding time in Book Delta (specifically, Chapters 10–14) of the *Physics*, whereby he defined *chronos* as a particular kind of *arithmos* (sections 219b1–2, 7–8), and, hence, as countable — for, no numbering or counting can take place unless a soul or intellect undertakes it (223a21–28). In addition, the generation of the 'soul' (*nafs*; *psukhê*; *anima*) was understood by Plato in the *Timaeus* (35b–36c) as being constituted out of numerical series (multiples of 2 and of 3). This opinion was explored in a quasi-arithmetical fashion by way of the following schema of portions, which was established in reference to seven numbers:

1

2 3

4 9

8 27

(Multiples of 2) (Multiples of 3)

68 The Pythagorean Philolaus (who was mentioned by Plato in the *Phaedo*, and by Aristotle in the *De Anima*) held that the soul was 'the harmony of the body'.

69 *Rasāʾil*, vol. 1, pp. 49, 75.

The epistle on arithmetic concludes with reflections on the nature of the soul,[70] and on its origin (*mabda'*) prior to becoming associated with the body, and in contemplating also its journey of return (*ma'ād*) to its source after departing in bodily death (*mawt*) from its embodied corporeality. In expressing their views on the soul in this context, and within an ecumenical appeal to the scriptures of Islam, Christianity, and Judaism, the Ikhwān also echo the antique Socratic injunction, as inspired by the Delphic inscription: 'Know thyself!' They also affirmed the associated maxim: 'If thou knowest thine essence [or 'knowest thyself'], then thou knowest God!'

Geometry[71]

The epistle on geometry (*handasa*; *jūmaṭrīyā*) is divided into twenty-seven chapters (*fuṣūl*), on topics such as: 'Types of Lines'; 'Epithets of Rectilinear Segments'; 'Names of Rectilinear Segments'; 'Types of Angles'; 'Types of Planar Angles'; 'Types of Curvilinear Segments'; 'Surfaces'; 'Rectilinear Polygons'; 'Visual Points Figurations'; 'Demonstration that the Triangle is the Origin of All Shapes'; 'Types of Surfaces'; 'Surveying'; 'The Human Need for Co-operation'; 'Mental [Intellective] Geometry'; 'On Imagining Distances'; 'The Reality of Distances in Mental Geometry'; 'Properties of Geometrical Figures'; 'Demonstrating these Properties'; 'The Benefits of this Art'. Here, we shall follow an approximately parallel subdivision of the epistle on geometry, suitable for the purposes of this study, to be noted with the sub-title 'Geometry' followed by a roman numeral.

Geometry I

The second epistle of the *Rasā'il Ikhwān al-Ṣafā'* on geometry (*Risāla fī al-jūmaṭrīyā*; transliterated from the Greek '*geômetria*') was inspired by the assimilation of Euclid's *Elements* within the

70 *Rasā'il*, vol. 1, pp. 76–77.

71 This part of the introduction presents updated geometrical analyses that complement the commentaries that were originally included in my chapter: 'Epistolary Prolegomena on Arithmetic and Geometry', in *The Ikhwān al-Ṣafā' and their 'Rasā'il'*.

mathematical circles of mediaeval Islamic civilization.[72] Euclid's objectives in the *Elements* (Books I–XIII)[73] were determined primarily with a view to systematizing the multiple relationships between figures into axiomatic forms,[74] which he regarded as the ideal way

---

72 According to the *Fihrist* (Index) of the bio-bibliographer, Ibn al-Nadīm, the first recorded translation of Euclid's *Stoikheia* into Arabic (*Kitāb Uqlīdis fī al-uṣūl*; also known as *Kitāb al-Arkān*) was prepared by al-Ḥajjāj ibn Yūsuf ibn Maṭar under the patronage of the caliph Ḥārūn al-Rashīd (r. 786–809). Another rendition was later commissioned to Ḥunayn ibn Isḥāq (known in Latin as Johannitius; ca. 808–873) by the caliph al-Ma'mūn. The first translation was known as *al-Hārūnī*, and the second was titled *al-Ma'mūnī*; a third version based on a revision of the latter was established by Thābit ibn Qurra, who also worked on Euclid's *Dedomena* (*al-Mu'ṭayāt* — a text connected mainly to Books I–VI of the *Stoikheia*). In addition, it is believed that the Latin translation of the *Elements* by Gerard of Cremona was based on Isḥāq's and Thābit's Arabic versions of the text (additional Arabic manuscripts were used by Adelard of Bath and Hermann of Carinthia), and these may have relied also on al-Ḥajjāj's rendition. In the Greek tradition, the most notable commentary on the *Elements* was by Pappus of Alexandria (fl. fourth century). For particulars, see Euclid, *The Thirteen Books of the Elements*, ed. Thomas L. Heath; Benno Artmann, *Euclid, the Creation of Mathematics* (New York: Springer–Verlag, 1999); Ian Mueller, *Philosophy of Mathematics and Deductive Structure in Euclid's Elements* (Cambridge, MA: MIT Press, 1981); Carmela Baffioni, 'Euclides in the *Rasā'il by Ikhwān al-Ṣafā*'', *Études Orientales* 5–6 (1990), pp. 58–68; Gregg de Young, 'The Arabic Textual Traditions of Euclid's *Elements*', *Historia Mathematica* 11 (1984), pp. 147–160; 'Isḥāq ibn Ḥunayn, Ḥunayn ibn Isḥāq, and the Third Arabic Translation of Euclid's *Elements*', *Historia Mathematica* 19 (1992), pp. 188–199; 'New Traces of the Lost al-Ḥajjāj Arabic Translations of Euclid's *Elements*', *Physis* 23 (1991), pp. 647–666.

73 The table of contents of Books I–XIII of Euclid's *Elements* may be summarized as follows: I 'Fundamentals of Plane Geometry: Theories of Triangles, Parallels, and Areas'; II — This book is titled inaccurately by some modern scholars as 'Geometric Algebra', but, rather, it interprets material attributable to algebra through geometry; III 'On Circles and Angles'; IV 'Constructions for Inscribed and Circumscribed Figures (Regular Polygons)'; V 'Theory of Abstract Ratios and Proportion'; VI 'Similar Figures and Geometric Proportions'; VII 'Fundamentals of Number Theory'; VIII 'Continued Proportions in Number Theory (Geometric Progressions)'; IX 'Number Theory Propositions'; X 'Classification of Incommensurables (Irrational Magnitudes)'; XI 'Solid Geometry'; XII 'Measurement of Figures'; XIII 'Regular Solids (On Constructing Regular Polyhedrons)'. The definitive edition of Euclid's *Elements* is preserved by the Teubner Classical Library in 8 volumes, with a supplement entitled *Euclides opera omnia*, ed. J. L. Heiberg and H. Menge (Leipzig: Teubner, 1883–1916).

74 The 'relationships' between geometrical figures are not established in this context in the so-called 'Euclidean space', given that such a concept and appellation was

to represent concrete physical bodies. The Euclidean axioms (*koinai doxai*; literally, 'common notions') and postulates, along with their supporting technical definitions, were presented as the foundational statements for the system of deductive logical inference, by virtue of which theorems were derived.

The Euclidean axioms may be designated as follows: Axiom 1: Things which are respectively equal to the same thing are necessarily also equal to each other (i.e., if $x = a$ and $y = a$, then $x = y$); Axiom 2: If equals are added to equals, the resulting wholes are equal (i.e., if $x = y$, then $x + z = y + z$); Axiom 3: If equals are subtracted from equals, the resulting remainders are equal (i.e., if $x = y$, then $x - z = y - z$); Axiom 4: Things that coincide with each other are equal (e.g., triangle *ABC* is equal to triangle *XYZ* if points *A*, *B*, *C* are superposed respectively on points *X*, *Y*, *Z*); Axiom 5: A whole is greater than any of its parts (i.e., if $x = y + z$, then $x > y$ and $x > z$). Moreover, the five postulates of Euclid's plane geometry can be summarized as follows: Postulate 1: A straight line segment can be drawn to join any two points; Postulate 2: Any straight line segment can be extended indefinitely in a straight line; Postulate 3: Given any straight line segment, a circle can be drawn that has the segment as its radius and one end-point as its centre; Postulate 4: All right angles are congruent; Postulate 5: If two lines are drawn which intersect a third in such a way that the sum of the inner angles on one side is less than two right angles, then the two lines inevitably must intersect each other on that side if extended far enough.

---

coined in relatively modern times, and is historically posterior to the geometry of figures as embodied in Euclid's *Stoikheia* (*Elements*). As noted in Euclid's *Data*, Proposition 55 (in correspondence with Proposition 25 of Book VI of his *Elements*), *khôrion* (which derives from *khôra*, approximating a designation of space in Greek) refers to 'area'— as in, 'if an area [*khôrion*] is given in form and in magnitude, its sides will also be given in magnitude'. The emergence of spatiality *qua* extension in geometry originated in the epistemic breakthrough associated with Ibn al-Haytham's (also known in Latinate versions of his name as Alhazen) mathematical definition of place in his *Qawl fī al-makān* (Discourse on Place); the Arabic critical edition and annotated French translation of this remarkable tract are both presented in Roshdi Rashed, *Les mathématiques infinitésimales du IXe au XIe siècle*, vol. 4 (London: al-Furqān Islamic Heritage Foundation, 2002), pp. 666–685.

The 'Fifth Postulate', also known as the 'parallel postulate' (*qaḍiyyat al-mutawāziyāt*), was problematic for geometers, given that it could not be proved as a theorem despite the attempts of many polymaths throughout history who believed that it could be decisively established on the basis of the other four postulates.[75] It is, however, worth noting in this context that the Ikhwān did not preoccupy themselves with demonstrations related to this question in their epistle on geometry; perhaps this was because of the complex and specialist mathematical content of this problem, which nevertheless had already been treated with great care and detail both by their predecessors and their contemporaries amongst the mathematicians.

In conjunction with the Fifth Postulate, Definition 23 in Book I of the *Elements* also addressed the question of 'parallelism' by stating that 'parallel lines [*parallêloi*; literally 'alongside one another'] are straight lines which, being in the same plane and being produced indefinitely [*eis apeiron*; literally 'without limit'] in both directions, *do not* meet one another in either direction' (italics added). Nevertheless, this Euclidean definition, which rests on the equidistance and directionality hypotheses of parallelism, would not have satisfied mathematicians and logicians, given that it proceeded by way of negation ('*do not* meet one another'), as well as lacking notional definiteness ('produced *indefinitely*').[76] To illustrate the content of Postulate 5 further, let us construct two straight lines, *AB* and *CD*

75 In the early 1700s, the Italian scholar Girolamo Saccheri (1667–1733) proposed negating the parallel postulate and then probing whether this results in a contradiction. However, the Russian mathematician Nicolai Lobachevsky, who developed the rudiments of 'non-Euclidean' geometry, was the one who accomplished the bold departure from Euclid's tradition; and his efforts were also anticipated by the Hungarian mathematician Janos Bolyai, and reinforced by the German mathematician Bernhard Riemann.

76 For a thorough geometrical evaluation of the problems posited by this Euclidean definition (I.23), principally in relation to the mathematical endeavours of Thābit ibn Qurra, and, in part, to the demonstrations of Ibn al-Haytham (*Sharḥ muṣādarāt Kitāb Uqlīdis fī al-uṣūl*), I refer the reader to Roshdi Rashed and Christian Houzel, 'Thābit ibn Qurra et la théorie des parallèles', *Arabic Sciences and Philosophy* 15 (2005), pp. 9–55. This article also includes Arabic critical editions and annotated French translations of two tracts composed by Thābit ibn Qurra which deal with Euclid's Fifth Postulate (*Fī burhan al-muṣādara al-mashhūra min Uqlīdis*).

(Figure 1, below), and let a third straight line, *EF*, intersect *AB* and *CD* respectively at points *G* and *H*; if the angle $\angle AGH$ added to the angle $\angle GHC$ results in a value < 180°, then lines *AB* and *CD* will meet if they are respectively extended far enough in the direction of *A* and *C*.

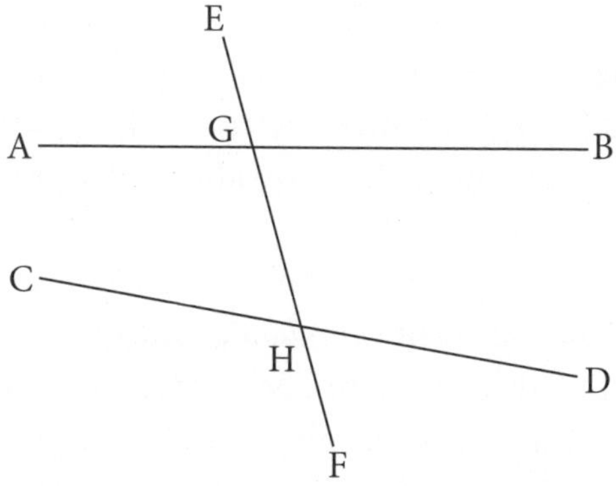

Figure 1

For parallel lines that are in one place, and produced indefinitely in both directions (directionality thesis), the distance between them is always the same (equidistance thesis); they neither converge nor diverge, and the perpendiculars to them are equal if drawn from one line to the other. Parallel lines remain equidistant and do not have a common point even at infinity. Attempts to prove this postulate were numerous. The most notable classical endeavours are those of Ptolemy and Proclus, and in the context of Islamic intellectual history, this was most prominently handled in the thirteenth century by the polymath Naṣīr al-Dīn Ṭūsī.[77]

77 In the seventeenth century, the effort to establish a proof was treated by John Wallis and Girolamo Saccheri (*Euclides ab omni naevo vindicatus*), to be followed in the eighteenth and nineteenth centuries by Johann Heintich Lambert and Adrien Marie Legendre. These paved the way for the development of the autonomous non-Euclidean geometry through the research of Schweikart, Taurinus, Gauss, Labochewsky, Bolyai, and Riemann.

Geometry II

Following Euclid, the Ikhwān held that the art of geometry begins with the point (*nuqṭa; sêmeion*), which is without parts and is one of the extremities of the line.[78] The line (*al-khaṭṭ; grammê*), which carries just one property, namely, length (*ṭūl*), was conceived as being a segment generated between two points, acting as the origin of the surface (*al-saṭḥ; epiphaneia*). The surface, in its turn, has two properties, length and width, and produces the geometric solid (*al-jism; stereos*) that carries the three properties of length, width, and depth (*dhū thalāthat abʿād; trikhê diastaton*).[79]

The Ikhwān classified lines as being straight (*mustaqīm*), curved/arched (*muqawwas*), or bent (*munḥanī*; namely, a line consisting of a straight and a curved segment). Moreover, they noted that lines could be equal (*mutasāwiyya*), parallel (*mutawāziyya*),[80] convergent (*mutalāqiyya*), in tangential contact with each other (*mutamāssa*), or intersecting (*mutaqāṭiʿa*). Lines could also have the additional properties of being perpendiculars, base-lines, chords, sides of polygons, hypotenuses, or diagonals.[81]

Progressing from lower-dimensional geometric entities, like lines, to higher-dimensional entities, like surfaces, the Ikhwān considered the variegated types of polygons that are generated from linear configurations, such as the triangle (trilateral figure), the square (the quadrangle, along with quadrilateral figures), the pentagon, hexagon, heptagon, octagon, nonagon, decagon, etc. It was also believed that, hypothetically, more shapes could be generated ad infinitum, in correspondence with the infinite progression of numbers in arithmetic.[82]

Furthermore, the Ikhwān classed surfaces (*suṭūḥ*) into three kinds: the planar (*musaṭṭaḥ*), the concave (*muqaʿʿar*), and the convex (*muqabbab*) — the latter two derived from conics. Moreover, surfaces

---

78 *Elements*, Book I, Def. 1, 3.

79 *Elements*, Book XI, Def. 1.

80 It is worth noting here that the Ikhwān did not elaborate on Euclid's Fifth Postulate, which preoccupied the polymaths of their age, as noted above in our section Geometry I.

81 *Rasāʾil*, vol. 1, pp. 81–84.

82 *Rasāʾil*, vol. 1, pp. 88–89.

enveloped solids as well as being the lower-dimensional entities from which the latter are generated; hence, a figure like the sphere is bound by one surface; the half-sphere is delimited by two surfaces; a quarter of a sphere, by three; a fiery solid (like a tetrahedron) by four triangular surfaces — or a solid (like a pyramid) delimited by four triangles and a square base; and the cube (*muka‘‘ab*) by six square-shaped surfaces.

The Ikhwān also made reference to four types of right-angled parallelepiped (*mutawāzī al-suṭūḥ*; i.e., a solid bound by parallelograms): (1) the cube (*muka‘‘ab*) was defined by them as a right-angled parallelepiped with three equal sides, hence having its length equal to its height, equal to its width; (2) the 'well-like shape' parallelepiped (*bi'rī*) has a height equal to its width, with each less than its length; (3) the 'brick-like shape' parallelepiped (*libnī*) has a height equal to its width, with each greater than its length; (4) the 'board-like shape' parallelepiped (*lawḥī*) has a height less than its width, which is less than its length.[83]

Geometry III

As stated in Plato's *Timaeus* (53a–55d), the four primary bodies — fire (plasma), air (gas), water (liquid), and earth (solid) — were all composed out of 'sub-atomic particles' having the shape of two right-angled triangles; namely, the scalene half-equilateral or the isosceles. Following this Platonic thesis, the trilateral figures were grasped by the Ikhwān as being the primary constituents of all figures (*al-muthallath aṣl kull al-ashkāl*);[84] it was similarly the case with the five Platonic solids: the tetrahedron (corresponding with fire), the octahedron (corresponding with air), the icosahedron (corresponding with water), the hexahedron (corresponding with earth), and the dodecahedron (corresponding with the visible cosmos).

In attempting to elucidate the properties of triangles, the Ikhwān based their definitions of angles (*zawāyā*) on Book I of Euclid's *Elements* (Def. 10–12). Thus, they noted that an angle could be planar (*musaṭṭaḥ*)

83 *Rasā'il*, vol. 1, pp. 94–95.

84 *Rasā'il*, vol. 1, p. 91.

or solid (*mujassam*), and that it could be right-angled (*zāwiya qā'ima*),[85] obtuse (*zāwiya munfarija*),[86] or acute (*zāwiya ḥādda*).[87] The Ikhwān then listed seven types of trilateral figures, as guided by Def. 20–21 of Book I of Euclid's *Elements*,[88] which may be summed up as follows: (1) The acute-angled equilateral triangle (*al-ḥādd al-zawāyā al-mutasāwī al-aḍlāʿ*); (2) The acute-angled isosceles triangle (*al-ḥādd al-zawāyā al-mutasāwī al-ḍilʿayn*); (3) The acute-angled scalene triangle (*al-ḥādd al-zawāyā al-mukhtalif al-aḍlāʿ*); (4) The right-angled isosceles triangle (*al-qā'im al-zāwiya al-mutasāwī al-ḍilʿayn*); (5) The right-angled scalene triangle (*al-qā'im al-zāwiya al-mukhtalif al-aḍlāʿ*); (6) The obtuse-angled isosceles triangle (*al-munfarij al-zāwiya al-mutasāwī al-aḍlāʿ*); (7) The obtuse-angled scalene triangle (*al-munfarij al-zāwiya al-mukhtalif al-aḍlāʿ*).[89]

On the different properties of trilateral figures, the Ikhwān evoked Pythagoras' theorem (without supporting explications),[90] in noting

85 Definition 10 of Book I of Euclid's *Elements* reads as follows: 'when a straight line set up on a straight line makes the adjacent angles equal to one another, each of the equal angles is right [*orthê*], and the straight line standing on the other is called a perpendicular [*kathetos*] to that on which it stands'.

86 Definition 11 of Book I of Euclid's *Elements* reads as follows: 'an obtuse angle is an angle greater than a right angle [*orthês*]'.

87 Definition 12 of Book I of Euclid's *Elements* reads as follows: 'an acute angle is an angle less than a right angle [*orthês*]'.

88 Definition 20 of Book I of Euclid's *Elements* reads as follows: 'of trilateral [*tripleuron*] figures, an equilateral [*isopleurôn*] triangle is that which has its three sides equal, an isosceles[*isoskelês*] triangle that which has two of its sides alone equal, and a scalene [*skalênos*] triangle that which has its three sides unequal.' As for Definition 21 of Book I of Euclid's *Elements*, it reads as follows: 'further of trilateral figures, a right-angled triangle is that which has a right angle, and obtuse-angled triangle that which has an obtuse angle, and an acute-angled triangle that which has its three angles acute'.

89 *Rasā'il*, vol. 1, pp. 104–106.

90 The Ikhwān allude to Pythagoras' theorem by evoking Proposition 47 in Book I of Euclid's *Elements* in their Epistle 5, on music. Euclid's *Elements* Book I, Prop. 47 reads as follows: 'In right-angled triangles, the square on the side subtending the right angle is equal to the squares on the sides containing the right angle'. As indicated earlier, this proposition was commonly associated with what is known as 'Pythagoras' Theorem', regarding the square on the hypotenuse. This theorem is known in populist expressions as 'the theorem of the nymph [or bride]' (*theorem tês numphês*) or 'the two-horned' (*dhū al-qarnayn*), on account of its geometric shape in the course of demonstration. See *The Thirteen Books of Euclid's Elements*, vol. 1, pp. 417–418.

that within any right-angled triangle, the square of the hypotenuse is equal to the sum of the squares of each of its two other sides. Hence, with a right-angled triangle *ABC*, where $\angle BAC = 90°$, we have: $BC^2 = AB^2 + AC^2$. The Ikhwān also indicated that for any triangle that has two acute angles, its third angle must be either 90° (right-angled; *zāwiya qā'ima*) or obtuse (*zāwiya munfarija*), because the sum total of all its angles is 180°.

Geometry IV

Regarding the definition of quadrilateral figures, the Ikhwān followed what was noted in Book I, Def. 22 of Euclid's *Elements*.[91] Hence, they pointed out that a square (*al-murabbaʿ*) is a quadrilateral figure that is equilateral and right-angled; that the oblong *qua* rectangle (*al-mustaṭīl*) is a quadrilateral figure that is right-angled but not equilateral; that the rhombus (*al-muʿayyan*) is a quadrilateral figure that is equilateral but not right-angled; that the rhomboid parallelogram (*al-shabīh bi'l-muʿayyan*) is a quadrilateral that is neither equilateral nor right-angled, and whose opposite sides and angles are equal; and that the trapezoid (*mukhtalif al-aḍlāʿ wa'l-zawāyā*) refers to the remainder of quadrilaterals.[92]

In reference to the properties of polygons, the Ikhwān noted that the side (*ḍilʿ*) of a hexagon (*musaddas*) is equal to half the diameter (i.e., equal to the radius) of the circle that contains it.[93] However, they did not provide a demonstration of this statement, which could be shown by way of a geometric construction (partly based on Proposition 1 of Book I of Euclid's *Elements*) as follows: let us construct two equal circles, C_1 and C_2 (with respective radii of R_1 and R_2, as shown in 'Figure 2' below), and let their circumferences pass through each others'

91 Definition 22 of Book I of Euclid's *Elements* reads as follows: 'Of quadrilateral [*tetrapleurôn*] figures, a square [*tetragônon*] is that which is both equilateral and right-angled [*orthogônion*]; an oblong that which is right-angled but not equilateral; a rhombus that which is equilateral but not right-angled; a rhomboid that which has its opposite sides and angles equal to one another but is neither equilateral nor right-angled. And call equilaterals other than these "trapezia"'.

92 *Rasā'il*, vol. 1, pp. 107–108.

93 *Rasā'il*, vol. 1, p. 108.

centres, O_1 and O_2, and also intersect at a point, Z; then, let each of the circles contain a hexagon, such as C_1 encircles hexagon H_1 with a corresponding side S_1, and C_2 encircles hexagon H_2 with a correlative side S_2. This construction means that:

$$R_1 = R_2 = O_1O_2 = O_1Z = O_2Z = S1 = S2,$$

where *S1* and *S2* are each subtended from centres O_1 and O_2 respectively at an angle of 60° to the line O_1O_2 (Figure 2, below).

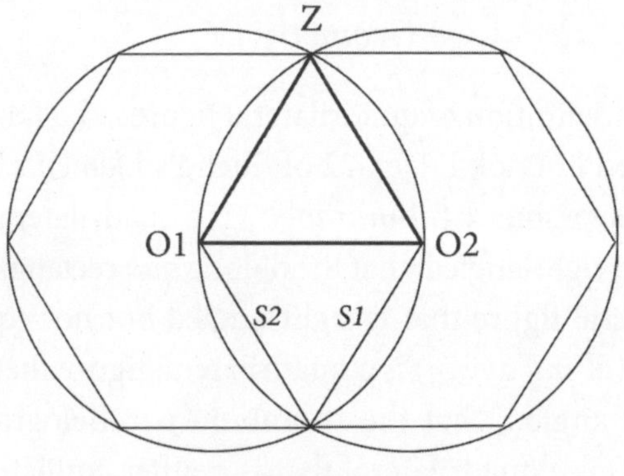

Figure 2

Geometry V

Based on the Pythagorean tradition related to figurate numbers (namely, numerical quantities represented by regular geometric configurations of equally spaced points),[94] the Ikhwān illustrated triangular numbers with ten marker dots (*al-muthallath min ʿasharat ajzāʾ*) as follows:

94 For instance, the series of triangular numbers (*ʿadad muthallath; arithmos trigônos*) begins with the following: 1, 3, 6, 10, 15, 21, 28, 36, 45, etc. In contrast, the so-called square numbers (*murabbaʿāt*; *arithmos tetragônos*) proceed in the form: 1, 4, 9, 16, 25, etc. Both of these series (triangular and square) increase following a pattern akin to the natural progression of numbers. Consequently, triangular numbers can be expressed with reference to the *n*th term as: $n(n + 1) / 2$. While square numbers, which are equal to the sum of two successive triangular numbers, are of the form n^2. In addition, we could note the following figurative numbers: Pentagonal numbers, like 1, 5, 12, 22, 35, 51, have the form: $n(3n - 1) / 2$. Hexagonal numbers, like 1, 6, 15, 28, 45, 66, take the form: $n(2n - 1)$. Heptagonal numbers, like 1, 7, 18, 34, 55, 81, are of the form: $n(5n - 3) / 2$. Octagonal numbers, like 1, 8, 21, 40, 65, 96, are of the form: $n(3n - 2)$. See J.

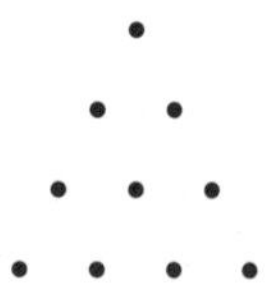

This diagram represented the so-called *tetraktys*,[95] which was a 'holy' symbol for the Pythagorean *mathêmatikoi* (i.e., Pythagoras' disciples). According to the Ikhwān, this figure described the procession of all numbers and corresponded with an onto-theological account of the manner in which God fashioned things (*al-ashyā'*) in the intellect (*al-'aql*), and how He manifested them in the soul (*al-nafs*) and in matter (*ṣawwarahā fī al-hayūlā*). Likewise, they held that this figure applied to the linguistic formation of letters (*al-ḥurūf*), and pointed to the theosophical significance of unity in arithmetic as a sign of God's Oneness (*waḥdāniyya*).[96] In ontological terms, and following a Neoplatonist interpretation, the tetraktys symbolized the embedded existential hierarchy of emanation from 'the One'. The unit/'number' 1 — a single dot in the diagram shown above — denoted unity, which correlated with 'the One' as the ultimate principle of reality, the number 2 — two dots in the diagram shown above — was associated with *Nous* (Intellect), the number 3 — three dots in the diagram shown above — was connected with *psuchê* (soul), and the number 4 — four dots in the diagram shown above — corresponded with the body and its constitution from the fourfold elements (earth, water, air, fire).

Geometry VI

The Ikhwān presented highlights from Euclid's *Elements*[97] regarding geometric entities associated with circles (*dawā'ir*), such as the centre

H. Conway and R. K. Guy, *The Book of Numbers* (New York: Springer–Verlag, 1996), pp. 30–62.

95 Marquet addresses the mystical significance of this figurate number in *Les 'Frères de la Pureté', pythagoriciens de l'Islam*, pp. 163–168.

96 *Rasā'il*, vol. 1, p. 54.

97 Definition 15 of Book I of Euclid's *Elements* reads as follows: 'a circle [*kuklos*] is a plane figure [*skhêma*] contained by one line such that all the straight lines falling upon it from one point among those lying within the figure are equal to one another'. Definition 16 of Book I of Euclid's *Elements* reads as follows: 'and

(*markaz*) and diameter (*quṭr*). They also elaborated on the chord (*watar*), the sagitta (*sahm*), the arc (*qaws*), the sine (*jayb mustawī*), and the versine (*jayb maʿkūs*), as well as making references to the positional relations between circles (concentric, tangential, intersecting).[98]

For instance, *al-watar* (the chord) is the straight line joining the two extremities/vertices of an arc (*al-khaṭṭ al-muqawwas*); as for *al-sahm* (the sagitta), it is the axis cutting both the chord and the arc into two equal segments (by being extended perpendicularly to the mid-point of the chord from the furthest point away from the chord situated on the arc). The sagitta is also the resultant of subtracting the apothem (i.e., the perpendicular dropped from the centre of a circle to the mid-point of a chord) from the radius of the circle, of which the arc is a segment. If the sagitta (*sahm*) relates to half the arc (*qaws*) it will be an inverted sine, namely a versine (*jayb maʿkūs*); while, if half the arc relates to half the chord, it will then be a 'straight' sine (*jayb mustawī*).[99]

To illustrate these geometric relationships, let us consider a unit circle with a centre O and then introduce a vertical sector chord AB that delimits the arc *AB* on that circle (Figure 3, below). Let us then project a perpendicular from O to chord AB such that it intersects with

the point is called the centre [*kentron*] of the circle'. Definition 17 of Book I of Euclid's *Elements* reads as follows: 'a diameter of the circle is any straight line drawn through the centre and terminated in both directions by the circumference [*periphereia*] of the circle, and such as a straight line also bisects the circle'. Definition 18 of Book I of Euclid's *Elements* reads as follows: 'a semicircle [*hemikuklion*] is the figure contained by the diameter and the circumference cut off by it. And the centre of the semicircle is the same as that of the circle'.

98 *Rasāʾil*, vol. 1, pp. 85–88. In addition, the Ikhwān named a variety of ellipsoids (derived as conic sections) that differ according to the proportions between their minor and major axes (carrying the following appellations: *bayḍī, hilālī, makhrūṭ ṣanawbarī, ihlīlajī, nīm khānjī, ṭablī, zaytūnī*); on this, see *Rasāʾil*, vol. 1, pp. 92–93. Besides the geometrical tradition of Euclid, and the editions of Theon of Alexandria, Pappus, or Diophantus, an influential line of transmission into Arabic geometry is also linked to the *Conica* of Apollonius of Perga (d. ca. 190); on this, see Apollonius of Perga, *Les coniques d'Apollonius de Perge*, tr. Paul ver Eecke (Bruges: Desclée De Brouwer, 1923). However, it is unclear whether the Ikhwān were aware of this tradition or of the *Sphaerica* of Theodosius of Bithynia (d. 90 BCE) and that of Menelaus of Alexandria (d. 130), or even whether they knew of the groundbreaking mathematical research of tenth-century polymaths like Ibn Sahl, al-Qūhī, and al-Sijzī.

99 *Rasāʾil*, vol. 1, p. 86.

its mid-point D, as well as intersecting with the arc *AB* at point C that is furthest away from the chord AB. Based on this figure, the radius of the circle is OC, the apothem is OD (namely, the perpendicular distance from the mid-point of the chord AB to the centre of the circle, O; this is also known as 'short radius' or 'in-radius'), and the sagitta (*sahm*) is DC (namely, the radius, OC, minus the apothem, OD). Let us also consider the angle $\angle AOC$ with a value (θ); then, the ordinary sine of ∠θ (i.e., vertical sine; *jayb mustawī*, or in Latin, *sinus rectus*) would be:

$\sin(\theta) = \frac{1}{2}$ chord AB = AD

While the cosine would be:

$\cos(\theta)$ = apothem OD.

And the *sinus versus* of ∠θ (also called flipped sine, versed sine or versine; *jayb ma'kūs*) would then be:

versin(θ) = sagitta DC.

Consequently, the sine is equal to half the chord; the cosine to the apothem; and the versine to the sagitta.[100]

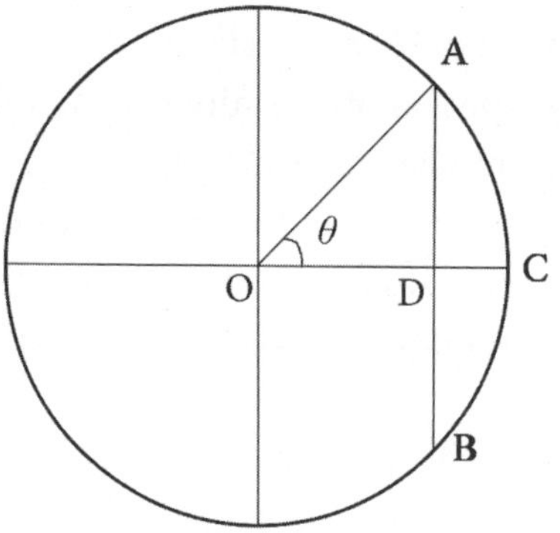

Figure 3

100 Just as a reminder concerning trigonometric ratios and reciprocals in reference to triangles: if we consider a right-angled triangle ABC, such as angle ACB = ∠C = 90°, we obtain the following trigonometric ratios: sin C = AC/AB, cos C = BC/AB, tan C = AC/BC. This also results in the reciprocal ratios: csc C = AB/AC, sec C = AB/BC, cot C = BC/AC (where sec C = 1/cos C, csc C = 1/sin C, and cot C = 1/tan C).

Geometry VII

The Ikhwān conceived geometry as a discipline that entailed knowing magnitudes (*maqādīr*) and dimensions (*ab'ād*; literally 'distances'), along with their properties. They divided this mathematical science (*'ilm*), which in their view underpinned all the applied sciences, into two branches: sensible geometry (*handasa ḥissiyya*) and intelligible geometry (*handasa 'aqliyya*). The former referred to sense perception, while the latter designated what is abstract (*mujarrad*) and conceivable only by the intellect. Furthermore, the Ikhwān held that the intelligible distances (*al-ab'ād al-'aqliyya*) were attributes of the (quantitative) sensory magnitudes (*al-maqādīr al-ḥissiyya*).[101] They also noted that the sensible (*ḥissiyya*) applications of this art were of expedient use to artisans, whilst the theoretical side assisted in understanding the influence that heavenly bodies and musical harmonics exercised on souls, as well as furnishing significant clues concerning the impact that the separate souls/intellects (*al-nufūs al-mufāriqa*) may have on corporeally embodied souls (*al-nufūs al-mutajassida*).[102]

Although the Ikhwān did occasionally conflate abstract mathematical propositions with empirical observations, or confuse them with the metaphorical and symbolic constructs of analogy, they nonetheless differentiated between the practical applications of geometry and its theoretical forms. This matter was articulated mainly in terms of the role assigned to the 'imagining' (postulating) of distances (*tawahhum al-ab'ād*).[103]

The phenomena of the sensible environment present geometers with various figures associated with material quantities. In this, sensible geometry, which is attested by means of physical objects, appears as an approximation of structurally abstract geometrical figures taken as invariant *idealities* (mathematical objects; *mathêmata*).[104]

101 *Rasā'il*, vol. 1, pp. 79–81.

102 *Rasā'il*, vol. 1, p. 113.

103 *Rasā'il*, vol. 1, pp. 101–103. It is worth noting that the process of abstraction in mathematics by means of imagining (postulating) proved to be a decisive factor in the geometrization of place by Ibn al-Haytham (Alhazen), in his *Qawl fī al-makān*, and its definition as *khalā' mutakhayyal* (imagined/postulated void).

104 Geometry opened up a growing stratification of mathematical primal idealities ('ideal' *qua* 'notional' objects), which were defined by their trans-temporal

In highlighting the practical merits of sensible geometry, the Ikhwān emphasized the connection between abstract geometry and its applications in the art of surveying (*al-misāḥa*), which is needed in building construction, quantity measuring, the apportioning of land, the distribution of irrigation channels, etc. However, they did not develop a method by which geometry could offer a theoretical grounding for the science of measurement, but, rather, they focused primarily on listing five scales or magnitudes (*maqādīr*) of linear measure (some being anthropometric), which (according to them) were deployed in Iraq by their contemporaries.[105]

In the attempt to realize an abstractive imagining (*tawahhum*) of pure mathematical figures, the Ikhwān made use of the application of (mechanical) motion to geometrical entities; perhaps this fact is an indication of the broad epistemic reception of the developments in mathematics that were achieved through the research of mathematicians in ninth- and tenth-century Islamic civilization.[106] In respect to this, the Ikhwān observed that a displaced point generates an imaginary line, and that a mobile line defines an imaginable surface, while a surface set in motion generates an imagined solid by its inclination in depth. And they showed that mathematical points, lines, surfaces, and solids are thus amenable to transformation via movement.

The imagined dimensions of height, length, and width are fully abstracted from physical entities in the intelligible type of geometry, unlike the application and manifestation of geometry in the sensible

---

constancy, their singularity, universality, exactness, and objectivity; as Hume observed: 'Although there never was a circle or a triangle in nature, the truths demonstrated by Euclid would forever retain their certainty and evidence.' See David Hume, *An Enquiry Concerning Human Understanding*, ed. L. A. Shelby-Bigge (Oxford: OUP, 1972), Section IV, Part 1, p. 25.

105 *Rasāʾil*, vol. 1, pp. 97–99. These scales included: (i) the finger (*al-iṣbaʿ*) = 6 capillaries (*shuʿayrāt*); (ii) the palm (*al-qabḍa*) = 4 fingers = 24 capillaries; (iii) the cubit (*al-dhirāʿ*) = 8 palms = 32 fingers = 192 capillaries; (iv) the doorway (*al-bāb*) = 6 cubits = 48 palms = 192 fingers = 1,152 capillaries; (v) the cord or rope (*al-ashl* or *al-ḥabl*; equivalent to the Greek *schoinion*) = 10 doorways = 60 cubits = 480 palms = 1,920 fingers = 11,520 capillaries. These measurements were also supplemented with surface-area magnitudes named *juraybāt*, *qufayzāt*, and *ʿushayrāt*.

106 Euclid avoided the application of motion in theoretical geometry, given that motion was primarily a phenomenon of physics.

realm. Imagination (*tawahhum*) is considered to be the source of the effective processes of abstraction that ground the study of theoretical geometry. This state of affairs describes an epistemic procession from focusing on knowable entities (*al-maʿlūmāt*) as objects of our sense faculties (*al-quwā al-ḥāssa*), to accounting for them by way of our imaginative faculty (*al-mutakhayyila*), and, through the imaginative faculty, grasping them via our cognitive capabilities (*al-mufakkira*), as well as preserving our knowledge of them in the memory (*al-ḥāfiẓa*).[107] Hence, the soul is not then in need of sense data or of the senses in order to derive the forms and be able to have the mystic promise of eventually taking a glimpse at the separate realities and archetypes.

The shift from sensory geometry to an intellective one constitutes a transition from practical arts (*ṣanāʾiʿ ʿamaliyya*) to theoretical ones (*ṣanāʾiʿ ʿilmiyya*),[108] as well as describing an epistemic shift from focusing on the sensible to placing emphasis on the intelligible (*min al-maḥsūsāt ilā al-maʿqūlāt*). Consequently, theoretical geometry studies the distances (*abʿād*) of solid figures as they are abstracted from corporeal entities and their sensible magnitudes (*maqādīr ḥissiyya*). It is thus a discipline that cultivates a capacity for abstraction and develops an ability to comprehend the intellective without recourse to the sensible.

The apprentice of mathematics is progressively trained to minimize reliance on the senses in conceiving entities that can be known (*al-maʿlūmāt*) or in attempting to contemplate the intelligible (*al-maʿqūlāt*). According to the Ikhwān, this pedagogical exercise opens up possibilities for acquiring 'higher forms of knowledge' that ultimately facilitate the study of theology, as well as training the intellect to hypothetically exit the realm of matter (*al-hayūlā*) in its reflections.[109]

107 *Rasāʾil*, vol. 1, pp. 103–104.

108 The Ikhwān addressed the practical arts in Epistle 7, and the theoretical sciences in Epistle 8; see respectively *Rasāʾil*, vol. 1, pp. 258–275 and 276–295.

109 The abstractive capabilities of the imagination were elaborated by Ibn al-Haytham (Alhazen), who established an ontological distinction between sensible beings (those apparent by means of *al-ḥiss*) and entities that exist through *al-takhayyul* ('the imagination'). Moreover, he held that the latter class exists by way of an ascertaining *taḥqīq* (in Latin, *certificatio*), while the perception of sensible entities is prone to error. He also added that the sensible does not exist in reality (*laysa huwa mawjūd ʿalā al-ḥaqīqa*), and that it is corruptible and unstable, while the imagined form is grasped according to its truth, and does

It is in this regard that arithmetic and geometry prime the apprentice for speculating on nature and the soul for studying theology. For this reason, the Ikhwān believed that arithmetic and geometry, along with their related measured proportions,[110] (theoretically) may all contribute to the constructive cultivation of the soul and the reformation of its ethics (*tahdhīb al-nafs wa iṣlāḥ al-akhlāq*).[111] However, the significance of mathematics for the Ikhwān was principally pedagogic rather than epistemic *per se*. By not exploring how mathematics assists in solving the problems of theoretical philosophy, they failed to give their epistemology a proper grounding in a mathematical-logical rationale, and, furthermore, by undermining the empirical impetus in research due to their mistrust of the senses, they mistook mystical analogism for rigorous explanation and sound acquisitions of scientific knowledge (*ʿilm*).

Having concluded their examination of geometry, the Ikhwān attempted to establish a synthesis between geometrical knowledge and arithmetic in the context of combining numbers in geometric figurate shapes (*numeri figurati*; figured numbers), in order to reveal arithmetical properties that cannot be derived without the aid of geometrical structuring.[112] These are embodied in various tables with square-units that are grouped in sets of 9, 16, 25, 36, 49, 64, and 81 and as multiples of 3, 4, 5, 6, 7, 8, and 9. They assign to these numerical-

not continuously change with the variation of whomsoever imagines it (*lā tataghayyar bi-taghayyur al-mutakhayyil lahā*). For an explanation of the role of imagination in mathematics in reference to Ibn al-Haytham's tract *Fī ḥall shukūk Kitāb Uqlīdis fī al-uṣūl*, see Rashed, *Les mathématiques infinitésimales du IXe au XIe siècle*, vol. 4, pp. 8–10.

110 The Ikhwān classed arithmetical proportion (*nisba ʿadadiyya*) as quantitative (*nisba bi'l-kammiyya*), while they took geometrical proportion (*nisba handasiyya*) to be qualitative (*nisba bi'l-kayfiyya*).

111 Ethics was also treated by the Ikhwān in Epistles 6 and 9; see respectively, *Rasāʾil*, vol. 1, pp. 242–257, 296–389. However, it is worthy of note that their reflections on the importance of mathematics in cultivating a moral inclination in their apprentices diverged from the classical precedents associated with Porphyry's *Sententiae ad Intelligibilia Ducentes* (*Aphormai pros ta noêta*), Plotinus' *Institutio Theologica* (*Stoikheiôsis theologikê*), or what centuries later emerged with Spinoza's *Ethica Ordine Geometrico Demonstrata*, whereby a philosophical tract is composed in a quasi-geometric fashion: stating definitions and axioms that carried 'metaphysical' content, followed by the deductions of 'demonstrative reasoning'.

112 *Rasāʾil*, vol. 1, pp. 109–112.

geometrical systems a talismanic (*ṭalasmāt*) significance that befits gnosis and the arts of magic (*siḥr*). The conclusion of the epistles on arithmetic and geometry points to the final Epistle 52 in their corpus. After highlighting the *telos* of arithmetic and geometry in their corpus, the Ikhwān return to the didactic sequence of the journey of knowledge by mentioning the epitome on astronomy that is to follow in their quadrivium course.

Technical Introduction

The Critical Edition and the Base Manuscript

The technical notes that I am including hereinafter have been partly informed by the analyses of the esteemed editors of the previously published volumes in this *Epistles of the Brethren of Purity* series, in connection with the assessment of the manuscripts and their use in the Arabic critical editions. This refers to the technical introductions in the volumes of *On Music* (pp. 1–15), *On Logic* (pp. 35–62), *The Case of the Animals* (pp. 57–59), and *On Magic, Part I* (pp. 69–84).[1]

As was the case with the earlier critical editions of this series, attempts to determine the filiation of the manuscripts and their genealogy, and also the effort to approximate a *stemma codicum*, all met with frustration. Owing to several reasons related to the circumstances under which the manuscripts were originally copied (which are discussed further below), so far no consensus has been reachable in terms of our editorial conclusions, including the outcomes of my own collation of the manuscripts. The difficulties are further accentuated given the differences in the contents and in the topics that have been treated within the various epistles that have been edited to date, along with the responses of the copyists of the manuscripts to the technical aspects of some of these tracts, which at times reveal a certain scribal incapacity to fully grasp the epistemic and conceptual bearings of the texts being copied, or of their internal nuances. The editorial

1 The bibliographical details of these publications are noted in the preliminary pages in this present volume, which describe the particulars of the *Epistles of the Brethren of Purity* series and its volumes published to date.

deductions reached hitherto concerning the connections between the manuscripts, and the lines of influence or the commonalities in terms of older sources, all show divergences in the conclusions of the various editors in relation to their respective epistles, with my own being no exception. Additional future inquiries need to be done in order to uncover the mysteries that surround the lineage of the manuscripts, if this aspect can indeed be realizable even by way of approximation.

The Arabic critical editions of Epistle 1: 'On Arithmetic' and Epistle 2: 'On Geometry' in this present volume have been principally based on the oldest manuscript, the Atif Efendi 1681 [ع],[2] which dates back to 1182 CE (AH 578), and contains the complete texts of the tracts on arithmetic and geometry without notable lacunae. It is unframed and set with ample proportionate margins. Its script is lettered with clarity and calligraphic elegance, whilst being fully vocalized, albeit also displaying minor grammatical errors in some of its vocalization accents. Moreover, this manuscript shows some instances of idiosyncrasy in style and content, which at times result in inconsistencies. This manuscript carries minimal notes in its margins, which incorporate selected variants in reference to certain terms that figure in the body of the text, in addition to introducing simple corrections. Each of the chapter headings in the manuscript has been clearly delineated with the term *faṣl* (chapter), and noted in red ink, whilst in most cases being left untitled. In this manuscript, Epistle 1: 'On Arithmetic' runs from folios 5a to 14b, and Epistle 2: 'On Geometry' from folios 15a to 23b, with each of these folios containing up to 25 lines when used to its full textual extent.

Even though this Atif Efendi manuscript is set as the base of the critical edition, it does not always offer the most reliable or convincing readings when compared with the other collated manuscripts, or when faced with the Dār Ṣādir printed Beirut edition of 1957. However, this manuscript is sound enough to be retained in most cases as the basis for establishing the body of the text, whilst subjecting it case-by-case to a thorough comparison with other readings from the remaining

2 The actual transliteration of the appellation of this manuscript in the Arabic script is *'Āṭif Afandī*.

manuscripts, and from the Beirut printed edition (the manuscript sources for which are unknown).

The Atif Efendi manuscript is introduced in the heading of the titles of both Epistles 1 and 2 with a subtitled eulogy that is noted in red ink, and that attributes the contents of the epistles to Sufi sources. This aspect emphasizes the suspicions that some scholars already hold regarding the possibility that this manuscript may have been affected by direct scribal interventions, which may have included intentional interpolations, or that the content of the manuscript has been altered through indirect channels of transcription from earlier sources, which may have distorted it by way of unchecked repetitions, scribal carelessness, or some forms of deliberate interference and tampering. Nonetheless, in the case of these two epistles, the deviations of the Atif Efendi from the other manuscripts remain ornamental and stylistic, without affecting the coherence and integrity of the mathematical contents, or their epistemic directives and conceptual correctness, with the exception of the calculation of the magnitudes of surveying as noted in Chapter 15 of Epistle 2, which resulted in some confusion.

The Full Set of Manuscripts and Sources

The critically edited Arabic text of Epistles 1 and 2 is based on the collation of several extant manuscript copies. The final Arabic text presented in this volume has been established through systematic comparisons between the various manuscripts that are noted below,[3] in addition to consulting the Dār Ṣādir printed Beirut edition of 1957:

(I) Bibliothèque nationale de France:
MS 2303 (1611 CE): [ر]
MS 2304 (1654 CE): [ز]
MS 6.647–6.648 (AH 695; Yazd): [د]

(II) Bodleian Library, Oxford:
MS Hunt 296 (not dated): [ج]

3 Besides these manuscripts, a much larger number of complete ones or fragments have been located in various libraries and collections, which are noted also in the 'Foreword' to this present volume; refer especially to pp. xx-xxi above.

MS Laud Or. 255 (not dated): [ح]
MS Laud Or. 260 (1560 CE): [خ]
MS Marsh 189 (not dated): [غ]

(III) El Escorial, Spain:
MS Casiri 895/Derenbourg 900 (1535–1536 CE): [س]
MS Casiri 923/Derenbourg 928 (1458 CE): [ش]

(IV) The Istanbul collections (mainly the Süleymaniye and associated libraries):
MS Atif Efendi 1681 (1182 CE): [ع]
MS Esad Efendi 3637 (ca. thirteenth century CE): [ن]
MS Esad Efendi 3638 (ca. 1287 CE): [أ]
MS Feyzullah 2130 (AH 704): [ف]
MS Feyzullah 2131 (AH 704): [ق]
MS Köprülü 870 (ca. fifteenth century CE): [ك]
MS Köprülü 871 (1417 CE): [ل]
MS Köprülü 981 (not dated): [و]
<MS Revan Kishk 1062 (AH 880)>[4]

4 The Istanbul Revan Kishk manuscript is not included in the original set that was given to the editors who are working on the *Epistles of the Brethren of Purity*. The reproduction of its contents was made available to us by courtesy of Professor Yahya Michot, who is a distinguished contributor to this book series and also a member of its Advisory Board. The transliteration of the appellation of this manuscript in its Arabic script is *Riwān Kishk*. It is dated as AH 880/1475 CE, and is part of the Istanbul collection. It contains Epistle 1 in folios 4 verso to 11 verso, and it covers Epistle 2 in folios 11 verso to 17 verso. Each of these folios can accommodate up to 29 writing lines if filled to its entire textual capacity. This manuscript comes perhaps the closest in content to the Tehran Mahdavī MS 7437 (AH 640), and yet this does not in itself indicate the existence of shared sources or links between them. The reproduction of this manuscript has been supplied on microfilm, and in a hard-copy printed format. The copies of the other Istanbul set manuscripts, along with the Tehran one, were all supplied on CD-ROM, which gave greater ease in reading due to the 'zooming' electronic functions. The remaining manuscripts were made available via microfilms and hard-copy printed formats, with some of these being reproduced in qualities that marked their presentation with some relative fuzziness in legibility, and resulted in some unease in terms of detecting all their words and vocalizations. All the copies of these manuscripts are preserved in the library special collections of the Institute of Ismaili Studies in London. However, it must be noted in this regard that under the copyright terms of the holders of the original manuscripts, all of these copies are only accessible for the purposes of the publication of the volumes

(V) Königliche Bibliothek zu Berlin:
MS 5038 (AH 600/1203 CE): [ب]

(VI) Tehran, The Mahdavī Collection:
MS 7437 (AH 640): [ط]

The Core Subset of Manuscripts

A core subset of manuscripts was used in establishing the editions of Epistles 1 and 2, whilst also consulting the other manuscripts from the broader set that has been listed above. The core subset of manuscripts has been determined on the account of antiquity, clarity of script, grammatical correctness, relative completeness, and reliability in reading. This selection has been also judged on the overall qualities of the manuscripts and their reproductions, also taking into account the respective condition of their epistles, which varies across the compendium between one manuscript and another. This core subset consisted of the following manuscripts,[5] which are described below in terms of some of their basic particulars.[6]

MS Atif Efendi 1681 (1182 CE): [ع]

As noted already above, this base manuscript offers the oldest complete text. It is written in a clear script with a neat hand. The text is unframed and it is well positioned with elegant measures on the folio, whilst having ample proportional margins, and a refined spacing in its lettering. The body of the text is fully vocalized, and the titles are noted in red ink, whilst the body of the text is set in black ink. The geometric figures in Epistle 2 are integrated within the body of the text

of the *Epistles of the Brethren of Purity* series, and that they are not available for consultation and for use in any other contexts outside the framework of this project.

5 The descriptions of the manuscripts of the core subset that are noted in this section follow the order in which they were assembled and collated in the editing process.

6 The description of the particulars of the other manuscripts figured in the technical notes of the editors of the earlier volumes in this series as noted above in the first footnote of this present section. There was no significant editorial need that necessitated their description again herein.

in small-scale drawings, mostly traced in red ink. The text carries minor notes and corrections in the margins by the hand of the same scribe.

Epistle 1: 'On Arithmetic' runs from folio 5a to 14b, and Epistle 2: 'On Geometry' from folio 15a to 23b, and each of its folios contains up to 25 lines of writing when used to its full textual capacity.

MS [Mahdavī] 7437 (AH 640): [ط]

The text is more or less complete, with minor lacunae in Chapter 21 of Epistle 2. It is written in a relatively clear script, albeit not with an elegant hand. The text is unframed and minimally vocalized. It carries many notes and corrections in the margins, which are placed in various directions (horizontally, vertically, and diagonally, or in-between the lines), whilst being lettered by a hand that is other than that of the scribe who copied the body of the text. The margins of the folios are very narrow, and the direction of the script leans towards the right side, set with an informal orthography. Unlike the other manuscripts, this one does not contain geometric figures in connection with Epistle 2.

This Tehran Mahdavī Collection manuscript incorporates in folios 27–28 a long passage in Epistle 2, which is not encountered in the other manuscripts, and is included in the 'Appendix to Epistle 2', in the original Arabic and in an English rendition (as noted in the table of contents of this present volume). This long passage is attributed within the manuscript to a certain 'scribe' (*nāsikh*) other than the copyist of the text. The content of the passage is linked to Book I of Euclid's *Elements*, and, at the end of Epistle 2, the manuscript carries the copyist's signature in its left margin, which can deciphered (despite relatively unclear handwriting) as follows: 'The weak servant, the one with the proper refined honours; son of Saʿd, of the people of Shahrābād; nicknamed, "The one famed for his staff".'

This manuscript contains Epistle 1 in folios 9 to 22, and Epistle 2 in folios 23 to 35. Moreover, each of its folios is divided into a left and a right segment, and it has in each of them up to 20 writing lines when used to its full textual extent.

MS Esad Efendi 3637 (ca. thirteenth century CE): [ن]

The text is incomplete, given that it reaches its conclusion at the end of Chapter 21 in Epistle 2, whereas all the other manuscripts in the core subset continue further till the end of Chapter 27 of that epistle. Of the remaining manuscripts, only the undated Oxford Bodleian MS Marsh 189 [غ] ends at the same location as [ن] in terms of its geometrical contents, however, it continues further by way of describing fourfold phenomena and the significance of the zodiac. It is perhaps worth noting here that, unlike all the other manuscripts, [غ] uses the term *kitāb* instead of *risāla*, and the word *dīwān* instead of *rasāil*.

Besides [ن] and [غ], the El Escorial MS Casiri 895/Derenbourg 900 [س] (1535–1536 CE), ends abruptly in Epistle 2; however, its ending is not demarcated in the exact same location as in [ن] or in [غ]. It also continues beyond the geometrical contents by way of expanding on honorific eulogies that celebrate the benefits of acquiring knowledge, and which seem to be unique to it.

The MS Esad Efendi 3637 [ن] also displays some affinities with the MS Esad Efendi 3638 [أ] (ca. 1287 CE), and yet these are perhaps not as pronounced as those found in the other epistles that have been edited and published so far in the series — for example, they conclude in different places at the end of Epistle 2.

The MS Esad Efendi 3637 [ن] is written with a legible, neat script. The text is unframed and lettered tightly, whilst being also partly vocalized. The titles in the text are written in red ink, and the margins of the folios are amply wide, with the script being gathered towards the folding middle. This manuscript carries some notes and corrections in the margins, and in-between the lines, by a hand that seems to be the same as that of the copyist of the body of the text. The contents display some similarities with at least one other manuscript, which may have been shared with the sources of [ع], and potentially with ones that may have been used in establishing the Dār Ṣādir printed Beirut edition.

This Esad Efendi 3637 manuscript has small-scale geometric figures in relation to Epistle 2, which are drawn in the body of the text.

This manuscript contains Epistle 1 in its folios 5 to 11, and it covers Epistle 2 in folios 12 to 18, with each of its folios divided into a left and a right segment, and each of these displaying up to 29 tight lines of writing when used to its full textual extent.

MS Feyzullah 2130 (AH 704): [ف]

The text is complete. It is written with a vocalized, clear, and neat script. The text is unframed and its titles are noted in red ink. The folios of the manuscript have narrow margins, and the geometric figures of Epistle 2 are mostly added in the margins in small scale rather than being part of the body of the text. This manuscript carries notes and corrections in the margins by a hand other than that of the scribe of the body of the text. It also incorporates additional passages in connection with Epistle 2 that correlate only with some of what was noted in [ل], but not to be found in other manuscripts. It may have consequently shared a common source with [ل], or its contents may have been available to the copyist of [ل]. In general, [ف] has been relatively unreliable in comparison with the rest of the manuscripts in the core subset (and this also applies to our assessment of [ل] as noted in its basic particulars below).

The [ف] manuscript contains Epistle 1 in folios 4 to 9, and it covers Epistle 2 in folios 9 to 14, and each of its folios is divided into a left and a right segment, with each of these containing up to 33 lines of writing when used to its full textual extent.

MS Köprülü 870 (ca. fifteenth century CE): [ك]

The text is complete. It is written with a clear script that is framed by golden lines. Each of its headings is set in an ornamental band that is drawn in blue ink with golden accents. The titles of the chapters are set also in navy blue ink. The text is vocalized, and it is placed on the folio in tight lettering, with cramped restrictions that occur particularly at the left side of the frame. The geometric figures of Epistle 2 are drawn to small scales in the margins and at times within the body of the text. In general, the contents of this manuscript may have shared a common older source with [ل], or they may have had a more direct influence on the copyist of [ل], though not through the same lineage as [ف].

This Köprülü 870 [ك] manuscript contains Epistle 1 in folios 45 to 50, and it covers Epistle 2 in folios 50 to 55, and each of its folios is divided into a left and a right segment, with each of these containing up to 31 tightly written lines when used to its full textual extent.

MS Köprülü 871 (1417 CE): [ل]

The text is complete. It is written in an elegant and very clear script that is also fully vocalized. The text itself was placed on the folio without a frame. Moreover, it carries numerous additions, notes, and corrections, which occur in every direction within relatively wide margins, all being drafted by a hand that is different from that of the copyist, and marked on most of the folios by the saying: 'reported by way of objection' (*bulligh muʿāraḍatan*).

This manuscript also carries the imprint of the circular seal of its owner: 'This is the property of the Vizier Abū al-ʿAbbās Aḥmad, son of the Vizier Abī ʿAbd'Allāh Muḥammad, who was known as Köprülü' (*'hādhih min waqf al-wazīr Abū'l-ʿAbbās Aḥmad ibn al-wazīr Abī ʿAbd'Allāh Muḥammad, ʿurifa bi-Kūbrīlā'*).

MS Köprülü 871 [ل] follows [ك], or else it was influenced by a common earlier source that was potentially shared with [ك]. In Epistle 2, [ل] contains a long passage that corresponds with what is noted in [ف] and is not encountered in any of the other manuscripts. In the folios of Epistle 2, it also contains the most clearly drafted geometric figures in comparison with the other manuscripts.

This [ل] manuscript contains Epistle 1 in folios 6 to 14, and it covers Epistle 2 in folios 20 to 29. Each of its folios is divided into a left and a right segment, with each of these containing up to 31 tightly written lines when used to its full textual extent. The gap that spans folios 15 to 19 between Epistle 1 and Epistle 2 contains a text that would correspond with Epistle 6, which has been classified in this context as being part of the quadrivium, and which is evidently misplaced in terms of the order of the epistles in this manuscript when compared with the other manuscript sources. In general, [ل] has been perhaps the least reliable in this core subset of manuscripts.

In addition to the core subset of manuscripts, we note the details of the following publication:

The Dār Ṣādir Beirut printed edition of 1957

Besides the manuscripts that were grouped in the core subset, we made use of the Dār Ṣādir Beirut printed edition of 1957, as edited therein by Buṭrus Bustānī. This published text includes Epistle 1 in pages 48 to 77, and Epistle 2 in pages 78 to 113.

Generally speaking, its technical mathematical contents seem to be more reliable than most of the manuscripts, even though it also required some corrections in arithmetic and geometry, as was the case with the other manuscripts, including the Atif Efendi and the remaining versions in the core subset.

This Dār Ṣādir edition has an additional and more complete section in Chapter 26 of Epistle 2 (running in its own pagination from page 110 to 112), which covers the tables with numbers beyond the sets of twenty-five cells. This edition also does not contain any of the digressions that figure in [ف] ,[ط], and [ل], and it goes beyond the lacunae and possibly abrupt endings of [غ],[ن], and [س]. It also offers a relatively more accurate calculation of the magnitudes that are used in the art of surveying as noted in Chapter 15 of Epistle 2 than what is found in the manuscripts.

Besides the Arabic sigla that were assigned to each of the manuscripts as noted in their complete list above, the Beirut Dār Ṣādir edition of 1957 has been given its own symbol [Ṣ] in order to note its variants in the apparatus of the Arabic critical edition.

Relations Between the Core Manuscripts

None of the core manuscripts takes the MS Atif Efendi 1681 [ع] as its archetypal source. The core manuscripts also differ amongst one another in terms of the readings they offer. This is especially the case in the context of Epistle 2, and more specifically in terms of the addition and omission of various short or long passages, and in presenting incompatible accounts of the calculation of the various magnitudes of surveying as set out in Chapter 15 of that same epistle.

It is difficult to speculate about the chronology of the redaction of the manuscripts, or to hypothesize about any *stemma codicum* that groups them together in a sequence. It remains the case that the determination of the genealogy of their filiations is a vexing matter in scholarship on the Brethren of Purity. Some of the corrections in the margins and in between the lines may signal certain elements of commonality between some of the manuscripts, as is also the case regarding the addition and omission of shared passages. Based on these criteria of analysis, it seems there are certain affinities between [ع] and [ن], and between [ن] and parts of [ف], and also between [ن] and certain passages in the Dār Ṣādir edition. There are also some similarities between [ف] and [ل], and consequently between some parts in [ن] and in [ل], and also separately between [ك] and [ل]. But [ن] has additions and omissions that demonstrate that its copyist may have had other sources at his disposal too. The manuscript [ط] seems to follow again another distinct tradition on its own, even though some aspects of its content show that it may have been based on an exemplar that has some indirect commonalities with the lineage of [ع].

Idiosyncrasies in the Manuscripts

The manuscripts have various linguistic idiosyncrasies and inconsistencies. They also reveal differences in their use of diverse honorific tiles, eulogies, exhortations, and in the endings of epistles. All the manuscripts display at least one of the following anomalies in their text: a reversal of the ordering of words; the usage of different cases, direct, indirect, or accusative; the use of feminine instead of masculine and vice versa; the use of singular instead of plural and vice versa; the elimination of the article; the deployment of varied expressions and terms to convey the same meaning; the use of different persons and tenses. To give some examples of such differences, we can note the following:[7]

- Variants occur in the opening lines, such as:

7 Refer also to the analyses of Professor Carmela Baffioni in the technical introduction of her volume on Epistles 10–14, *On Logic*, pp. 35–62.

أيّدك الله وإيّانا بروح منه

versus

أيّدنا الله وإيّاك بروح منه

- The same applies to the differences in sayings, like:

واعلَم أيها الأخ البارّ الرحيم

versus

واعلَم يا أخي

- Inconsistencies occur also in these forms:

هو versus هي

الأعداد الصحيح or العدد الصحيحة

- Or in the use of الياء instead of ئ , like in:

البسائط versus البسايط

- The order of terms is reversed in some of the manuscripts, for example:

العلوم والمعارف والرياضات

versus

الرياضات والمعارف والعلوم

- Differences occur in eulogies and in references to the Divine, like in the following appellations:

تعالى / جل جلاله / عزّ وجلّ / سبحان عظمته / جلّ إسمه / جل ثناؤه

- Some minor or major changes in the sequences of chapters, along with the introduction at times of chapters with titles, and at other times by simply writing 'chapter', in the form:

فصل

Such idiosyncrasies in the manuscripts may have been due to several factors, like the carelessness of scribes and their later copyists (or of those who attempted to correct them in the margins and in-between the lines) and the confusion that arises as a result of omitting the diacritical points or misusing them, in addition to errors in vocalization. There are also instances of errors in concordance between nouns and adjectives, and between pronouns and nouns, in addition to spelling mistakes, or errors in the transposition of letters, or in the inversion of the ordering of words, as well as blunders arising as a consequence of unchecked repetitions, or else as a result of intentional scribal expression of uniqueness in style and flair. In other cases, the variations may have resulted from tampering and interpolation, or even from the incapability of the scribes to fully comprehend and analyse the technical contents of selected epistles, which then necessitated the intervention of a specialist in the given field in question.

Method

The editing method consisted of a close and systematic collation of the extant manuscripts and the identification of the core subset that was described above. The MS Atif Efendi 1681 [ع] was then compared against each of the manuscripts that constituted the core of the edition, as well as the Dār Ṣādir printed version. The readings of Atif Efendi were retained within the body of the text whenever possible, in the case that this did not result in confusion of meaning or yield grammatical and idiomatic errors. At times, the terms used in Atif Efendi were placed in the critical apparatus instead of retaining them in the body of the text, in view of favouring more reliable and appropriate variants. Overall, the editing method was not mechanical but selective, with a focus on selection criteria that addressed issues such as: mathematical correctness, stylistic forms, proper grammar, nuances in the scribal idioms, and other clear and significant deviations in meaning — in addition to simply correcting mistakes in writing. Moreover, words or expressions that occurred only in the Atif Efendi manuscript were also recorded as additions in the critical apparatus. I have indicated the titles of the chapters as they were found in the Dār Ṣādir edition, which in most cases were confirmed in part by various manuscripts.

The texts of the Arabic critical edition and the annotated English translation show the beginning of each of the folios of the MS Atif Efendi 1681 [ع] in the body of the text. References to the pagination of the Ṣādir edition [Ṣ] are also noted in the body of the text. At times, certain sentences and passages are placed between <…> for the clarity of demarcating their content as a main segment that has been omitted or added from a given manuscript and that did not figure in other sources elsewhere.

I followed similar procedures with the annotated translation. In terms of approach, I tried as much as possible to retain the closest rendition that approximated the intended meaning, the style, and the idiom of the original Arabic text, without resulting in literal translations that may lead to confusion or readings that lacked eloquence. The fidelity to the original was sustained whilst trying to present the content in a sound and clear English rendering. A consistency in translating terms was retained throughout, with slight variations in the selection of synonyms being adopted at times, in order to have a more nuanced and non-repetitive or less monotonous English text.

Regarding Epistle 1, the annotated English translation that is presented in this volume was also compared against an earlier rendering of this tract that was at the time based on the Ṣādir edition of 1957. This earlier translation was established by Bernard R. Goldstein, and published in 1964, in Volume 10 of the journal *Centaurus* (pp. 135–159), under the title 'A Treatise on Number Theory from a Tenth-Century Arabic Source'.

Referencing

In terms of the referencing signs, the beginning of each of the folios of the Atif Efendi 1681 [ع] manuscript is indicated in the body of the texts of the annotated English translation and of the Arabic edition of both Epistle 1: 'On Arithmetic' and Epistle 2: 'On Geometry'. For example, at verso folio 5 we have (fol. 5b) noted in the English, and |ع ٥ ظ| in the Arabic (whereby ع indicates that it is the Atif Efendi manuscript, the number ٥ shows that it is folio 5, and ظ notes that it is verso); or in terms of recto folio 6, it takes the form: (fol. 6a) in the English, and |ع ٦ و| in the Arabic (herein و showing that it is recto).

The pagination of the Ṣādir edition is also noted in the body of the text within square brackets; for example, this takes the form [p. 48] in the English, and the reference [٤٨] in the Arabic.

Additional directives, specifically related to the Arabic, are indicated in the first footnotes of the critical apparatus of the edited Arabic texts of each of the two epistles in this present volume.

(fol. 5a) [p. 48]

Epistle 1
On Arithmetic

[Being the first epistle from the first section of the *Epistles of the Brethren of Purity*, on the Propaedeutical and Mathematical Sciences]*

(fol. 5b) *In the name of God, the Compassionate, the Merciful*

Chapter 1[1]

Know, O righteous and compassionate brother, may God aid us and you with a spirit from Him,[2] that it is the method of our noble Brethren, may they be aided by God, to study all the sciences of existent beings [*al-mawjūdāt*] that are in this world, be they substances, accidents, abstract entities, simple or compound, and to inquire into their principles and the quantity of their species, kinds, and properties, and into their arrangement and order, as they are at present, as well as into the process of their originating and growing out of one cause and of one origin, by one Creator, may His loftiness be exalted! They rely in demonstrating these [things] on numerical analogies and geometric proofs, similar to what the Pythagoreans used to do. Therefore, we had to situate this present epistle before all our other tracts, and to mention in it significant things that belong to the science of number

---

* The title in the Atif Efendi MS (fols. 5a–b) runs as follows: 'The First Epistle from the First Section on the Propaedeutical [and Mathematical] Sciences, on "Arithmetic", among the Fifty-One Epistles of the Brethren of Purity for the Purification of the Soul and the Clarification of Morals as a Part of a Sufi discourse'. (It is worth noting in this context that the insertion of a phrase that attributes the *Epistles* to a Sufi source was discussed elsewhere, in *The Ikhwān al-Ṣafā' and their 'Rasā'il'.*) This introductory line is then completed by the following eulogy: '*In the name of God, the Compassionate, the Merciful, O Lord, make our way easy. Praise to God! For nothing turns out fine except when initiated by praises to Him. Every being endowed with speech, or remaining silent, is God's worshipper. The minds of the people of knowledge are humbled when witnessing the Loftiness of His Majesty. Peace upon the best of His creations, the Prophet Muhammad, and on the members of his household, and the saved amongst his companions and entourage, and the righteous amidst the worshippers, and the people of his own* Umma.' The Ṣādir edition differs in the last four lines of this eulogy by evoking the 'hearts' instead of the 'minds' in the following manner: '*He is truly the One to whom the hearts reach out in the furthest of their limits in His Greatness and point at it. Praise to God, and peace upon His worshippers whom He has elected.*'

1 The numbering of chapter divisions throughout the text is an editorial interpolation, not to be found in the same form in the manuscripts.

2 This initiatory phrase occurs throughout the *Epistles of the Brethren of Purity* in its full form or in shorter versions, and at times in an inverted ordering of some of its terms. Its most common form is a rendering of the Arabic: '*i'lam ayyuhā'l-akh al-bār al-raḥīm, ayyadaka Allāhu wa-iyyānā bi-rūḥin minhu*'. The term *al-bār* can be also rendered as 'virtuous', 'pious', or 'reverent'; it has been noted herein as 'righteous', and it corresponds to the French word 'bienfaisant'.

and its properties, which is called 'arithmetic' [*arithmāṭīqī*]. We did so in terms of a preface or an introduction in order that the pathway might become easier for students to acquire the wisdom that is called 'philosophy', and that such acquisition may become simpler for novices in the study of the propaedeutic sciences [*al-'ulūm al-riyāḍiyya*].

So we say that the beginning of philosophy is the love of the sciences, the middle of it is the knowledge of the true nature of existent beings [*al-mawjūdāt*] by virtue of human ability, and its end is speech and action that is in accord with knowledge.

Chapter 2

The philosophical sciences are of [p. 49] four kinds: the first kind is the propaedeutic sciences, the second is the logical sciences, the third is the natural sciences, and the fourth is the theological sciences. The propaedeutic sciences are of four kinds: the first kind is arithmetic, the second is geometry, the third is astronomy, and the fourth is music. Music is the knowledge of the composition of sounds, and the principles of melodies are derived from it. Astronomy is the science of the stars by means of proofs that are recorded in the book [Ptolemy's] *Almagest*. Geometry is the science of mensuration by means of proofs that are recorded in the book of Euclid.[3] Arithmetic is the study of the properties of numbers and the qualities of the existents that conform to them, following what Pythagoras and Nicomachus recorded in this regard.[4] One begins the study of the philosophical sciences with the

3 Referred to here is the *Stoikheia* of Euclid of Alexandria, which was translated into Arabic as *Kitāb Uqlīdis fī al-uṣūl* (also known as *Kitāb al-Arkān*) by al-Ḥajjāj ibn Yūsuf ibn Maṭar under the patronage of the caliph Ḥārūn al-Rashīd (r. 786–809). Another rendition was later established by Ḥunayn ibn Isḥāq, with a third version revised by Thābit ibn Qurra. See Euclid, *The Thirteen Books of the Elements*; Ian Mueller, *Philosophy of Mathematics and Deductive Structure in Euclid's* Elements; Carmela Baffioni, 'Euclides in the *Rasā'il by Ikhwān al-Ṣafā*"; Gregg de Young, 'The Arabic Textual Traditions of Euclid's *Elements*'; Gregg de Young, 'Isḥāq ibn Ḥunayn, Ḥunayn ibn Isḥāq, and the Third Arabic Translation of Euclid's *Elements*'; Gregg de Young, 'New Traces of the Lost al-Ḥajjāj Arabic Translations of Euclid's *Elements*'.

4 This is contained in the *Arithmêtikê eisagôgê* (*Introductio Arithmetica*; *al-Madkhal ilā 'ilm al-'adad*) of Nicomachus of Gerasa. Nicomachus' *Introduction to Arithmetic* was translated into Arabic by Thābit ibn Qurra, under the title

propaedeutic sciences, and the first of these abstract sciences is the study of the properties of numbers because it is the easiest science to acquire, then mensuration, to be followed by astronomy, (musical) composition, the logical sciences, the natural sciences, and, finally, the theological sciences.

The first thing about which we will speak on the science of numbers is by way of an introduction (fol. 6a) or a preface.

Expressions [*al-alfāz*] point to certain meanings that are the objects of names, and the expressions are the names. The most general expression or name is the saying 'thing' [*al-shay'*], and a thing may be one or more than one. The 'one' [*al-wāḥid*] is said in two ways: in its proper usage [*bi'l-ḥaqīqa*; literally, 'in a real way'], and by way of metaphor [*al-majāz*]. In its proper usage it is a thing that cannot be partitioned or divided. And everything that cannot be divided is one when looked upon from the aspect by which it cannot be divided. If you wish, you may say: '"One" is that in which there is nothing else but itself, insofar as it is one'.

As for 'one' in metaphor, it is every aggregate [*jumla*; multitude] that is considered a unity. So, for example, ten is called a 'unit', and a hundred is called a 'unit', and a thousand is called a 'unit'. One is the epitome of oneness, like black is the epitome of blackness; and oneness is the attribute of being one, as blackness is the quality of being black.

Chapter 3

Plurality [*al-kathra*] is an aggregate of ones, and the first of the plural numbers is two, then three, then four, five, and so on, ad infinitum. Plurality is of two kinds, numbers and that which is numbered. The difference between them is that a number is the quantity of the form [*ṣūra*] of things in the mind of the counter [*fī nafs al-ʿādd*], [p. 50] while the numbered are things themselves. As for reckoning [*al-ḥisāb*],

---

Kitāb al-Madkhal ilā ʿilm al-ʿadad; refer to this text in *Arabische Übersetzung der Arithmêtikê Eisagôgê des Nikomachus von Gerasa*; see also *Nicomachi Geraseni Pythagorei Introductionis arithmeticae*; *Introduction to Arithmetic*; *Introduction arithmétique*.

it is the putting of numbers together [*al-jam*ʿ; i.e., addition] and their separation [*al-tafrīq*; subtraction].

Numbers are of two kinds, whole numbers [ʿ*adad ṣaḥīḥ*; i.e., positive integers] and fractions [*kusūr*]. One [*al-wāḥid*; i.e., the unit 1], which precedes two [i.e., the number 2], is the source and principle of all numbers, and from it all the numbers are generated, both whole and fractional, and they may be reduced to it again.[5]

The whole numbers are generated by augmentation [*tazāyud*], and the fractional numbers by division [*tajazzu*ʾ]. An example of this is what I stated regarding the generation of whole numbers:[6] when another one is adjoined [*uḍīfa*; added] to one, it is said that they are two; and when another one is adjoined to the two of them, the aggregate is called three; and when another one is adjoined to them, it is called four; and when one is adjoined to them, it is called five. Similarly, the whole numbers are generated by increasing them one by one, ad infinitum; and this is their form [*ṣūra*]:

1 2 3 4 5 6 7 8 9

Numbers are reduced to one as follows: if one is taken from ten, then nine remains; and if one is taken from nine, eight remains; and when one is taken from eight, seven remains; and similarly ones are taken away until only one remains. But nothing can be removed from one because (by definition) a part cannot be taken from it. So now it has been shown how the whole numbers are generated from one and how they are reduced to it.

Chapter 4

Fractional numbers are generated [*nushū*ʾ; also 'emerge'] from one as follows: the whole numbers are put in their natural order [*al-naẓm al-ṭabī*ʿ*ī*], namely, one, two, three, four, five, six, seven, eight, nine,

---

5 The literal meaning is 'they may be analysed back [*yanḥall*] into it'.

6 The copyist of the Atif Efendi manuscript varies the stylistic aspect of the text by utilizing sometimes the first person singular 'I', and at other times as the first person plural 'we'; this may indicate the possibility that the epistles were not compiled by a single scribe. In any case, the style of writing accentuates the ambiguities still surrounding the question concerning authorship.

ten; and one is pointed out from every aggregate. It is then clear how fractions are obtained from one. If one is pointed out from an aggregate of two, it is called a half; (fol. 6b) and if one is pointed out from an aggregate of three, it is called a third; and if one is pointed out from an aggregate of four, it is called a fourth; and if one is pointed out from an aggregate of five, it is called a fifth; and similarly if it is pointed out from an aggregate of six, it is called a sixth, and so forth for the seventh, eighth, ninth, and tenth. Moreover, if one is pointed out from an aggregate of eleven, it is called one part [*juz'*] in eleven; and from twelve, a half of a sixth; and from thirteen, one part in thirteen; and from [p. 51] fourteen, a half of a seventh; and from fifteen, a third of a fifth; and one may consider [*yaʿtabir*] the rest of the fractions according to this pattern. So now, based on this measure, it has been shown how the fractional numbers, as well as the whole numbers, are generated from one, and how one is the origin of both [sets] of them, and this is their form:

2	3	4	5	6	7	8	9
half	third	fourth	fifth	sixth	seventh	eighth	ninth

10	11	12	13	14	15
tenth	eleventh	twelfth	thirteenth	fourteenth	fifteenth

Know, O brother, that the whole numbers are arranged [*ruttib*] in four ranks: units [*āḥād*; ones], tens, hundreds, and thousands. The units are [the numbers] from one to nine, the tens from ten to ninety, the hundreds from one hundred to nine hundred, and the thousands from one thousand to nine thousand.

Twelve single words [in Arabic] suffice to encompass all the names of the numbers, namely, [the numbers] from one to ten are covered by ten words; and one word, 'hundred'; and one word, 'thousand'; so there are twelve single words in all. The other names [for numbers] are derived from these, or combined from them, or they are a repetition of them. For example, twenty is derived from ten, thirty from three, forty from four, fifty from five, sixty from six, and so on. Combinations such as 'two hundred', 'three hundred', 'four hundred', and 'five hundred', are constituted from the word 'hundred' joined with respective unit numbers. Similarly, 'two thousand', 'three thousand', and 'four thousand' are combinations of the word 'thousand' with the terms used for the unit numbers, the tens, and the hundreds, in such a way that one says: 'five thousand', 'seven thousand', 'twenty thousand', 'one hundred thousand', etc., and this is their form:

1 2 3 4 5 6 7 8 9[7]

10 20 30 40 50 60 70 80 90

[p. 52]

100 200 300 400 500 600 700 800 900

1,000 2,000 3,000 4,000 5,000 6,000 7,000 8,000 9,000

10,000 20,000 30,000 40,000 50,000 60,000 70,000 80,000 90,000

100,000 200,000 300,000 400,000 500,000 600,000 700,000 800,000 900,000

The units are 1 2 3 4 5 6 7 8 9; the tens are 10 20 30 40 50 60 70 80 90; the hundreds are 100 200 300 400 500 600 700 800 900; the thousands are 1,000 2,000 3,000 4,000 5,000 6,000 7,000 8,000 9,000.

Chapter 5

Know, O brother, may God aid you and us with a spirit from Him, that the existence of numbers in the four ranks of units, tens, hundreds, and thousands is not a matter that necessarily follows from the (fol. 7a)

7 The Ikhwān calligraphically represented numbers through the 'abjad' sequence (*ḥurūf al-jumal*), as was traditional before the introduction of the Arabic numerals conventionally used in their day. The alphabetic letters were set in simple calligraphic forms or in composites of two or more letters, as shown in Chapter 4 of the Arabic edition of this volume.

nature of numbers, as [is the case with] the existence of even and odd numbers, whole and fractional numbers, which follow one another. But, rather, it is a conventional matter that the philosophers have laid down by their own will. They did this so that numbers would conform to the arrangement of natural things, for most natural things were established by the Creator, exalted be His praise, in fourfold [*murabbaʿāt*] orders.[8] For example, there are: (a) four natures [*ṭabāʾiʿ*], as heat, cold, dampness, and dryness; (b) and four elements [*arkān*], as fire, air, water, and earth; (c) and four humours [*akhlāṭ*], as blood, phlegm, and the two biles, yellow bile and black bile; (d) and four seasons [*azmān*], of the spring, summer, autumn, and winter; (e) and four winds [*riyāḥ*], the east wind, the west wind [p. 53], the south wind, and the north wind;[9] (f) and four orientations [*jihāt*], upward, downward, right, and left; (g) and four *cardines* [*awtād*], the ascendant, descendant, *medium coeli* and *imum coeli*;[10] (h) and four [sub-lunar] beings [*mukawwanāt*] of metals, plants, animals, and humans. Consequently, one finds that most natural things come as fourfold.

Chapter 6

Know, O brother, that these natural things come in fours by the graceful intention of the Creator, exalted is His name, and by the exigencies of His wisdom. The categories of natural things thus conform to the spiritual things that are above natural beings and are incorporeal, for things that are beyond the natural entities are [also] set in four ranks. The first of them is the Creator, exalted be His praise; then, under Him, the Active Universal Intellect [*al-ʿaql al-kullī al-faʿʿāl*]; then, under it, the Universal Cosmic Soul [*al-nafs al-kulliyya al-falakiyya*]; and under it, Prime Matter [*al-hayūlā al-ūlā*]; and all these are not corporeal.

8 This follows the fourfold schema of the antique Greek tradition that is associated with the legacies of Hippocrates and Galen.

9 These are *al-ṣabā*, *al-dabūr*, *al-janūb*, and *al-shimāl*.

10 These four *cardines* (*awtād*) are noted in Atif Efendi as *al-zīr*, *al-muthnā*, *al-muthlath*, *al-yamm*; whilst in the Ṣādir edition they figure as *al-ṭāliʿ*, *al-ghārib*, *watad al-smāʾ*, *watad al-arḍ*. These designate the first, seventh, tenth, and fourth houses respectively.

Know, O brother, may God aid you and us with a spirit from Him, that the relation of the Creator, exalted be His praise, to all existent beings [*mawjūdāt*] is like the relationship of the unit one [to the other numbers]; and the relation of the [Universal] Intellect to existents is like the relation of the number two [to the other numbers]; and the relation of the [Universal] Soul to existents is like the relation of the number three [to the other numbers]; and the relation of Prime Matter to the universe is like the relation of the number four [to the other numbers].

Chapter 7

Know, O brother, may God aid you and us with a spirit from Him, that every number has its units [*āḥād*], its tens, its hundreds, and its thousands or what exceeds them, ad infinitum. The origin of all of them is the set [of numbers] from one to four: 1, 2, 3, and 4; the remaining numbers are composed and generated from them, since they are the source of all the numbers. You see this when you add one to four, the total is five; and when you add two to four, the total is six; and when you add three to four, the total is seven; and when you add one and three to four, the total is eight; and when you add two and three to four, the total is nine; and when you add one and two and three to four, the total is ten.

This is the rule for the [p. 54] rest of the numbers, the tens, the hundreds, and the thousands, and what exceeds them, ad infinitum. Similarly, the elements of writing are four, and the rest of the letters are compounded from them, since words are composed from the letters, as we shall explain later. Consider it [*iʿtabirha*], and you will find that what we say is true and correct. Let those who wish to know how God, exalted is His name, invented things in the Universal Intellect, and how He brought things into existence in the [Universal] Soul, and how He formed them in [Prime] Matter, consider then what we have discussed in this chapter.

Chapter 8

Know, O brother, may God aid you (fol. 7b) and us with a spirit from Him, that the first thing which the Creator, exalted is His name, invented and innovated from the light of His unity was a simple substance called 'Active Intellect', as He made two arise from one, by repetition. Then He made the Universal Soul arise from the light of the [Universal] Intellect, as He made three from the adding of one to two. Then He fashioned Prime Matter from the movement [*ḥaraka*] of the [Universal] Soul, as He generated four by adding one to three. He then made the rest of the created beings from Prime Matter, and He arranged it by the intermediary of the [Universal] Intellect and Soul, as He made the rest of the numbers from four by adjoining to it what precedes it, as we noted in the examples above.

Chapter 9

Know, O brother, may God aid you and us with a spirit from Him, that when you reflect upon what we have said regarding the composition and generation of numbers from the one that is prior to the number two, you will find it the clearest proof of the uniqueness of the Creator, exalted be His praise, and of the process of His creation and invention of things. For, although the existence of numbers and their composition can be conceived from the one, as we explained earlier, nothing essential to it is changed, and it is indivisible. Similarly, although God, exalted in His loftiness, is the one who created all things from the light of His unity, and generated them, and made their beginning, and made them grow, they have their existence, duration, completeness, and perfection through Him, and nothing essential to Him has changed in His unity prior to His act of creation, as we will explain in the epistle 'On the Intellective Principles'.[11] We already informed you that the relation [*nisba*] of the Creator, exalted is His name, to all existents is analogous to the relation of the [number] one to the numbers; as one is the origin

11 This being 'Epistle 33' (*Risāla fī al-mabādi' al-ʿaqliyya*), as the second epistle from the third section of the *Epistles of the Brethren of Purity*, on the psychical sciences (*al-ʿulūm al-nafsāniyya*).

of the numbers and that which generates them, their beginning and their end, similarly God, exalted and lofty as He is, is the cause of all things and their Creator, their beginning [p. 55] and their end. God cannot be compared or likened to anything in His creation, and the one, which accounts for the generation of all numbers, cannot be divided, nor can it be compared to any other number. God knows all things and their essences [*māhiyyāt*]. Hence God is exalted in grandeur and magnificence over what the unjust say.

Chapter 10

Know, O brother, that the ordered ranks of the numbers are four, according to most people, as we mentioned already, but the Pythagoreans place them in sixteen ranks, and this is their table:

Units [*āḥād*; ones]: 1
Tens [*ʿasharāt*]: 10
Hundreds [*miʾāt*]: 100
Thousands [*ulūf*]: 1,000
Ten thousands [*ʿasharāt ulūf*]: 10,000
Hundred thousands [*miʾāt ulūf*]: 100,000
Millions [*ulūf ulūf*]: 1,000,000
Ten millions [*ʿasharāt ulūf ulūf*]: 10,000,000
Hundred millions [*miʾāt ulūf ulūf*]: 100,000,000
Billions [*ulūf ulūf ulūf*]: 1,000,000,000
Ten billions [*ʿasharāt ulūf ulūf ulūf*]: 10,000,000,000
Hundred billions [*miʾāt ulūf ulūf ulūf*]: 100,000,000,000
Trillions [*ulūf ulūf ulūf ulūf*]: 1,000,000,000,000
Ten trillions [*ʿasharāt ulūf ulūf ulūf ulūf*]: 10,000,000,000,000
Hundred trillions [*miʾāt ulūf ulūf ulūf ulūf*]: 100,000,000,000,000
Quadrillions [*ulūf ulūf ulūf ulūf ulūf*]: 1,000,000,000,000,000

Know, O brother, may God aid you and us with a spirit from Him, that the fractions have many ranks [p. 56], and every whole number has one part, two parts, or several parts. For example, twelve has a half (fol. 8a), a third, a fourth, a sixth, and a twelfth, and similarly for twenty-eight, and so on for the other numbers. But, although the ranks and divisions of fractions are numerous, their scheme is in a descending order, and each rank is smaller than the previous one. All of them are included in ten words: one word, which is general and ambiguous, and nine words, which are special and fixed. Among the fixed nine words is one word without [etymological] derivation [from its whole number], and that is a 'half' [*al-niṣf*], and eight words which are derived: a third [from three], a fourth [from four], a fifth [from five], a sixth [from six], a seventh [from seven], an eighth [from eight], a ninth [from nine], and a tenth [from ten]. The word that is general and ambiguous is a 'part' [*juz'*] because one in eleven is called 'a part in eleven', and similarly for thirteen, seventeen, and so on for what is of their likeness amongst the numbers. The remaining expressions, which are used in designating the fractions, they get formed by combinations from the aforementioned ten words. For example, one in twelve is called 'a half of a sixth', and one in fifteen is called 'a fifth of a third', and one in twenty is called 'a half of a tenth'. The rest of the significations of the fractions are similarly understood as the adjoining of one of them to another.

Know, O brother, that these two kinds of numbers continue in quantity ad infinitum. Whole numbers start with the smallest quantity, [the number] two, and continue to increase without limit. As for the fractions, they begin with the largest [discrete] quantity, a half, and diminish without limit. So both of them begin at a fixed point, but have no ending point to limit them.

Chapter 11
On the Properties of Numbers

Know, O brother, that every number has one or more special properties, meaning the particular qualities of a described object that nothing else other than itself shares with it. The special property of the [unit] one is that it is the source of all the numbers, as we explained above, and that it generates all the numbers, odd and even. A special property of [the number] two is that it is absolutely [*muṭlaqan*] the first whole number, and that it generates half of the numbers [p. 57], the even numbers as opposed to the odd numbers. A special property of [the number] three is that it is the first odd number and it generates a third of the numbers, some odd, some even. A special property of [the number] four is that it is the first perfect square [*ʿadad majdhūr*]. A special property of [the number] five is that it is the first circular [*dāʾir*] number, also called 'spherical' [*kariyy*]. A special property of [the number] six is that it is the first perfect [*tāmm*] number. A special property of seven is that it is the first complete [*kāmil*] number. A special property of [the number] eight is that it is the first perfect cube [*mukaʿʿab*]. A special property of [the number] nine is that it is the first odd perfect square [*ʿadad fard majdhūr*], and it is the last of the rank of units [*āḥād*; 'ones']. A special property of [the number] ten is that it is the first number of the rank of tens [*ʿasharāt*]. A special property of [the number] eleven is that it is the first deaf [*aṣamm*] number. A special property of [the number] twelve is that it is the first abundant [*zāʾid*] number.

In general, a special property of any number is that it is half the sum of its adjacent numbers [*niṣf ḥāshiyatayh majmūʿatayn*], and if its adjacent numbers are added together, their sum will be (fol. 8b) twice the given number. For example, one of the numbers adjacent to five is four and the other is six, their sum is ten, and five is half of it; and a similar measure [*qiyās*] applies to the rest of the numbers — and this is their form:

1 2 3 4 5 6 7 8 9

[The unit] one has only one adjacent number, which is [the number] two; and one is half of it, and it is twice one.

We say that one is the source and generator of the numbers, because when *one* is removed from existence, all the numbers are removed with it, but when the numbers are removed from existence, one is not removed. We say that two is the first whole number because numbers are a plurality [*kathra*] of ones, and the first plurality is two. We say that three is the first odd number because two is the first number, and it is even, and three, being adjacent to it, is odd. We say that it generates a third of the numbers, some odd and some even, because it comes after [*tatakhaṭṭā*; literally, 'surpasses'] these two numbers, and it can be counted the third from them. This third number will sometimes be even and sometimes odd.[12]

We say that four is the first perfect square [*ʿadad majdhūr*]. This is the case because it is the product of [the number] two multiplied by itself. And any number that is multiplied by itself is a [square] root, and the product is a perfect square.

We say that [the number] five is the first circular [*dā'ir*; recurrent] number, because when it is multiplied by itself [p. 58] it returns to itself, and if that number is multiplied by itself, it again returns to its essence, and so on forever. So, for example, five times five is twenty-five, and if this number is multiplied by itself, the product is 625, and if this number is again multiplied by itself, the product is 390,625, and if this number is multiplied by itself, the product is another number ending in twenty-five. So, do you not see how five conserves itself and whatever derives from it eternally, whatever it may reach? And this is their form:

5 25 625 390,625

As for [the number] six, it is similar to five in this sense, but it is not self-abiding [*mulāzima*] as five is. Its prolongation is of this form:

6 36 1,296

Six times six is thirty-six; six returns to itself, and thirty appears. When thirty-six is multiplied by itself, the product is 1,296 — six appears again, but not thirty. So it is evident that six conserves itself but not what is derived from it. But five conserves itself and what derives from it eternally and forever.

It was said that a special property of the number six is that it is the first perfect [*tāmm*] number; namely, if the divisors of a number add up to itself, it is called a perfect number, and six is the first of them. Six has a half, which is three, and a third, which is two, and a sixth, which is one, and if these divisors are added up, the sum is equal to six. No

12 This whole paragraph between <…>, from 'We say that one is the source and generator of the numbers…' to '…This third number will sometimes be even and sometimes odd', is omitted from the Atif Efendi manuscript (fol. 8b), and a version of it figures on p. 57 of the Ṣādir edition. The variants across the manuscripts in this passage have been accounted for in the Arabic critical edition.

number before six has this property, but after it is the number twenty-eight, then 496, and, after that, 8,128, and all are perfect numbers. And this is their form:

6 28 496 8,128

It was said that seven is the first complete [*kāmil*] number because seven combines in itself the meanings of all the [preceding] numbers. For all the numbers are even or odd; two is the first even number, and four is the second. Odd numbers also have a first [p. 59] and a second. Three is the first odd number, and five is the second. If the first odd number is added (fol. 9a) to the second even number, or the first even number is added to the second odd number, the sum is seven. So, for example, if you add two, which is the first even number, to five, which is the second odd number, then the sum is seven. Similarly, if you add three, which is the first odd number, to four, which is the second even number, then the sum is seven. And if [the unit] one, which is the source of all numbers, is taken with six, which is a perfect number, then the sum is seven, which is a complete number. And this is their form:

1, 2, 3, 4, 5, 6, 7

This is a special property of seven, which no other number before it possesses, and it has other special properties, which we will discuss when we consider the fact that existent beings are constituted in accordance with the nature of numbers.

It was said that eight is the first perfect cube [*muka‘‘ab*] because of the following argument: If any number is multiplied by itself, it is called a [square] root and the product of two of them is a perfect square [*majdhūr*], as we explained before; however, if the perfect square is multiplied by its [square] root, then the product is called a perfect cube. Two is the first number, and if it is multiplied by itself the product is four, which is the first perfect square. And if this perfect square is multiplied by its [square] root, which is two, then the product is eight. Hence eight is the first perfect cube.

Eight is the first solid [*mujassam*] number, because there cannot be a solid body without interlocked surfaces, and there cannot be a surface without mutually adjoining lines, and there cannot be a line without ordered points, as we will explain in the epistle 'On Geometry' [i.e.,

Epistle 2]. The shortest line consists of two points and the narrowest surface consists of two lines, and the smallest solid body consists of two surfaces, so the conclusion from these premises [*muqaddimāt*] is that the smallest solid body has eight parts. One of them is a line, which has two parts. If a line is multiplied by itself, the resulting form is a surface that has four parts, and if the surface is multiplied by one of its lengths, it will have depth from it, so then there will be eight parts in all, two of length, two of width, and two of depth; and this their form:

2, 4, 8

It was said that nine is the first odd perfect square [*majdhūr*] because three times three is nine and neither seven nor five nor three consist of a perfect square.

[p. 60] Ten is clearly the first number of the rank of tens [*ʿasharāt*], as one is the first in the rank of units [*āḥād*]; and this is clear without the necessity of commentary [*sharḥ*]. It has another special property similar to a property of the unit one, namely, that it only has a single number adjacent to it, namely twenty, and ten is half of it, as we explained in the case of one, which is half of two.

It was said that eleven is the first deaf [*aṣamm*] number because it has no fractional part with a name of its own, but a part is called one part in eleven or two parts in eleven. And each number that has such an attribute is called 'deaf', like thirteen, seventeen, and the like of numbers; and this is their form:

11, 13, 17, 23, 29, 31, 37, 41, 43, 47, 53, 59, 61, 67, 71, 73, 79, 83, 89, 91

It was said that twelve is the first (fol. 9b) abundant [*zāʾid*] number, because if the sum of all the divisors of a number are added up and are greater than it, it is called an 'abundant number', and twelve is the first such number. It has a half which is six, and a third, which is four, and a fourth, which is three, and a sixth, which is two, and a twelfth, which is one. If these divisors are added up, the total is sixteen, which exceeds twelve by four.

So, in general, every whole number has a special property peculiar to itself, but we will omit their mention to avoid superfluous lengthiness in our accounts.

Chapter 12

Know, O brother, may God aid you and us with a spirit from Him, that numbers are divided into two divisions, whole numbers and fractions.[13] Whole numbers are divided into two subdivisions, even numbers and odd numbers. An even number is any number that can be divided into two halves which are whole numbers, while an odd number is any number that [p. 61] exceeds an even number by one or that falls short of an even number by one. The generation of even numbers begins from the number two, continuing by repetition without end as is seen in this form:

2, 4, 6, 8, 10, 12, 14, 16, 18, 20

The generation of odd numbers begins from the unit one, to which two is adjoined continually, ad infinitum; and their form is as follows:[14]

3, 5, 7, 9, 11, 13, 15, 17, 19

Even numbers are divided into three kinds: powers of two [*zawj al-zawj*], pairs of odd numbers [*zawj al-fard*], and pairs of pairs of odd numbers [*zawj zawj al-fard*]. The powers of two are all numbers that may be divided into two equal halves of whole numbers, which in turn may be so divided, continuing until the process of dividing reaches one. For example, sixty-four: half of it is thirty-two, and half of that is sixteen, and half of that is eight, and half of that is four, and half of that is two, and half of that is one. And the generation of these numbers begins with two, which is multiplied by two, and the product is multiplied by two, and so forth, continuing ad infinitum [*bālighan mā balagh*].

Whoever wishes to understand this thoroughly ought to double the squares of the chess-board, because he will always remain within the powers of two, and these numbers have other special properties that Nicomachus explained at length in his book [*Arithmetic*], and we will paraphrase a part of it, as when he says: Let these numbers be set in

13 As already explained above.

14 The Brethren's account is ambivalent here, given that they construe one as a unit and not as a number per se, and that all numbers are generated from it; hence the odd numbers start ultimately from the number three, just as the even numbers begin from the number two.

their natural order, which is one, two, four, eight, sixteen, thirty-two, sixty-four, and so on, ad infinitum. One of their special properties is that if one multiplies the two extreme terms, the product will be equal to the mean term [*al-wāsiṭa*] multiplied by itself, if there is only one mean term; or if there are two mean terms [p. 62], the product of the extreme terms is equal to the product of the two mean terms. For example, let sixty-four be the last term of the series and one the first. This series has one mean term, which is eight, so I say: if one is multiplied by sixty-four, or two times thirty-two, or four times sixteen, the product is equal to eight times eight, and this their form:

1, 2, 4, 8, 16, 32, 64

(fol. 10a) If one adds to it another rank so that there will be two mean terms, then I say: if one multiplies the two extreme terms, it will be equal to the product of the two mean terms. For example, if 128 is multiplied by one, or sixty-four by two, or thirty-two by four, the product will be equal to the product of sixteen times eight. And this is their form:

1, 2, 4, 8, 16, 32, 64, 128

These numbers have another special property. If one adds the numbers of the series starting with one and ending arbitrarily, the sum will be one less than the next number of the series. For example, take one, two, and four — the sum is smaller than eight by one. And if eight is added to it, the sum is smaller than sixteen by one. And if sixteen is added to it, the sum is smaller than thirty-two by one. Similarly, you discover [with all] the ranks of these numbers, however great, and this is their table:

1, 2, 4, 8, 16, 32, 64, 128, 256

Each of these numbers is equal to half the number that is immediately higher than it.

The pairs of odd numbers are all numbers that can be divided in half once, but do not lead to one by division, such as six, ten, fourteen, eighteen, twenty-two [p. 63], or twenty-six. All of these examples are numbers that can be divided once, but do not lead to one. These numbers are obtained by multiplying every odd number by two, and this is their table:[15]

6, 10, 14, 18, 22, 26, 30, 34, 38, 42, 46

Pairs of pairs of odd numbers include all numbers that may be divided in half more than once, but do not lead to one by division, such as twelve, twenty, twenty-four, twenty-eight, and similar numbers, and this is their table:

12, 20, 24, 28, 36, 44, 52, 60, 68

These numbers are generated by multiplying a pair of odd numbers by two, once or many times. And these numbers have other special properties whose mention we will omit, fearing [them] to be redundant.

Odd numbers are divided into two subdivisions, prime [*awwal*] numbers and composite [*murakkab*] numbers. Composite numbers are of two kinds, those that are associated [*mushtarak*] with one another, and those that are relatively prime [*mutabāyin*]. The distinction is this: the prime numbers include all numbers, together with one, which are not generated by another number, such as three, five, seven, eleven, thirteen, seventeen, nineteen, twenty-three, and the like of numbers. The special property of these numbers is that they have no fractional part other than the one named after them.[16] So, three has no fractional part except a third; five has no fractional part except a fifth, and similarly seven has no fractional part except a seventh; and so on for eleven, thirteen, and seventeen. In general, all the deaf numbers (fol. 10b) cannot be generated except by one, and the name of their fractional parts is derived from them.

The composite numbers include all the numbers that are generated by another number, excluding one, such as nine, twenty-five, forty-nine, eighty-one, etc. And this is their table [p. 64]:

9, 25, 49, 81, 121

15 The numbers are not shown on p. 63 of the Ṣādir edition; they are only noted in part through the use of words instead of numerals, and then represented by Arabic alphabet letters. The same applies to numbers that are mentioned on p. 64 of the Ṣādir edition.

16 A prime number is that which is measured by a unit alone; Euclid, *Elements* VII, Definition 11; Nicomachus, *Arithmetic* I.II.2; Aristotle, *Posterior Analytics* II.13.96a36.

Two numbers are *associated* [*mushtaraka*] with one another if both of them are generated by the same number, excluding one. For example, nine, fifteen, and twenty-one are associated because three generates all of them [i.e., they are all countable by sets of threes]. Similarly, fifteen, twenty-five, and thirty all are generated by five. These numbers, and those like them, are said to be associated in connection with the number that generates them.

Two numbers are *relatively prime* [*mutabāyina*] if two different numbers other than one generate them, but what generates one of them does not generate the other, such as nine and twenty-five. Three generates nine but does not generate twenty-five, whereas five generates twenty-five, but does not generate nine. So these numbers and others like them are called 'relatively prime'.

Chapter 13

Know, O brother, may God aid you and us with a spirit from Him, that every odd number has the special property that if it is divided into two parts, in any way, one of the parts will be even and the other will be odd; and every even number has the special property that if it is divided in any way, the parts will be either both odd or both even, and this is their table:

	Odd			Even	
10	11	1	9	10	1
9	11	2	8	10	2
8	11	3	7	10	3
7	11	4	6	10	4
6	11	5	5	10	5

Chapter 14
On Perfect, Deficient, and Abundant Numbers[17]

Know, O brother, may God aid you and us with a spirit from Him, that numbers may be divided [p. 65] into three kinds, by considering them from another point of view, such as perfect [*tāmm*], abundant [*zā'id*], and deficient [*nāqiṣ*].

A *perfect* number is any number whose divisors add up to itself,[18] such as six, twenty-eight, 496, and 8,128. If the divisors of each of these numbers are added up, the sum will be equal to itself. There is only one perfect number in each rank of the numbers: six in the units, twenty-eight in the tens, 496 in the hundreds, and 8,128 in the thousands. This is their form:

6 28 496 8,128

An *abundant* number is any number whose divisors add up to more than itself, such as twelve, twenty, sixty, and the like of numbers. Half of twelve is six, and a third of it is four, and a fourth of it is three, and a sixth of it is two, and a twelfth of it is one; all these divisors add up to sixteen, which is more than twelve.

A *deficient* number (fol. 11a) is any number whose divisors add up to less than itself, such as four, eight, ten, and the like of numbers. Half of eight is four, and a fourth of it is two, and an eighth of it is one; the sum of them equals seven, which is less than eight. The rest of the deficient numbers are subject to the same rule.

Chapter 15
On Amicable Numbers

Know, O brother, may God aid you and us with a spirit from Him, that from another point of view the numbers may be divided into two subdivisions; one of them is called 'amicable' [*mutaḥābba*] numbers. This appellation designates any two numbers, one abundant, and one deficient, such that the sum of the divisors of the abundant number

17 This chapter heading was placed in error within p. 64 of the Ṣādir edition at a paragraph higher than where it should be.

18 A perfect number is that which is equal to its own parts; Euclid, *Elements* VII, Definition 22; and *Elements* IX, Proposition 36; Nicomachus, *Arithmetic* I.16.

is equal to the deficient number, and the sum of the divisors of the deficient number is equal to the abundant number. For example, consider 220, which is an abundant number, and 284, which is a deficient number. The sum of the divisors of 220 is equal to 284 [p. 66], and the sum of the divisors of the latter number is equal to 220. So these numbers, and others like them, are called 'amicable' and there are [only] a few of them. And this is their table:

Abundant number:	*220*
a half of it	110
a fourth of it	55
a fifth of it	44
a tenth of it	22
an eleventh of it	20
a twentieth of it	11
a twenty-second of it	10
a forty-fourth of it	5
a fifty-fifth of it	4
a hundred and tenth of it	2
a two hundred and twentieth of it	1
Total sum	284

Deficient number:	*284*
a half of it	142
a fourth of it	71
a seventy-first of it	4
a hundred and forty-second of it	2
a two hundred and eighty-fourth of it	1
Total sum	220

Chapter 16
On the Duplication of Numbers

Know, O brother, that one of the special properties of numbers is that they increase by duplication [*taḍʿīf*; duplation] and by addition [*ziyāda*] without limit. That happens in five ways:(i) firstly, in their natural order of: 1 2 3 4 5 6 7 8 9 10 11 12, and so on, ad infinitum; (ii) secondly, in the order of even numbers: 2 4 6 8 10 12 14, and so on, ad infinitum; (iii) thirdly, in the order of odd numbers: 1 3 5 7 9 11 13 15 17, and so on, ad infinitum; (iv) fourthly, in terms of subtraction [*al-ṭarḥ*], and by following any of the preceding methods of calculation; and (v) fifthly, in terms of multiplication [*al-ḍarb*], which we shall explain later.

Chapter 17
On the Special Properties of the Types [of Numbers]

Know, O brother, may God aid you and us with a spirit from Him, that each type of the numbers has many properties that have been recorded in the book of *Arithmetic* [of Nicomachus] in detail, but we will restate parts of it in this chapter.

We say that [p. 67] one of the special properties of the natural order of numbers is that the sum from one to any arbitrary number is equal to the product of one more than the last number multiplied by half of the last number.[19]

For example, when we say: what is the sum of the numbers from one to ten as per their natural order? The measure of this is that we add one to ten and multiply it by half of ten, and we get fifty-five; or that we multiply five by itself, which is twenty-five. We then multiply five by the other 'half', which is six [11 – 5 = 6], and we get thirty. Their sum [25 + 30] is fifty-five. And this is its solution and the rule that was sought.

(fol. 11b) The order of even numbers is that of: one, two, four, six, eight, ten, twelve, etc., ad infinitum. One of the properties of this order is that the sum is always odd. Moreover, the sum of one to any arbitrary number, following a natural ordering, is equal to the product of half

19 I.e., $\Sigma n = \frac{1}{2}n\,(n + 1)$.

of this times one more than the other half of this number, adding one to the total.

<For example, when we say to you: what is the sum of the numbers from one to ten according to the even order? You take half of ten and add one to it, then you multiply it by the other half, and you add one to the total, and this is thirty-one, and so on similarly for the rest of the numbers.

The order of odd numbers is that of: one, three, five, seven, nine, eleven, etc., ad infinitum. One of its properties is that when these numbers are added according to their natural order, there are two [types of] sums, one even, and the other odd, which, following after the other, continue ad infinitum. And all of the sums [of odd numbers] will be perfect squares [*majdhūrāt*]. Moreover, when they are added according to their natural order from one to any arbitrary number, no matter how far it reaches, the sum is equal to half of the last number rounded off [algebraically] [*majbūran*; literally, by way of restoration] to the next whole number and then squared>.[20]

For example, when we say: what is the sum from one to eleven according to the order of odd numbers? Its solution is that you take half of eleven, and you then round it off algebraically and multiply it by itself, such as that the result is thirty-six. Its form is also that of taking half the number, which is five and a half, and rounding it off to six; then multiply it by itself, which equals thirty-six, and that is its solution, so take this as a rule.

Chapter 18

[p. 68] Know, O brother, may God aid you and us with a spirit from Him, that the meaning of multiplication [*ḍarb*] is the duplication [*taḍʿīf*; duplation] of one of the numbers by the number of the units of one in the other number, as for example, when we inquire: how

20 The two passages between <…> (from 'For example, when we say to you…' to '…the last number rounded off to the next whole number and then squared') have been noted in the margin of the Atif Efendi manuscript on fol. 11b, which correlates with p. 67 in the Ṣādir edition.

much is three times four? And, its meaning is: how much is the sum of three taken four times?

Know, O brother, that numbers are of two kinds, whole numbers and fractions, as we explained before. Moreover, the multiplication of numbers is of two kinds, simple and compound.

Simple multiplication is of three kinds: (i) a whole number by a whole number, like two times three, or three times four, etc.; (ii) a fraction by a fraction, like a half times a third or a third times a fourth, etc.; and (iii) a whole number times a fraction, like two times a third, or a third times four, etc.

Compound multiplication is also of three kinds: (i) a fraction and a whole number times a whole number, like two and a third times five, etc.; (ii) a whole number and a fraction times a whole number and a fraction, like two and a third times three and a fourth etc.; and (iii) a whole number and a fraction times a fraction, like two and a third times a seventh.

Chapter 19
On Whole Numbers

Know, O brother, that the multiplication of whole numbers is of four kinds, and that there are ten categories [*abwāb*] for the multiplication of all of them. The four ranks of numbers are units, tens, hundreds, and thousands. And the ten categories are units times units [where] one of them is one and ten of them are ten; units times tens [where] one of them is ten and ten of them are a hundred; units times hundreds [where] one of them is a hundred and ten of them are a thousand; units times thousands [where] one of them is a thousand and ten of them are ten thousand; and these are four categories.

As for tens times tens, one of them is a hundred and ten of them are a thousand; and [with] tens by hundreds, one of them is a thousand and ten of them are ten thousand; and [with] tens by thousands, one of them is ten thousand and ten of them are [p. 69] a hundred thousand; and these are three categories.

As for hundreds times hundreds, one of them is ten thousand and ten of them are a hundred thousand; and [with] hundreds (fol. 12a) times thousands, one of them is a hundred thousand and ten of them are a million; and these are two categories.

As for thousands times thousands, one of them is a million and ten of them are ten million, and this is one category. So, there are ten categories in all, and this is their table:

units x units; units x tens; units x hundreds; units x thousands;

tens x tens; tens x hundreds; tens x thousands;

hundreds x hundreds; hundreds x thousands;

thousands x thousands.

Chapter 20
On Multiplication, Square Roots, Perfect Cubes, and the Terms Used by the Algebraists and Geometers, and their Meanings

So we say [that] for any two numbers, whatever pair would they be, if one of them is multiplied by the other, then the product is called a rectangular [*murabbaʿ*][21] number. But if the two numbers are equal, the product is called a perfect square [*murabbaʿ majdhūr*],[22] and the two numbers are called the square roots [*jadhrayn*] of this number. For example, if the number two is multiplied by the number two, then the product is four; or three times three is nine; or four times four is sixteen. Four, nine, and sixteen, and other similar numbers are all called perfect squares; while two, three, and four are called square roots, since two is the square root of four, three is the square root of nine, and four is the square root of sixteen. One considers the rest of the perfect squares according to this pattern, and the square roots are as follows:

21 The literal Arabic term for 'rectangular' would be *al-mustaṭīl*, but here, the word *al-murabbaʿ*, which means literally 'square', is itself used by the Brethren of Purity to distinguish the rectangular number from the perfect square, namely, what they refer to as '*al-murabbaʿ al-majdhūr*'. The rectangular number is not a perfect square; it is a '*murabbaʿ ghayr majdhūr*'.

22 Definition 18 of Book VII of Euclid's *Elements* notes that 'a square number is equal multiplied by equal, or a number that is contained by two equal numbers'.

2	3	4	5	6	7	8	9
4	9	16	25	36	49	64	81

[p. 70] If one multiplies any number by any other number, then the product of them that is not a perfect square [*murabbaʿ ghayr majdhūr*] is called a rectangular [*murabbaʿ*] number, and the two different numbers are called its factors, and they are called the sides of this rectangle, which is the term that the geometers use. For example, two times three, or three times four, or four times five, and so on with the likes of these. The product of these numbers that are multiplied is called a rectangle that is not a [perfect] square [*murabbaʿ ghayr majdhūr*].

Chapter 21
Concerning the Rectangular Number

When any rectangular [*murabbaʿ*] number (fol. 12b), whether a perfect square or not, is multiplied by any number whatever it may be, the product is called a solid [*mujassam*] number, but if the number is a perfect square [*murabbaʿ majdhūr*], and it is multiplied by its square root [*jadhr*], then the product is called a perfect cube [*mujassam mukaʿʿab*].[23] For example, if four, which is a perfect square, is multiplied by two, which is its square root, then the product is eight. Similarly, if nine, which is also a perfect square, is multiplied by three, which is its square root, then the product is twenty-seven. Likewise, if sixteen, which is a perfect square, is multiplied by four, which is its square root, then the product is sixty-four. Hence, eight, twenty-seven, and sixty-four, and similar numbers, are called perfect cubes. A perfect cube is a solid such that its length, its width, and its depth are equal. It has six square faces [*suṭūḥ murabbaʿāt*] whose sides are equal and perpendicular to each other; and it has twelve parallel edges, eight solid [*mujassama*] angles, and twenty-four plane [*musaṭṭaḥa*] angles.

If a perfect square is multiplied by a number less than its square root, the product is called a diminished solid number [*ʿadad mujassam*

23 Definition 19 of Book VII of Euclid's *Elements* notes that 'a cube is equal multiplied by equal and again by an equal, or a number that is contained by three equal numbers'.

libnī], which is a solid whose length and width are equal but whose height [*simk*; literally, 'thickness'] is less than they are. It has six faces whose sides are parallel and perpendicular [*mutawāziya wa qā'imat al-zawāya*]; [p. 71] but it has only one pair of opposite faces that are squares and whose sides are equal and perpendicular, and four faces that are elongated [and whose sides are unequal]. It has twelve edges, every pair of which are parallel; eight solid angles; and twenty-four plane angles.

If a perfect square is multiplied by a number greater than its root, the product is called an augmented solid number [*'adad mujassam bīrī*], such as if four, which is a perfect square, is multiplied by three, which is greater than its square root, then the product is twelve. Similarly, if nine is multiplied by four, which is greater than its square root, then the product is thirty-six. Hence, twelve, thirty-six, and similar numbers are augmented solid numbers; and an augmented solid is one whose height is greater than its length and width. It has six rectangular faces: one pair of opposite faces are rectangles whose sides are equal and perpendicular, and four oblong [*mustaṭīla*] faces whose sides are parallel and perpendicular. It has twelve edges, and every pair of these is equal and parallel. It also has eight solid angles and twenty-four plane angles.

If any rectangular number which is not a perfect square is multiplied by its shorter side, then the product is called a diminished solid; and if it is multiplied by its longer side, the product is called an augmented solid; and if it is multiplied by a number smaller than both of them or greater than both of them, the product is called a free solid [*mujassam lawḥī*].[24] For example, if twelve, which is a rectangular number and not a perfect square, one of its sides being three and the other (fol. 13a) four, is itself multiplied by three, then the product is thirty-six, which is a diminished solid number. If it [twelve] is multiplied by four, then the product is forty-eight, which is an augmented solid number; and if it [twelve] is multiplied by a number less than three or more than four, it is called a free solid [number]. A free solid is one whose length is

24 These diminished, augmented, and free solid numbers (as *libnī*, *bīrī*, and *lawḥī* solid numbers, respectively), correspond with geometric solids that carry the same appellations, which will be described also below in Chapter 13 of Epistle 2.

greater than its width, and its width is greater than its height. It has six faces, every pair of which is equal and parallel; twelve edges, every pair of which is parallel; eight solid angles, and twenty-four plane angles.

[p. 72]

Chapter 22
On the Properties of Perfect Squares

We say that, for every perfect square, if one and twice its square root are added to it, then the sum is a perfect square.[25] If a perfect square is diminished by twice its square root, and also [decreased] by one, then the remainder would be a square.[26]

For every two perfect squares that follow each other, if the square root of one of them is multiplied by the square root of the other and a fourth [¼] is added to it, the total will be a perfect square. For example, if the square root of four, which is two, is multiplied by the square root of nine, which is three, the product is six, to which is added a fourth, totalling six and a fourth, and its square root is two and a half.[27] The product of two and a half times itself is six and a fourth, whose square root is two and a half. For every two perfect squares that follow each other, if the square root of one of them is multiplied by the square root of the other, the product is the geometric mean [*wasaṭ*] between them, and the three numbers are in one proportion. For example, four and nine are perfect squares whose roots are two and three: two times three is six, and four is to six as six is to nine. The other cases follow the same pattern.

Chapter 23
On the Propositions from the Second Book of Euclid's *Elements*[28]

Given any two numbers, if one of them is divided into any number of parts, then the product of the two numbers is equal to the product of

25 I.e., $x^2 + 2x + 1 = (x + 1)^2$.
26 I.e., $x^2 - 2x + 1 = (x - 1)^2$.
27 I.e., [$\sqrt{6¼} = 2½$].
28 This chapter contains the first ten propositions from Book II of Euclid's *Elements*, supplied here in their textual forms, whilst being furthermore supported by

the one that was not divided, times all the parts of the number that was divided, one part after the other. For example, take ten and fifteen, and let fifteen be divided into three parts: seven, three, and five; then we say:

1. The product of ten times fifteen is equal to the product of ten times seven plus ten times three plus ten times five [as the law of 'distributivity'].[29]

2. [p. 73] Let any number be divided into [unequal] parts arbitrarily, then the product of this number by itself is equal to the product of this number multiplied by all its parts.[30] For example, let ten be divided into two parts, seven and three, then I say: the product of ten times itself is equal to the product of ten times seven plus ten times three.[31]

3. Let any number be divided into two [unequal] parts, then we say: the product of this number times one of its parts is equal to the product of this part times itself plus the product of the two parts.[32] For example, let ten be divided into two parts, three and seven, then we say: the product of ten times seven is equal to the product of seven times itself plus three times seven.[33]

4. (fol. 13b) Let any number be divided into two [equal] parts, then we say: the product of this number times itself is equal to the product of each part times itself plus twice the product of the two parts.[34] For example, let ten be divided into two parts, seven and three, then we say:

---

algebraic equations. The propositions (*masā'il*) figure within the *Elements* in a geometrical context, and their content is expressed in a textual form, supported by geometric demonstrations. Each of the ten propositions will be noted in its original form within its corresponding footnote below.

29 Prop. 1 reads as follows in Book II of Euclid's *Elements*: 'If there be two straight lines, and one of them be cut into any number of segments whatever, the rectangle contained by the two straight lines is equal to the rectangles contained by the uncut straight line and each of the segments'; i.e., $a(b + c + d + \ldots) = ab + ac + ad + \ldots$

30 Prop. 2 reads as follows in Book II of Euclid's *Elements*: 'If a straight line be cut at random, the rectangle contained by the whole and both of the segments is equal to the square on the whole'.

31 I.e., $(a + b)(a + b) = a\,(a + b) + b\,(a + b)$; or $(3 + 7)(3 + 7) = 3(3 + 7) + 7(3 + 7)$.

32 Prop. 3 reads as follows in Book II of Euclid's *Elements*: 'If a straight line be cut at random, the rectangle contained by the whole and one of the segments is equal to the rectangle contained by the segments and the square on the aforesaid segment'.

33 I.e., $b\,(a + b) = ab + b^2$; or $7(3 + 7) = 3 \times 7 + 7^2$.

34 Prop. 4 reads as follows in Book II of Euclid's *Elements*: 'If a straight line be cut

the product of ten times itself is equal to the product of seven times itself plus three times itself plus twice seven times three <since the product of seven times itself, plus three times itself is fifty eight; then twice seven times three is forty two, which when added to fifty eight results in one hundred, which is equal to ten times itself>.[35]

5. Let any number be divided into two halves, then into two different parts; the product of one of the different parts times the other, plus half the difference between them multiplied by itself is equal to the product of half of the number times itself.[36] For example, let ten be divided into two halves, then into two unequal parts, three and seven. Now we say that the product of seven times three plus half the difference between them, which is two, times itself is equal to the product of five times itself.[37]

6. Let any number be divided into two halves, then add something to it. We say that the product of this number together with its increment, times that increment plus half of the number times itself is equal to the product of half of that number together with the increment times itself.[38] For example, let ten be divided into two halves, then add two to it. We say that the product of twelve times two plus five times itself is equal to the product of two plus five together times itself.[39]

7. [p. 74] Let any number be divided into two parts, then we say that the product of that number times itself plus the product of one of its parts times itself is equal to twice the product of that number

---

at random, the square on the whole is equal to the squares on the segments and twice the rectangle contained by the segments'.

35 I.e., $(a + b)^2 = a^2 + b^2 + 2ab$; or $(7 + 3)^2 = 7^2 + 3^2 + 2(3 \times 7)$.

36 Prop. 5 reads as follows in Book II of Euclid's *Elements*: 'If a straight line be cut into equal and unequal segments, the rectangle contained by the unequal segments of the whole together with the square on the straight line between the points of section is equal to the square on the half'.

37 I.e., $[(a + b)/2]^2 = ab + [(a - b)/2]^2$; or $[(3 + 7)/2]^2 = 3 \times 7 + [(7 - 3)/2]^2$; or $25 = 21 + 4$.

38 Prop. 6 reads as follows in Book II of Euclid's *Elements*: 'If a straight line be bisected and a straight line be added to it in a straight line, the rectangle contained by the whole with the added straight line and the added straight line together with the square on the half is equal to the square on the straight line made up of the half and the added straight line'.

39 I.e., $a\,(x + a) + (x/2)^2 = [(x/2) + a]^2$; or $2(10 + 2) + 5^2 = (5 + 2)^2$; or $24 + 25 = 49$.

times that part plus the product of the other number times itself.[40] For example, let ten be divided into two parts, seven and three. Then we say that the product of ten times itself plus seven times itself is equal to the product of twice ten times seven plus three times itself.[41]

8. Let any number be divided into two parts, then add one of the parts to the original number. We say: the product of all that [the number plus the part] times itself is equal to four times the product of that number times the part, plus the other part times itself.[42] For example, let ten be divided into two parts, seven and three, then add three to it. Now we say: the product of thirteen times itself is equal to the product of ten times three taken four times, plus the product of seven times itself.[43]

9. Let any number be divided into two unequal parts, then the sum of the product of each of them times itself is double the product of half of that number times itself, plus the product of half the difference of what is between the two numbers times itself.[44] For example, let ten be divided into two halves, then into two unequal parts, three and seven. We say that the product (fol. 14a) of seven times itself plus three times itself is twice the product of five times itself together with the product of two [which is half the difference between the two parts] times itself.[45]

10. Let any number be divided into two halves, then add some increment to it. Now the product of that number with its increment

---

40 Prop. 7 reads as follows in Book II of Euclid's *Elements*: 'If a straight line be cut at random, the square on the whole and that on one of the segments both together are equal to twice the rectangle contained by the whole and the said segment and the square on the remaining segment'.

41 I.e., $(a + b)^2 + b^2 = 2b(a + b) + a^2$; or $(3 + 7)^2 + 7^2 = 2 \text{ x } 7(3 + 7) + 3^2$; or $100 + 49 = 140 + 9$.

42 Prop. 8 reads as follows in Book II of Euclid's *Elements*: 'If a straight line be cut at random, four times the rectangle contained by the whole and one of the segments together with the square on the remaining segment is equal to the square described on the whole and the aforesaid segment as on one straight line'.

43 I.e., $(2a + b)^2 = 4a(a + b) + b^2$; or $(10 + 3)^2 = 4 \text{ x } 3(10) + 7^2$; or $169 = 120 + 49$.

44 Prop. 9 reads as follows in Book II of Euclid's *Elements*: 'If a straight line be cut into equal and unequal segments, the squares on the unequal segments of the whole are double of the squares on the half and of the square on the straight line between the points of section'.

45 I.e., $a^2 + b^2 = 2[(a + b)/2]^2 + 2[(a - b)/2]^2$; $7^2 + 3^2 = 2[(7 + 3)/2]^2 + 2[(7 - 3)/2]^2$; $49 + 9 = (2 \text{ x } 25) + (2 \text{ x } 4)$; $49 + 9 = 58$.

times itself plus the product of the increment times itself is twice the product of half the number with the increment times itself together with the product of half the number times itself.[46] For example, let ten be divided in half, then add two to it. [p. 75] We say that the product of twelve times itself plus the product of two times itself is twice the product of seven times itself together with the product of five times itself.[47]

Chapter 24
On the Science of Number and of the Soul

Know, O righteous and compassionate brother, may God aid you and us with a spirit from Him, that the philosophers have put the study of the science of numbers before the study of the rest of the propaedeutic sciences because this science is potentially embedded in everyone, and humans ought to reflect [on it] with their reasoning powers alone, without taking examples from another science — but from it one takes examples for everything else that can be known.

The examples that we illustrated through Indic numerals in this treatise are for the beginner students whose mental powers are rather weak, but for those who are sharp-witted, these examples are not necessary if they reflect [on these matters].

Know, O righteous and compassionate brother, may God aid you and us with a spirit from Him, that one of our goals in [composing] this epistle is what we explained in its beginning, and the other goal is to bring attention to the science of the soul [*'ilm al-nafs*] and [to provoke] the incitement to the knowledge of its essence. For, when an understanding, intelligent person studies the science of numbers and reflects upon the quantity of its species, the divisions of its several

46 Prop. 10 reads as follows in Book II of Euclid's *Elements*: 'If a straight line be bisected, and a straight line be added to it in a straight line, the square on the whole with the added straight line and the square on the added straight line both together are double of the square on the half and of the square described on the straight line made up of the half and the added straight line as on one straight line'.

47 I.e., $(a + x)^2 + x^2 = 2[(a/2 + x)^2 + (a/2)^2]$; $(10 + 2)^2 + 2^2 = 2[(5 + 2)^2 + 5^2]$; $144 + 4 = 2(49 + 25)$; $148 = 2 \times 74$.

branches, and the special properties of these several branches, this person knows that all of them are accidental and have their being and existence in the soul. So the soul is an essence, because accidents do not have existence other than in essence, and cannot exist except through it.

Chapter 25
On the Goal of the Sciences

Know, O brother, may God aid you and us with a spirit from Him, that the goal of the wise monotheistic [*muwaḥḥidūn*] philosophers in studying the propaedeutic sciences [*al-ʿulūm al-riyāḍiyya*], and the training of their apprentices in them, is indeed the passage [p. 76] to the natural sciences. The goal of studying the natural sciences is the rise and ascent from it to the theological sciences, which are the highest goal of the philosopher and the aim to which they are ascending with true knowledge.

The first step of the study of the theological sciences is the knowledge of the essence of the soul [*jawhar al-nafs*] and the search for its source, where it was before its fastening to the body; and inquiry into its return to the source [*al-maʿād*], where it will be after its separation from the body, which is called death; and inquiry into the manner of reward for the good people, and how it will be in the world of spirits [*ʿālam al-arwāḥ*]; and inquiry into the lot of evil doers and how it will be in the afterlife [*dār al-ākhira*]. Moreover, another quality that humans are recommended (fol. 14b) to acquire is the knowledge of their Lord, and there can be no means of knowing Him except after knowing oneself; as God, the Exalted, has said, *Who forsakes the religion of Abraham but he who is ignorant of himself*,[48] meaning by it one who is ignorant of the soul [*jahl al-nafs*]. And as it is said, 'He who knoweth himself knoweth his Lord'.[49] And it has been said, 'If he informs you of himself, he informs you of his Lord'. Every scholar is bound to study the science of the soul and to get to know its essence and cultivation. God, exalted

48 Qur'an 2:130.

49 This is a weak (unauthenticated) hadith which was commonly evoked by the Sufis, and it may have had a distant connection with the Socratic injunction from the temple of Apollo at Delphi: 'Know thyself!'.

may He be, has said, *And by the soul, and He who fashioned it, and who taught it its sin and its piety, he who keeps it pure will be happy, and he who corrupts it will be disappointed.*[50] And God said in the story of al-ʿAzīz's wife in the narrative of Joseph [peace be upon him],[51] '[...] *The soul enjoins unto evil, save that whereon my Lord has had mercy'.*[52] And God has said, *As for him who fears to stand in the presence of his Lord and forbids his soul from low desires, surely Paradise will be (his) abode.*[53] And God has said, *On the Day (of Judgement) when every soul will come pleading for itself.*[54] And God has said, *O soul that art at rest: Return to your Lord completely satisfied, and be amongst my servants and enter into my Paradise.*[55] And God has said, *God takes the souls at the time of their death and as for those that die not, (he takes them) during their sleep.*[56]

Thus there are many verses and proofs in the Qur'an on the existence of the soul and on its changeable conditions, and they are decisive against anyone who denies the existence of the soul.[57]

[p. 77] When those philosophers who used to discuss the science of the soul before the descent of the Qur'an, the New Testament [Gospels], and the Torah, inquired into the science of the soul with the natural talents of their pure minds, they deduced the knowledge of its essence by the conclusions of their reasoning. This induced them to compose philosophical books that were mentioned previously in the beginning of this first epistle.[58] But because of extensive discourse in them and their transmission from language to language, one cannot understand their meaning or know the goal of their authors. The understanding

50 Qur'an 91:7–10.

51 The term 'al-ʿAzīz', which means 'the exalted one', is the official court title of the Pharaoh's chief chamberlain and captain of the guards, Potiphar (Genesis 37:36), who was a eunuch and childless. He was married to Zulaykhā, who was nominally his wife, and who may have been a virgin (*Tafsīr al-Jalālayn* XII.21).

52 Qur'an 12:53.

53 Qur'an 79:40–41.

54 Qur'an 16:111.

55 Qur'an 89:27–30.

56 Qur'an 39:42.

57 The Atif Efendi manuscript specifically mentions the 'Khurramiyya' (a quasi-Mazdakite Zoroastrian sect), whilst the Ṣādir edition points at the 'astronomers' (*al-munajjimūn*), most likely meaning by this the materialists.

58 There seems to be some confusion here over chronologies and dating.

of the meaning of these books is closed to those who inspect them and their authors' goals trouble those who examine them. We have taken the core of their meaning and the highest goals of their authors, and we have presented them as briefly as possible in fifty-two epistles,[59] of which this is the first. Its other companions amongst the epistles follow it in fidelity, and you find them according to the order of the numbers, if God, most exalted, wills it!

Praise to the One who grants us reason, and guides us on the straight path.

The first epistle, from the first part of the propaedeutic mathematical sciences, which comprises thirteen epistles, has been completed. Praise to God, Lord of the world. May God bless His apostle, Muhammad the Prophet, and the members of his family who are righteous, and may He surely grant them peace!

---

59 The copyist of the Atif Efendi manuscript, and the scribes of other manuscripts, enumerated the epistles as fifty-one, which points to discrepancies and nuances within the classificatory systems that were adopted in compiling these tracts. This is discussed in *The Ikhwān al-Ṣafāʾ and their 'Rasāʾil'*; see especially the 'Prologue' and Chapter 2. It is also noted in the critical apparatus of the Arabic text in this present volume.

(fol. 15a) [p. 78]

Epistle 2
On Geometry

[Being the second epistle from the first section of the *Epistles of the Brethren of Purity*, on the Propaedeutical and Mathematical Sciences]*

(fol. 15b) *In the name of God, the Compassionate, the Merciful*

Chapter 1[1]

Know, O righteous and compassionate brother, may God aid us and you with a spirit from Him,[2] that we have completed the epistle on numbers in arithmetic, and we explicated the properties of numbers with sufficient effort, and we progressed from that tract to this present one, which is the second in the epistles of the propaedeutic sciences, and constitutes an introduction to the science of geometry. So, we say: know, O righteous and compassionate brother, may God aid you and us with a spirit from Him, that the sciences, which the ancients used in cultivating their offspring and in training their apprenticeship, consist of four genres: the first of these are the propaedeutic sciences, the second are the logical sciences, the third are the natural sciences, and the fourth are the theological sciences.

The propaedeutic sciences are also of four kinds: the first is arithmetic, which studies numbers, the quantity of their genus, and their properties and kinds, together with the attributes of their types. The principle of this science is the [number] one that is prior to the [number] two. The second [of the propaedeutic sciences] is geometry,

---

* The title in the Atif Efendi MS (fols. 15a–b) runs as follows: 'The Second Epistle on Geometry, which is mensuration' [from the First Section on the Propaedeutical and Mathematical Sciences], among the Fifty-One Epistles of the Brethren of Purity for the Purification of the Soul and the Clarification of Morals as a Part of a Sufi Discourse'. (It is worth noting in this context the insertion of this phrase that highlights the attribution of the *Epistles* to a Sufi source, as it was also the case with Epistle 1 on arithmetic. This question was discussed elsewhere in *The Ikhwān al-Ṣafāʾ and their 'Rasāʾil'*, ed. El-Bizri.) This introductory line is then completed by the following eulogy: '*In the name of God, the Compassionate, the Merciful, O Lord, make our way easy*'; it is also noted in the Ṣādir edition: '*Praise to God, and peace upon His worshippers whom He has elected. God is goodness, for they should not worship anyone else besides him*'.

1 On fol. 15b, and after the *basmala*, it is noted again that this is an: 'Introductory Epistle on the Science of Mensuration (Geometry), which belongs to a total of Fifty-One Epistles of the Brethren of Purity; may God increase their [retinue] in Purifying the Soul and Reforming Morals'.

2 This initiatory phrase occurs throughout the *Epistles of the Brethren of Purity* in its full form or in shorter versions, and at times with slight changes in the order of the words. It is a rendering of the more common phrase in Arabic: '*iʿlam ayyuhāʾl-akh al-bār al-raḥīm, ayyadaka Allāhu wa-iyyānā bi-rūḥin minhu*'. The term *al-bār* can be also rendered as 'virtuous', 'pious', or 'reverent'; it has been noted herein as 'righteous'.

which is the science that inquires about magnitudes, distances, and the quantity of their kinds, along with the properties of their types. The origin of this science is the point, which is the extremity of the line, namely, as its end. The third [of the propaedeutic sciences] is astronomy, namely the science of the stars that inquires about [p. 79] the composition of the heavenly spheres, the constellations of the zodiac, the enumeration of the planets, [the investigation] of their natural properties, and their significance and indications with regard to existent entities in this world and in relation to the Sun and its movement. The fourth [of the propaedeutic sciences] is music,[3] which inquires about the harmonic compositions and ratios that exist between entities that differ in substance and sustain opposite potencies. The origin of this science lies in the equality of ratios, like the ratio of equivalence and qualification, such as the ratio of three to six, which brings the same result as the ratio of two to four.

The definition of 'logic' is that it is a science that discloses what is unknown in terms of conceptions and assents [*al-taṣawwurāt wa'l-taṣdīqāt*] by way of data that rest on its own principles.[4] The logical sciences inquire into the properties of existent things as they are pictured in the mind's thoughts. Their origin is the intellect [*al-ʿaql*] and the soul [*al-nafs*].[5]

As for the natural sciences, they study the substance of bodies in terms of knowing what happens to them as a result of accidents. The principle of this science lies in movement and rest.

Regarding the theological sciences, these investigate the forms that are abstracted from matter in this world. The origin of this science emerges from knowing God, may He be exalted in majesty, and [having knowledge] about the substance of the soul [*jawhar al-nafs*], the angels,

---

3 The four kinds of propaedeutic sciences correspond therefore to the quadrivium: arithmetic, geometry, astronomy, and music. See *On Music*, ed. and tr. Owen Wright; and *Epistles of the Brethren of Purity. On Astronomy and Geography: An Arabic Critical Edition and English Translation of Epistles 3 & 4*, ed. and tr. Jamil Ragep et al. (New York–London: OUP–IIS, forthcoming).

4 See *On Logic*, ed. and tr. Carmela Baffioni.

5 The Ṣādir edition notes 'substance' (*al-jawhar*) instead of 'the intellect and the soul' (*al-ʿaql wa'l-nafs*). Refer to other variants in the critical apparatus of the Arabic edition.

demons, and jinn, and the manner in which bodies are for them, as disembodied spirits, three dimensional.[6] The principle behind knowing this science is that of knowing the substance of the soul.[7]

For each of these sciences we have composed an epistle that is akin to an introduction and offers a set of premises. The first of these [epistles], which is prior to this [present tract], deals with numbers. We explicated in it some of the properties of numbers, the quantity of their kinds, and the manner they emerge from the [unit] one that is prior to the [number] two. So, in this [present] epistle, we seek to note and explain the principle origin of geometry, which is that of the three magnitudes, the quantity of their kinds and the properties of their types, and the qualification of the manner in which they are constituted from the point that lies at the end of the line and serves the science of geometry like the [unit] one does in the science of arithmetic.

(fol. 16a) Know, O righteous and compassionate brother, may God aid you and us with a spirit from Him, that geometry consists of two types, the intellective and the sensible. The sensible [type of geometry] consists of knowing magnitudes and what belongs to them in terms of properties when one of them is added to another. And this pertains to what is visible by the sense of sight and is tangibly sensed by touch. It is also what can be known and grasped, since what is visible by the sense of sight [p. 80] is the line and the surface, along with the solid that has [three] dimensions, and what belongs to them in terms of weights. This is akin to how weightiness, as that which is in the weighty, cannot be known but through the intellect, and weight is the designation [*ʿayn*] of the weighty.

Magnitudes [*maqādīr*] are of three kinds: lines, surfaces, and solids. [Their] geometry is applied in all the arts of manufacturing since every maker who considers magnitudes within his art, prior to labouring on its objects, is involved in one of the modes of engaging intellective geometry. This [type of geometry] consists of knowing

6 The statement that 'bodies are for them three-dimensional' seems to be a repeated phrase that is copied out of context. It does not seem to be meaningful in the way it occurs in the text after the mention of 'disembodied spirits'. It is therefore unclear in this setting whether the 'disembodied spirits' have some sort of *images* that are three-dimensional without being corporeal qua material.

7 The study of theology presupposes the knowledge of psychology as a prerequisite.

about dimensions [*abʿād*], and of what belongs to them in terms of properties when added to one another. It deals with what is imagined in the mind by way of thinking, and it is divided into three kinds: length, width, and breadth. And these intellectually conceivable [*ʿaqlī*; mental] dimensions are the descriptive attributes of the sensible magnitudes. This is the case given that the line is one of the magnitudes, and that it has only one property, which is length. As for the surface, it has a second magnitude, and it has two properties, namely, length and width. The solid has a third magnitude that has the three properties of length, width, and depth.

Know that an inquiry about these dimensions as abstracted from bodies is the art of the philosophizers [*al-mutafalsifīn*].[8]

We begin by describing sensible geometry, since it is more immediate to the comprehension of the apprentices, so, we say that the sensible line, which is one of the magnitudes, originates from the point, as we have demonstrated in the epistle on the properties of numbers [Epistle 1: 'On Arithmetic'], in terms of how the one [unit 1] is the origin of numbers. This is the case, since, if the sensible point

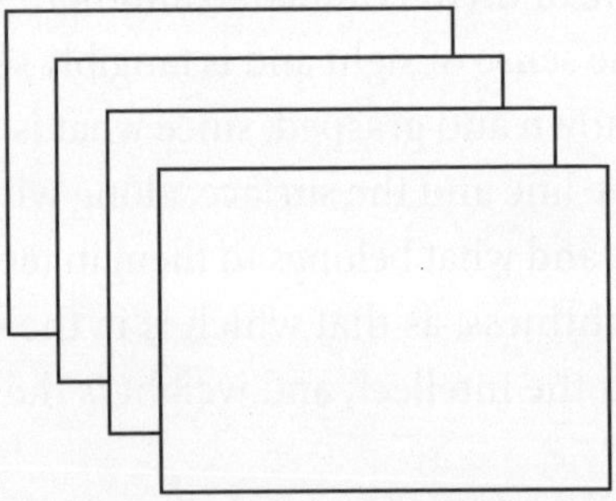

[*al-nuqṭa al-ḥissiyya*] is ordered in a sequence, then the line appears to the sense of sight, such as in this [form]:

....

We do not claim that this point [*al-nuqṭa*; *sêmeion*] is an entity that does not have parts [i.e., an atom, as classically conceived], but that

8 The Ṣādir edition notes the term 'investigators' (*muḥaqqiqīn*) instead of 'philosophizers' (*mutafalsifīn*).

the conceptual point [*ʿaqliyya*] is that which does not have parts [i.e., cannot be partitioned and is indivisible].[9]

We also say that the line [*al-khaṭṭ*; *grammês*] is the origin of the surface [*al-saṭḥ*; *epiphaneia*], like the point is the origin of the line, as it is the case with the [unit] one being the origin of [the number] two. And [the number] two is the origin of even numbers, as we demonstrated before in the epistle on numbers [Epistle 1: 'On Arithmetic']. If lines are ordered contiguously, then the surface appears to the sense of sight, like in this [form]:

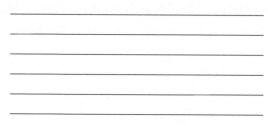

[p. 81] We say, moreover, that the surface is the origin of the solid [*al-jism*; *stereos*], like the line is the origin of the surface, and the point is the origin of the line,[10] and the [unit] one is the origin of [the number] two, and [the number] two along with the [unit] one are the two origins of odd numbers, as we demonstrated before [in the epistle 'On Arithmetic']. If surfaces have been accumulated on top of one another, then the depth of the solid body appears to the sense of sight, like in this [form]:

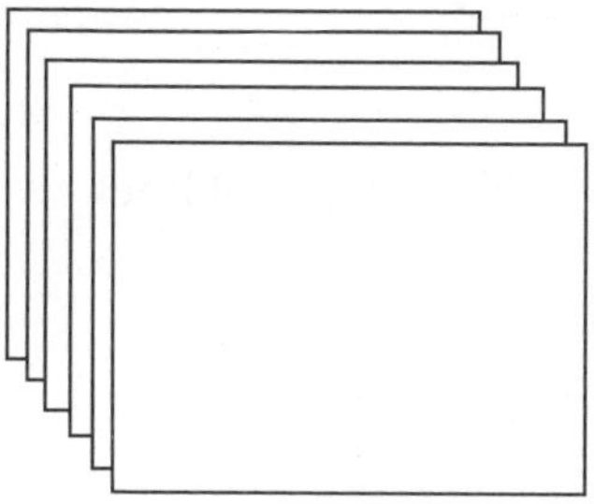

9 This statement corresponds with Definition 1 in Book I of Euclid's *Elements*, which reads as follows: 'The point is that which has no parts [*sêmeion estin oû meros outhen*]'.

10 With such definitions we determine also the prior by means of the posterior: a *point* as the extremity of a line, a *line* as the edge of a surface, a *surface* as the limit of a solid.

Chapter 2
On the Kinds of Lines

We say that lines are of three types. The first [type of line] is straight [rectilinear], like the one that is drawn with a ruler, such as that which appears in the following form:

The second [type of line] is arched [*muqawwas*],[11] like the one that is traced with a compass [*birkār*], such as in this [form]:

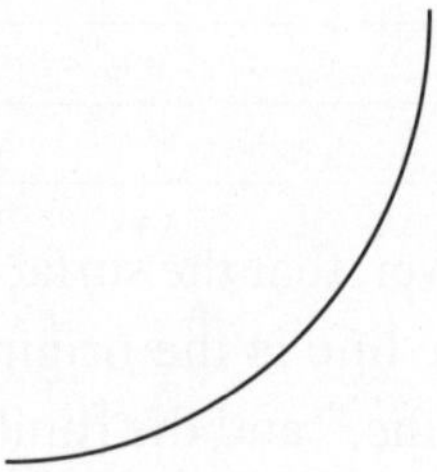

The third [type of line] is (fol. 16b) bumpy [*munḥanī*];[12] it is a composite of the two [other lines], like in this [form]:

And these constitute the three types of lines.

11 By this is meant 'curved lines', and yet, for a rendering that is closer to the Arabic *muqawwas*, I have opted to use throughout this translation the term 'arched', but designating by it curvature.

12 It has a bump or waviness; literally, it has an *inclination* along a curvature, or that traces a smooth *deviation* from being straight. Hence it is named *munḥani*, meaning 'inclined' or 'deviating'.

[p. 82]

Chapter 3
On the Appellations of the Straight Lines[13]

We say that if the straight lines are grouped with one another, they will then be either equal [*mutasāwiya*], parallel [*mutawāziya*], convergent [*mutalāqiya*], contiguous [*mutamāssa*], or intersecting one another [*mutaqāṭiʿa*]. The equal [lines] have equivalent lengths, such as in this [form]:

The parallel [straight lines] are such that if they were in a single surface plane, and they were extended in either direction indefinitely, they would never meet [in either direction];[14] and this is like the following [form]:

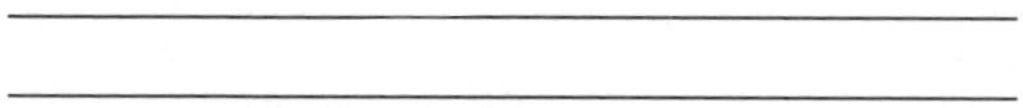

The convergent [lines] are the ones that meet in one of their directions, and they determine a single angle; such as in this [form]:

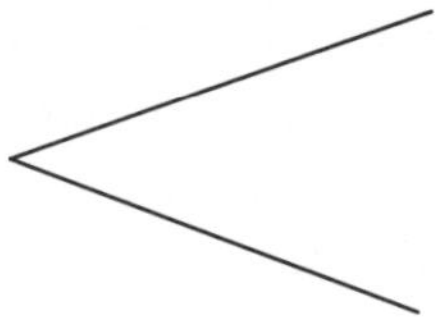

13 'Titles' is the literal translation of the Arabic *alqāb* instead of 'appellations' or 'epithets'.

14 This corresponds with Definition 23 and Postulate 5 in Book I of Euclid's *Elements*.

The contiguous [lines] produce two angles, or a single one, like in this [form]:

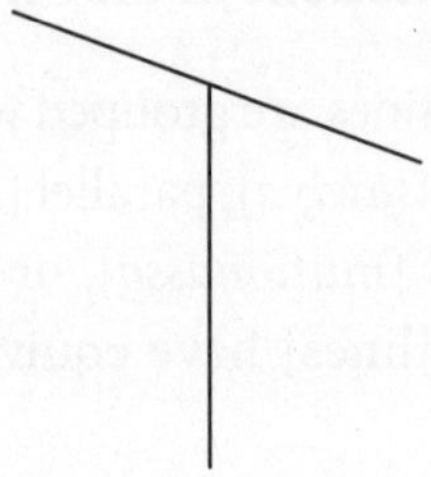

[p. 83] The intersecting [lines] produce four angles, such as this [form]:

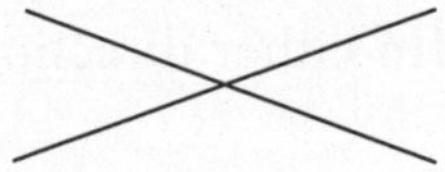

And these are the appellations of the straight lines.

Chapter 4
On the Names of the Straight Lines

If a straight line falls onto another in a rectilinear manner without inclination [to any side] in its extremities, such line is then called a 'perpendicular' [*al-qā'im al-'āmūd*], and [the line] onto which it is perpendicular is named 'the base' [*al-qā'ida*]; like in this [form]:

Each line that is opposite to a given angle is called a chord [*watar*] of that angle that faces it, like this:

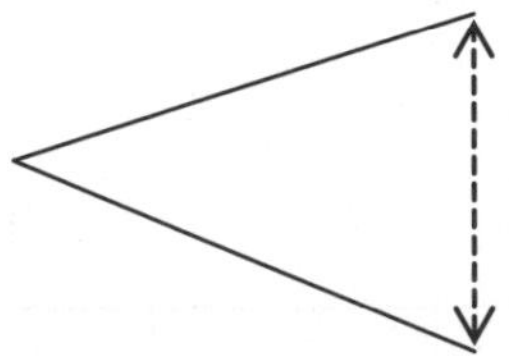

<If the extensions of two lines delimit an angle [*zāwiya*], then these [lines] are called the 'legs' [*al-sāqatān*] of that angle; as in this [form]:

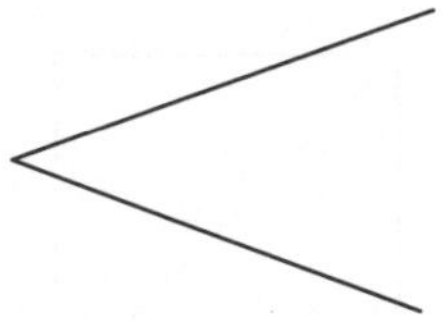

If a straight line falls onto another, and that [straight] line and the one onto which it falls have an inclination [*mayl*] to one of their sides, then this results in having two angles, one of which is larger than the other and is called 'obtuse' [*munfarija*], whilst the one that is smaller is named 'acute' [*ḥādda*]. And every straight line that subtends a given angle is referred to as the 'chord' of that angle; like this>:[15]

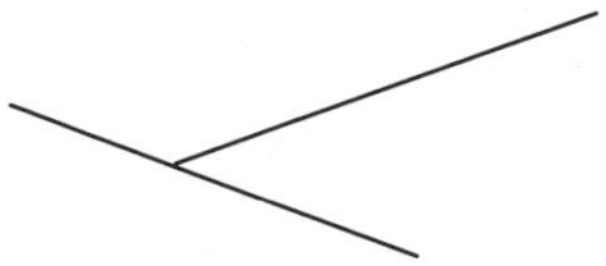

15 The two passages and two figures which are included between <...> in the body of the text have been omitted from the Atif Efendi manuscript, but they are noted in the Ṣādir edition and in some of the manuscripts as shown in the critical apparatus of the Arabic text.

[p. 84] If lines delimit a given surface [*saṭḥ*], they are then called the sides of that surface; such as in this [form]:

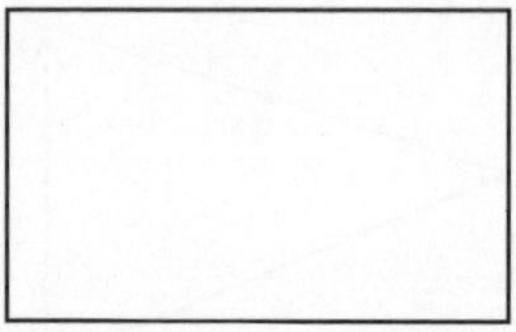

Every line that is extended from one angle of [a given square] and reaches another [of its angles], is called the 'diagonal of the square'; such as in this [form]:

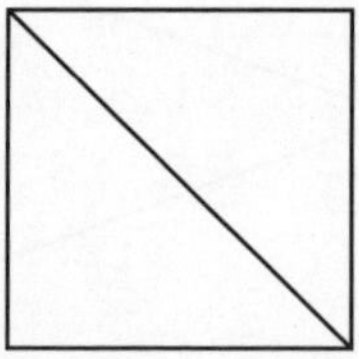

Each line that is extended from the angle of a triangle and falls onto its opposite side [*al-ḍilʿ al-muqābil*], in such a way that it is perpendicular to it, would then be called 'the trajectory of a falling-stone'[16] [*masqaṭ al-ḥajar*], or, as it is also named, 'perpendicular' [*al-ʿāmūd*; orthogonally upright];[17] as for the line on which this perpendicular falls, it is referred to as the 'base' [*al-qāʿida*]. And this is its example:

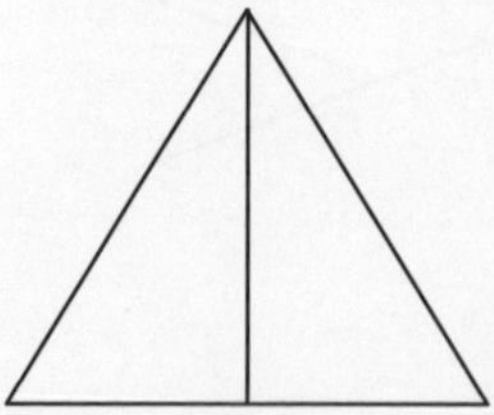

And these are the names [*asmāʾ*] of the straight lines.

16 This designates the perpendicularity/verticality of a plumb line.

17 This designates a perpendicular angle bisector within that given triangle.

[p. 85]

Chapter 5
On the Kinds of Angles

We say that angles are of two types, planar [*musaṭṭaḥa*] and solid [*mujassama*]. The planar [angles] are subtended by two lines that do not extend into each other rectilinearly, namely such as in this [form]:

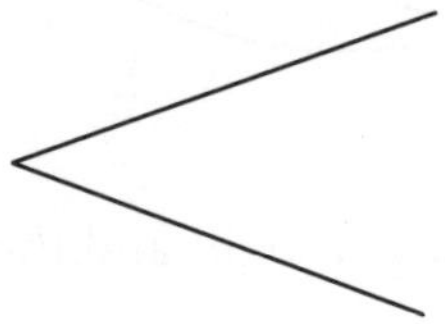

As for the solid [angles], they are subtended by three lines, and each pair of these does not extend into each other rectilinearly; such as in this [form]:

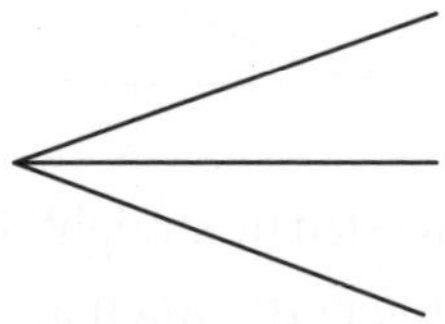

Chapter 6
On the Types of Planar Angles in Other Varieties

The directionality [*jiha*] of lines is of three types:

It is either that of two straight lines, such as in this [form]:

Or of two arched [*muqawwas*] lines, such as in this [form]:

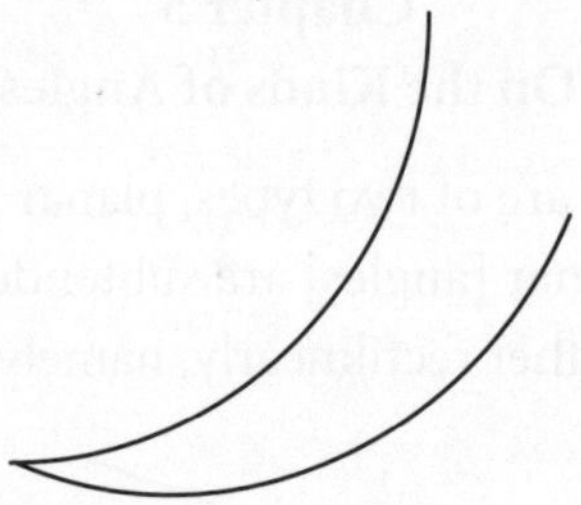

Or of one of them being straight and the other arched, such as in this [form]:[18]

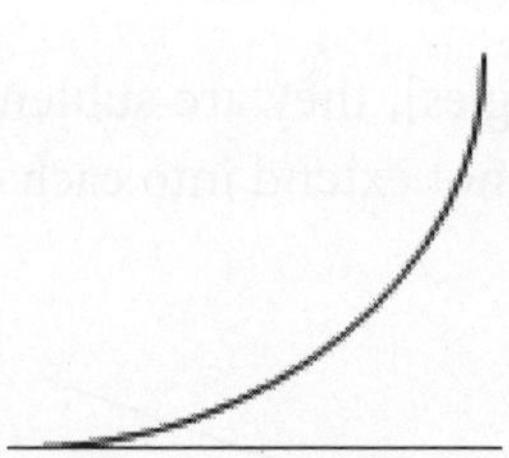

The angles that are subtended by straight lines differ in their qualities according to three kinds: the perpendicular, the obtuse, and the acute. The perpendicular is the one that results from one straight line falling onto another straight line in an orthogonally upright manner [*qiyām mustawī*], such that the angles that are produced from either of its two sides are equal. Each one of these [angles] is called a 'right angle' [*zāwiya qā'ima*], such as in this [form]:

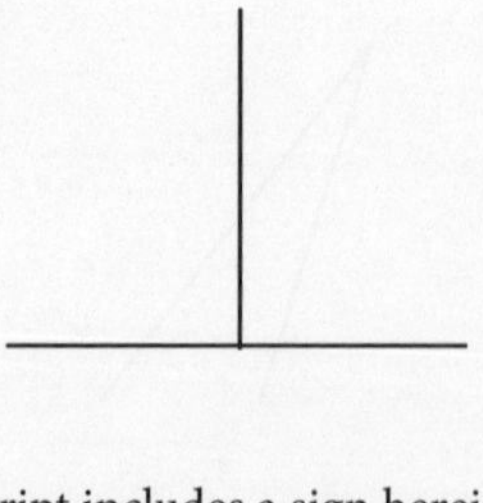

18 The Atif Efendi manuscript includes a sign herein that a new untitled chapter (*faṣl*) begins, however, it only covers a very short passage before the start of Chapter 7 below, and it seems to be irrelevant and incoherent to include it as a new untitled chapter heading in this context.

[p. 86] If a line falls onto another in a (fol. 17a) non-orthogonal manner, the angles that are produced from either of its two sides will be unequal. The one of them that is greater than a right angle [*akbar min'al-qā'ima*] will be called 'obtuse' [*munfarija*], whilst the one that is lesser than a right angle [*aṣghar min'al-qā'ima*] is named 'acute' [*ḥādda*]. The total sum of their [magnitudes] will be equal to two right angles, since the acute angle is lesser than a right angle by the same magnitude [*miqdār*] by which the obtuse angle is greater than a right angle.[19] This is its example:

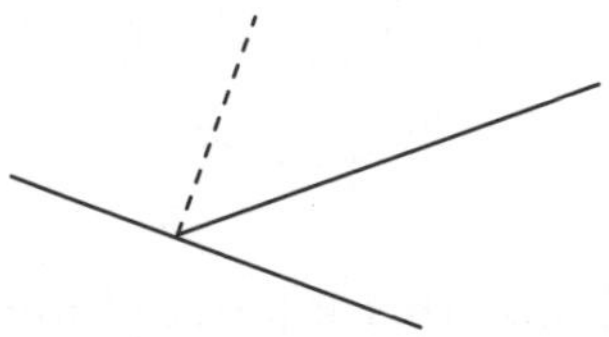

And these constitute the enumerations of the types of angles.

Chapter 7
On the Types of Arched Lines

We say that the arched lines are of four kinds. Some constitute the circumference of a circle, others consist of a semicircle, whilst other kinds have a curvature that exceeds the extendedness of a semicircle, or yet others are lesser [in the extendedness of their curvature] than a semicircle.[20]

The centre [*al-markaz*] of a circle [*al-dā'ira*] is the point within it that is [*dākhilaha*] in the middle [*wasaṭ*] of the circle, and from where all the lines extending out to the circumference would be equal. The diameter [*al-quṭr*] of a circle is the straight line that cuts the circle into two halves [semicircles]. The chord is the straight line that connects the two extremities of an arched line [*al-khaṭṭ al-muqawwas*]. The

19 Let 'a' be an acute angle, 'o' an obtuse angle, and 'r' a right angle, then: $r - a = o - r$.

20 All of these figures are illustrated in Chapter 8 below. These instances of repetition may be for didactic purposes.

sagitta [*al-sahm*] is the straight line that divides the chord and its subtended arch into two halves, and if it were to be added to half the arch, it would then be called a 'versine' [*al-jayb al-maʿkūs*], and if half the chord were to be added to half the arch, it would then be called a 'straight sine' [*al-jayb al-mustawī*].

The arched lines [*al-muqawwasāt*] that have their circumferences in parallel formations are concentric, such as this [form]:

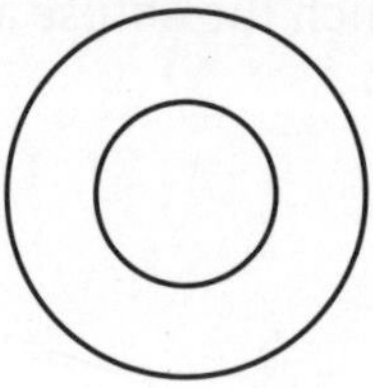

[p. 87] The arched lines that intersect [*mutaqāṭiʿa*] have distinct centres, such as is the case with this [form]:

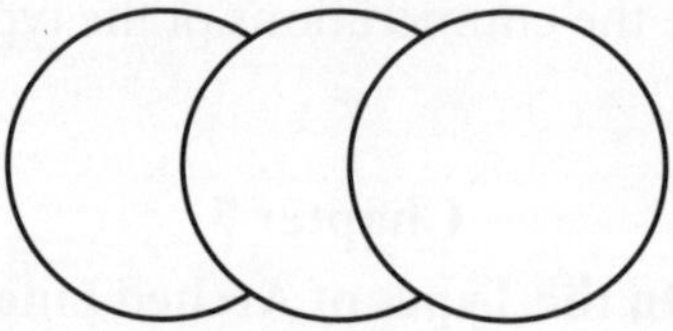

The arched lines that are contiguous [*mutamāssa*] with one another are ones that touch each other from the inside or from the outside without coming to an intersection, such as is the case with this [form]:

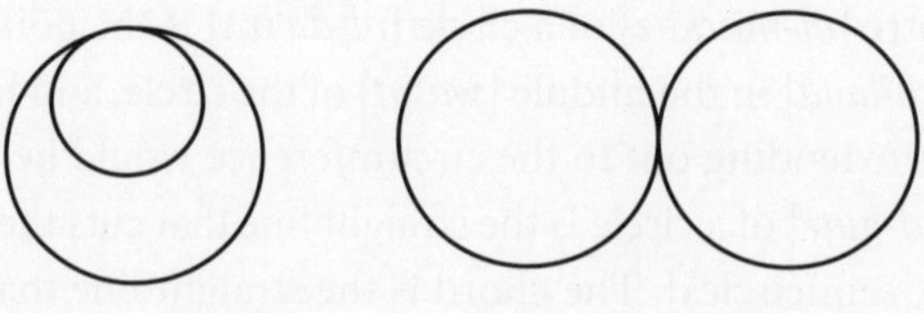

As for the bumpy [*munḥaniya*] lines [that are composites of arched and straight lines], we left them out of our inquiry since they are not used [in specific applications]. Let everyone know that! The mentioning of lines is concluded.

Chapter 8
On Surfaces

We say that a surface is a figure that is surrounded by a single line or by multiple lines. A circle [*dāʾira*] is a figure that is surrounded by a single line as such:

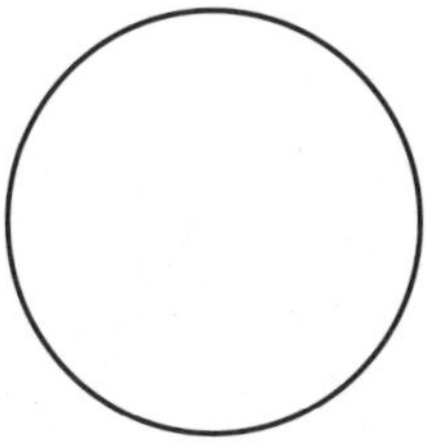

Inside [the circle] there is a point such that any straight lines issuing from it in any direction, and which would end [by intersecting its circumference], would be equal to one another.

A semicircle constitutes a figure that is surrounded by two lines, one arched, which is half the circumference, and the other straight, which is the horizontal diameter of the circle, such as in this [form]:

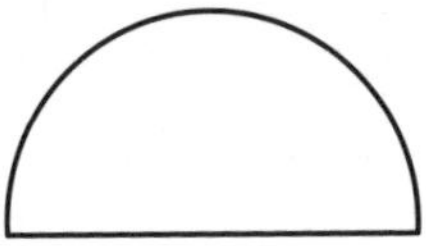

[p. 88] <A segment of a circle [*quṭʿat dāʾira*] is a figure that is delimited by a straight line and an arched line that is part of a circle's circumference, and that is either greater than the semicircle or lesser than it, as we explained and noted earlier>.[21]

21 Chapter 8 repeats some of what has been partly indicated in Chapter 7 above on the circle, the semicircle, etc., whilst including herein some illustrative geometric figures. The passage that is included between <…> has been omitted in the Atif Efendi manuscript.

Chapter 9
On Rectilinear Figures and their Kinds

We say that the first of the rectilinear figures [those delimited by straight lines][22] is the triangle [*al-muthallath*] that is surrounded by three equal [straight] lines, and it has three angles, such as in this [form]:

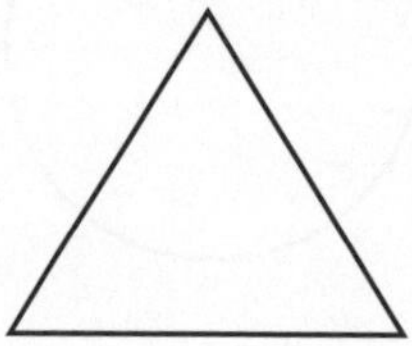

The square [*al-murabbaʿ*] is that which is delimited by four [equal] straight lines and has four right angles, like this [form]:

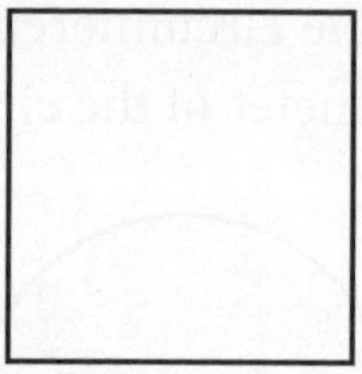

22 These are regular polygons (trilateral, quadrilateral, multilateral) with a certain number of edges and vertices, enumerated herein from three to seven in terms of the number of their edges and vertices by way of illustrating how these polygons are generated and how they progress as does the progression of numbers as delineated in the arithmetic epistle. These rectilinear figures are described by Euclid in Book I of his *Elements*, in Definition 19, as follows: 'Rectilinear figures are those which are contained by straight lines, trilateral figures being those contained by three, quadrilateral those contained by four, and multilateral those contained by more than four straight lines'. These are planar figures with shapes that consist of straight-line edges and multiple angles-cum-vertices, which constitute a closed multilateral form.

The pentagon [*al-mukhammas*] is a figure that is delimited by five [equal straight] lines and has five angles, like this [form]:

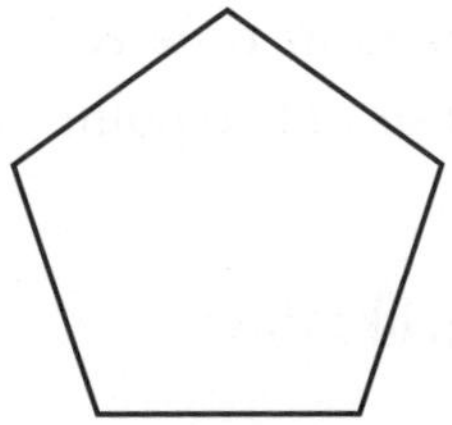

The hexagon [*al-musaddas*] is a figure that is delimited by six [equal straight] lines and has six angles, like this [form]:

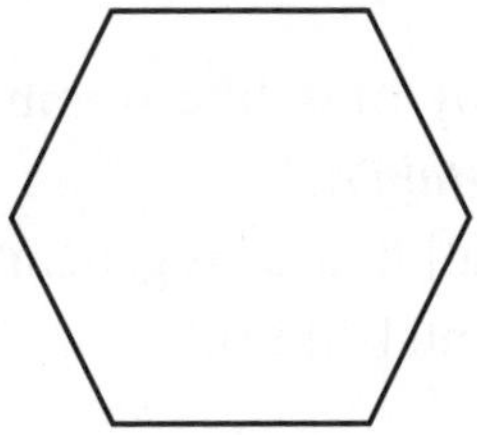

[p. 89] Then the heptagon [*al-musabbaʿ*] is a figure [that is delimited by seven equal straight lines and has seven angles], like this form:

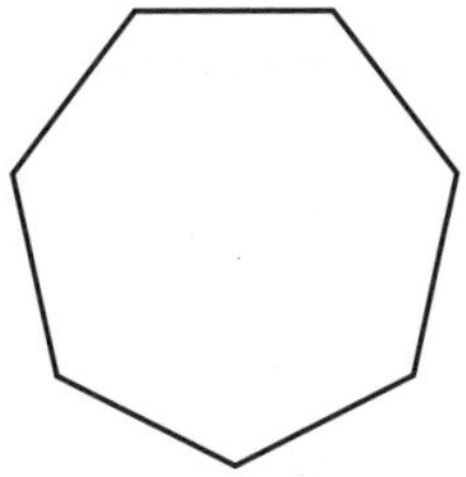

(fol. 17b) And it is by following this pattern [*qiyās*] that the [rectilinear] figures are generated in the same way as numbers.[23]

23 This progresses ad infinitum, though without being able to name all of them.

Chapter 10
On Points [as Visible] by the Sense of Sight

We have shown that the extension of lines appears to the sense of sight from the point, if points are ordered in a sequence [*intaẓamat*].[24]

The shortest line consists of two points, like this:

. .

Then from three [points], like this:

. . .

Then from four [points], like this:

. . . .

Then from five [points], like this:

. . . . .

It then increases one by one, like numbers increase in natural progression [*al-naẓm al-ṭabīʿī*].[25]

The smallest [trilateral] figure [as generated visually from points] is of three parts [*qua* points], like this:

.

. .

Then follows the one of six parts [as a trilateral figure], like this:

.

. .

. . .

Then follows the one with ten parts [as a *tetraktys*], like this:

.

. .

. . .

. . . .

24 This expresses the Pythagorean *skhêmatographein* (namely, the penchant to represent things with figures).

25 The Atif Efendi manuscript includes a sign here that a new untitled chapter (*faṣl*) begins; however, it seems more relevant and coherent in this context not to interrupt the flow of this whole section by including a new untitled chapter heading.

[p. 90] Following this pattern [*qiyās*], the figures increase, and they are generated as such, like numbers are increased by additions in their natural progression.

As for the quadrilateral figures, the first amongst them appears in four parts, like this:

After it [the figure] with nine parts [quadrilateral], like this:

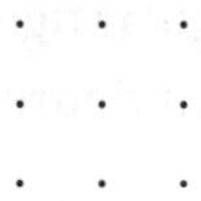

After it [the figure] with sixteen parts [quadrilateral], like this:

After it [the figure] with twenty-five parts [quadrilateral], like this:[26]

Following this pattern, the squared figures [quadrilateral] ever increase, like numbers get added up in the natural progression of odd numbers [*al-afrād*], and they are all perfect squares [*majdhūrāt*].

26 The three quadrilaterals, which are respectively delimited by 9, 16, and 25 points, correspond also with three tables that figure at the end of this present epistle, which also include sets of square-cells as 9, 16, and 25. See also Chapter 26 below.

[p. 91]

Chapter 11

Showing that the Triangle is the Origin of All Figures

We say that the triangular figure [*al-shakl al-muthallath*; a trilateral polygon] is the origin of all the rectilinear figures.[27]

This is the case just as the [unit] one is the origin of all numbers, and the point is the origin of lines, and the line is the origin of surfaces, and the surface is the origin of solids, as we demonstrated before. This is the case, given that if a triangular figure were to be added to another figure that is like it, the resultant figure from their composition would be a square, like this [form]:

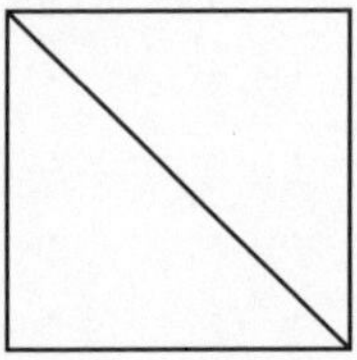

<If another triangular figure is added to both of them [namely, to the two triangles that formed the square], then the product is a five-sided figure. If an additional triangular figure is added, the resultant is a six-sided figure. And if yet another triangular figure is added to them, the resulting figure is a heptagon, which is of this [form]>:[28]

27 The view that the triangle is the origin of all rectilinear figures is embodied in Plato's physical-geometrical theory as explicated in the *Timaeus* 54c–55c. It is stated therein that all the elements (fire, air, water, earth) are solids bound by surfaces that are constituted from triangles, such as one half of an equilateral triangle and the isosceles. See Plato, *Timaeus*, ed. and tr. R. G. Bury, Loeb Classical Library, Vol. IX (Cambridge, MA.: Harvard University Press, 1929; 8th rep., 1999). Refer also to Definitions 20–21 of Book I of Euclid's *Elements* on trilateral figures.

28 The shape that would result from the Brethren of Purity's description would not be a heptagon as such but rather a seven-sided irregular figure. The accounts that are presented herein between <…> are inaccurate and vague in the form they appeared in the text. This is the case, since:

If a triangle is added to the two original triangles that formed the square, then the resultant figure would be four-sided and not a pentagon, twisted or not, such as in this form:

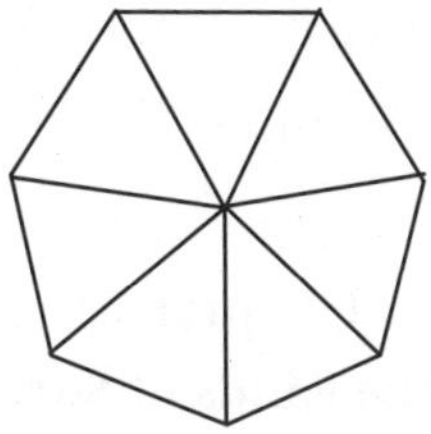

Following this pattern, all rectilinear figures [multilateral polygons] with numerous [delimiting straight] lines [edges] and with multiple angles [vertices] are generated. If trilateral figures are conjoined with one another, they can increase ad infinitum, just as numbers are generated from [the units of] ones [*āḥād*], if these were to be added to one another, ad infinitum, as we demonstrated earlier [in Epistle 1].

It has been shown that all rectilinear figures are composed from trilateral figures, and that solids are composed from surfaces, and that surfaces are composed from lines, and that lines are composed from points, all being like numbers that are composed from [the units of] ones.

[p. 92] The point in the art of geometry is like the [unit] one in the art of arithmetic. Just as the [unit] one has no [component] part

---

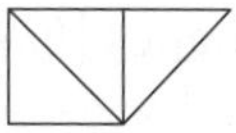

Moreover, if another triangle is added to the above four-sided figure that is composed from three triangles, then the resultant shape would be four-sided again and would not be a hexagon, twisted or not, such as in this form:

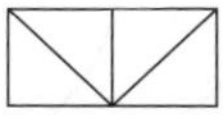

The account presented in the text could have been accurate if the Brethren described each set of triangles as being composed from respective segments (with respective angles) that diagonally bisect the pentagon into five equal trilateral figures, or that divide the hexagon into six equal trilateral figures, such as in these two respective forms:

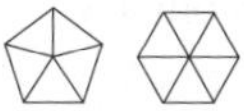

[*lā juzʾ lahu*; i.e., it is indivisible], the conceptual point [*al-nuqṭa al-ʿaqliyya*][29] also has no [component] part.[30]

Chapter 12
On the Kinds of Surfaces

When considered in qualitative terms [*min jihat al-kayfiyya*], surfaces are diversified according to three kinds: [rectilinearly] planar [*musaṭṭaḥ*], concave [*muqaʿʿar*], and convex [*muqabbab*].[31] The planar is similar to the surface of [flat] boards [*alwāḥ*], (fol. 18a) the concave is like the curved hollowness of pans [*qaʿr al-awānī*], and the convex is akin to the exterior of domes [*ẓahr al-qibab*].

Of the shapes of surfaces, there are those that are egg-like [*bayḍī*], such as in this [form]:

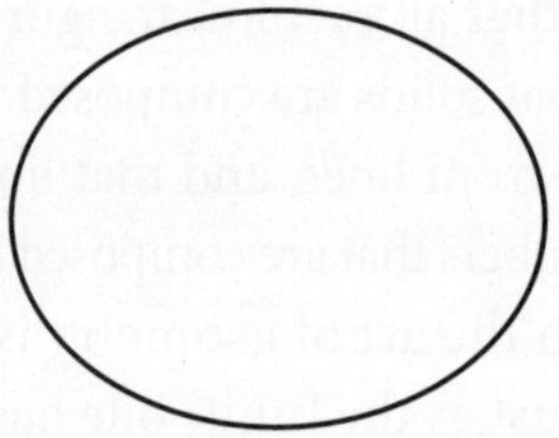

There are also crescent-like shapes [*hilālī*; like lunes], such as in this [form]:

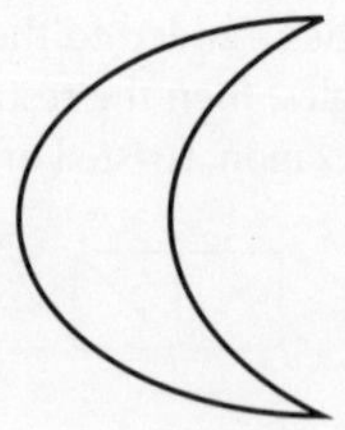

29 In other words, this *point* belongs to the domain of intellective geometry, as opposed to the approximation of a physical demarcation of a given point in the sensible realm of applied geometry.

30 Euclid, *Elements* I, Def. 1.

31 The concave and convex sections are derivable from conics.

There is also the cone that is shaped like a pine nut [*al-makhrūṭ al-ṣanawbarī*], such as in this [form]:

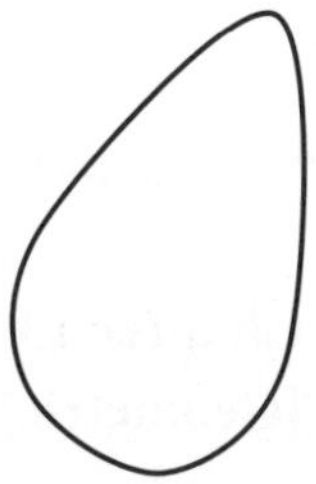

There is also the oval-like shape [*ihlīlijī*; named in Arabic after the fruit *al-ihlīlij* (from '*al-lawz al-hindī*' meaning 'Indian almond')]:

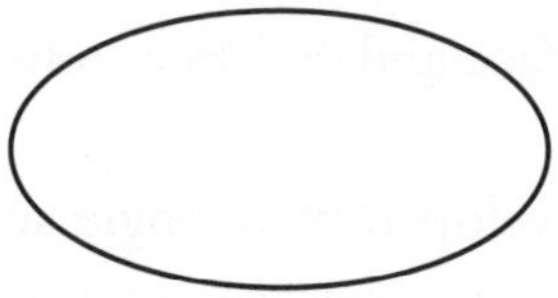

[p. 93] And the elliptical shape [*nīm khānjī*], which is of this [form]:

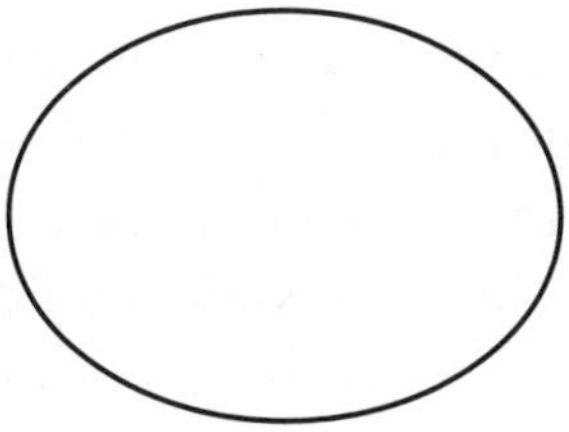

And the drum-like shape [*al-ṭablī*; being bi-concave], such as in this [form]:

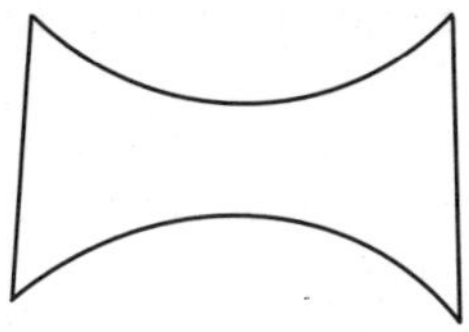

Or the olive-like shape [*al-zaytūnī*; as bi-convex], such as in this [form]:

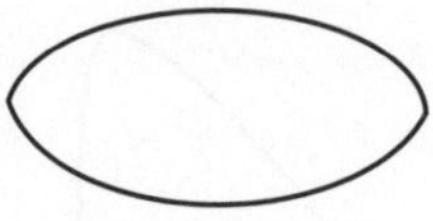

Chapter 13
On the [Geometric] Solids

We say that surfaces constitute the limits of solid bodies [*ajsām*], and that the limits of surfaces consist of lines, and that the limits of lines are points. This is the case since every line has to begin from a given point and end in another. [Likewise,] surfaces start from a given line and end in another or more than one other [line], and [likewise] solids begin with a given surface and end in another or in more than one other [surface].

Some solids are enveloped by a single surface, like the sphere. Others are surrounded by two surfaces, such as the hemisphere, which is delimited by a convex surface [*muqabbab*] and a plane-circular one [*mudawwar*]. Some other solids are delimited by three surfaces, like a quarter portion of a sphere. Moreover, there are solids surrounded by four trilateral surfaces [formed as a tetrahedron or pyramid], which are called 'fiery' [*nārī*] shapes.[32] There are also solids that are delimited by five surfaces, and yet others are furthermore surrounded by six quadrilateral surfaces. This latter kind of solids [that is delimited by six surfaces] encompasses: the cube [*al-muka''ab*] and [the cuboids qua parallelepipeds]: the *bi'rī* [shaped like a square-based water-well], the *libnī* [shaped like a sun-dried mud-brick], and the *lawḥī* [shaped like a board].[33]

The cube is a solid whose length is equal to its width, and its width equal to its depth. It has six square surfaces that have equal edges, with

32 This corresponds with Plato's geometric modelling of the four elements as delineated in the *Timaeus* 55b–c: the fire molecule being formed as a tetrahedron, air as an octahedron, water as an icosahedron, earth as a cube, with all being constituted in their surfaces from composites of half equilateral and isosceles triangles.

33 These were also treated in arithmetic, in Chapter 23 of Epistle 1, above.

[p. 94] eight solid right-angled vertices, twenty-four planar angles, and twelve equal edges, with each set of four among them being parallel; and this is its form:

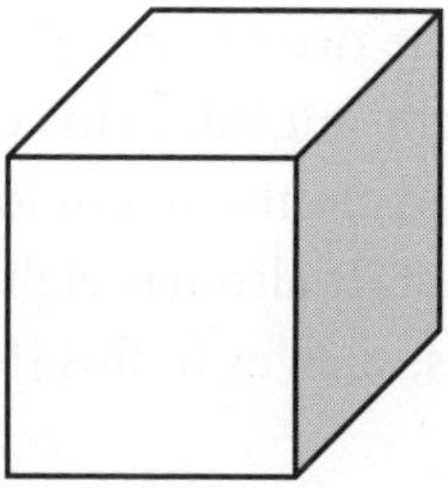

The *bi'rī* cuboid [as an augmented parallelepiped that looks like a squared-section water-well][34] has its length equal to its width, whilst its depth is larger than these. It has six quadrilateral surfaces, two of which are parallel and with equal edges, and right-angled, whilst its four other surfaces are narrower and oblong [and parallel in pairs], with equal edges [in opposite pairs], and right-angled. It has twelve edges, four of which are long, equal, and parallel, and the remaining eight edges are short, equal, and parallel [in opposite pairs]. It also has eight solid angles, and twenty-four planar angles [and its form is the following]:

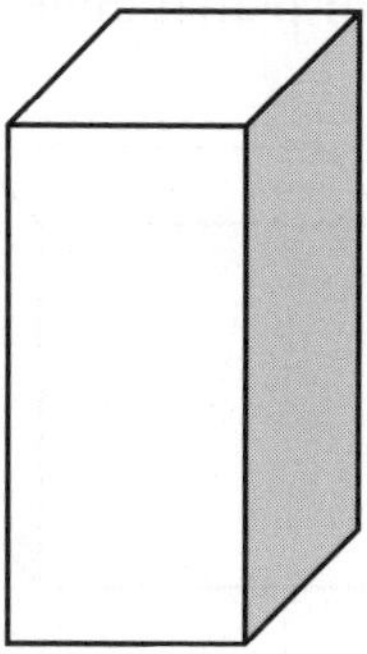

34 The *bi'rī*, *lawḥī*, and *libnī* parallelepipeds were associated with solid numbers in Chapter 21 of Epistle 1, and they were designated respectively by the appellations 'augmented', 'free', and 'diminished'.

The *lawḥī* cuboid [as a free parallelepiped that looks like a flat board] is that whose length is greater than its width, and its width is greater than its depth. It has six quadrilateral surfaces. Two of these are elongated and stretched out whilst being parallel, with equal edges, and right-angled, whilst the two other surfaces are short and narrow, with equal edges, and right-angled. [This cuboid] has twelve edges, four of which are long, and another set of four being shorter, and the last four being the shortest. It also has eight solid angles, along with twenty-four planar angles, such as in this [form]:

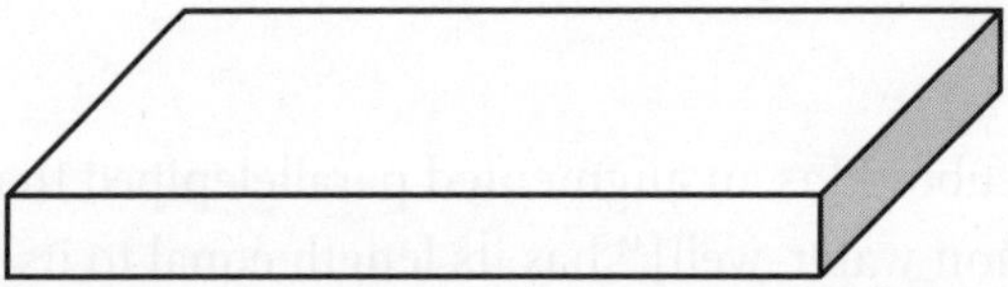

[p. 95] The *libnī* cuboid [as a diminished parallelepiped that looks like a sun-dried mud-brick] is that whose length is like its width, whilst its depth is less than both of them. It has six quadrilateral surfaces, two (fol. 18b) of which are broad and parallel, with equal edges, and right-angled, whilst the other four are narrow and oblong, with equal edges, and also right-angled. [This cuboid] has twelve edges; four of which are short, equal, and parallel, with the other eight being long and equal, with four sets amongst them being parallel. It also has eight solid angles and twenty-four planar angles, such as in this [form]:

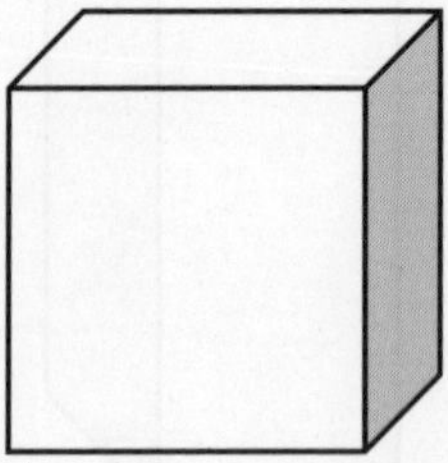

As for the spherical solid [*al-jism al-karīyy*], it is that which is surrounded by a single surface and carries within itself a given point. All straight lines that are stretched out from that point to the [inner]

surface of the sphere are equal. This point is called 'the centre of the <sphere>'. If the sphere rotates, two opposite points on its [outer] surface remain at rest [with respect to that revolution], and these are called 'the poles [*quṭb*] of the sphere', such as in this [form]:

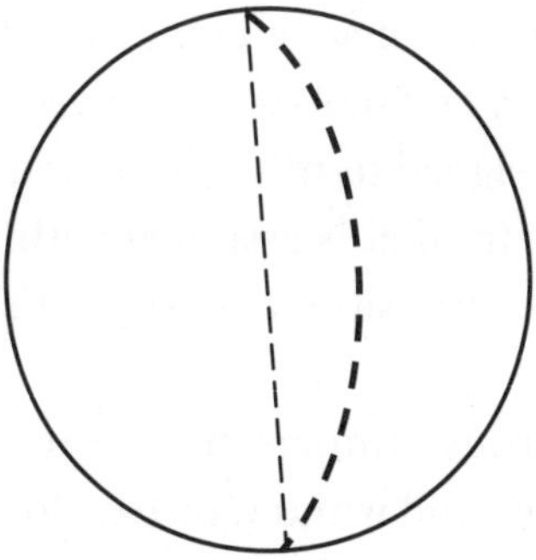

If a straight line connects these [two poles of the sphere] in such a way that it also passes by the centre of the sphere, it is then called the 'axis [*miḥwar*] of the sphere'; and if a [straight] line connects a point [on the surface of the sphere] to another [opposite] point [on the surface of the sphere, in such a way that it also passes by the centre of the sphere], it is then an axis.[35]

Chapter 14[36]

As we have noted this aperçu on the origin of sensible geometry, by way of offering an introduction and postulations, and as we have also said that this science is needed by manufacturers, then we ought now to demonstrate such [a proposition]. This [application of geometry] is that which one reckons [*taqdīr*][37] prior to labouring [*'amal*], since every artisan [*ṣāni'*] configures the [relations of] solids with respect to

35 The insertion of terms between [...] herein address the lacunae in the original text, which otherwise would have offered the following inaccurate and potentially misleading reading: 'if a line stretches from a point to another, it is then an axis'

36 This is an untitled chapter heading noted within the body of the text of the Atif Efendi manuscript, and it is relevant to include it herein given that it offers as smoother textual transition.

37 The term *taqdīr* figures here in the context of reckoning the three-dimensional

one other and composes them. Such a maker ought first to determine the place [*makān*], namely, the position in which the composition of solids is to be undertaken, then the time at which to begin [p. 96] its compositional making.

The [maker is faced with certain aspects to consider]:[38] is the emplacement [*imkān*; i.e., of such composites][39] achievable or not? What are the machinery and tools that are needed in its making? How would the parts be designed to fit together and get configured? This is the [kind of] geometry that is assumed within the arts of making, which deal with the compositional design of solids with respect to one another.

Know that numerous animals engage in fabrications that are inherent to their nature, and without instruction. This is the case with the bees in taking up their dwellings, since they construct their hives in compacted, curved shapes like piled-up shields [*al-attrāss*] laid out on top of one another. They then puncture the hives with openings that are hexagonal in the shapes of their edges and angles, with what this denotes in terms of the workings of a perfected wisdom, given that the property of this [hexagonal] figure is that it is larger than the square or the pentagon. They also use these [hexagonal shapes] to fit them without looseness, in such a way that air is not admissible within, in order not to corrupt the honey or make it mouldy. And this is the example [of their form]:[40]

---

magnitude ('magnitude' itself being termed in Arabic *miqdār*). In other usages, *taqdīr* may be rendered into the English as 'taking account of' or 'considering'.

38 The insertion of terms between [...] again aims to facilitate the flow in reading, and to address the lacunae in the original text.

39 This term is used here to designate geometrical emplacement, and it derives from the Arabic word that refers to place (Greek: *topos*), namely, *makān*.

40 A long passage has been added to the text at this point. This constituted a diversion that only occurred in the MS 7437 [ط] from the Mahdavī Collection, Tehran (fols. 27 and 28). The contents of this passage are geometrical, and they allude to what is noted in Book I of Euclid's *Elements*. The English translation of this passage is contained in the 'Appendix to Epistle 2'. The transcribed Arabic version of this passage has been noted in an appendix to the critical edition, and I did not include it in the body of the Arabic text or in its critical apparatus.

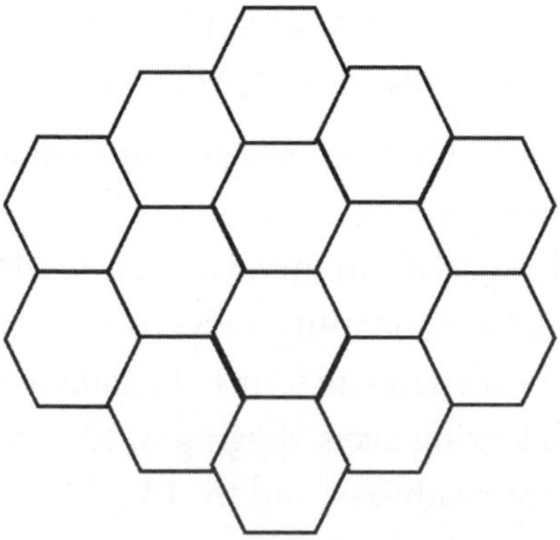

Consider also the spider that weaves its web in the corners of the house or the walls (fol. 19a) in order that it might shelter it from the piercings of the wind or from being torn apart. As to how it does this weaving, it stretches its threads rectilinearly, while the stitching lines are woven in a curvilinear way — all with ease in the making. And this is its example:

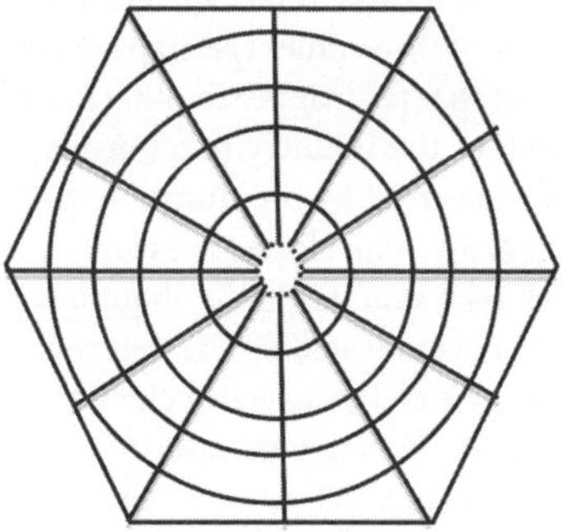

[p. 97] Some people generate a given art from their own proper talent [*qarīḥa*] and intelligence, without being preceded in this by anyone else, whilst most makers acquire their art by way of demonstration and learning under mentors.

Chapter 15
On [the Art of] Surveying

Know, O brother, may God aid you and us with a spirit from Him, that the art of geometry is integral to all the [applied practical] arts of making [*al-ṣanāʾiʿ*], especially in surveying [*al-misāḥa*], since it is an art that is needed by builders, quantity-surveyors, farm-workers, lords of hamlets, and owners of estates, in terms of regulating their transactions in connection with levying taxes, digging waterways, laying down the measure-markers for roadways, and the like.[41]

Know that the magnitudes that are used in surveying the lands in Iraq are of five scales: the cord [*al-ashl*], the doorway [*al-bāb*], the arm [*al-dhirāʿ*], the fist [*al-qabḍa*], and the finger [*al-iṣbaʿ*].[42]

---

41 The latter refers specifically to working on roads or pathways that extend the length of *barīdāt*, namely, the plural of the so-called *barīd*, which is conceptually equivalent to a milestone. The *barīd* is a distance travelled by a message bearer. It is determined by the extent of a short journey of travel to deliver a letter. Literally, the *barīd* designates the mail, the postman, and the post-service. Each *barīd* equals two *parasangs* (*farsakhān*), and each parasang (*al-farsakh*) is measured by the duration of time it takes to cross a stretch of land on a mule's back, which extends about three miles (perhaps also as a distance that separates two villages that are not that far from each other). The measure of the parasang is related to the march of the infantry in an army, whereby the extent of the distance it crosses is measured by the duration of time it takes following the pace of the marching army. The closest measure to this mode of determining long distances is the Greek *stadion* that is about 600 feet. The parasang is over 30 stadia, and the *barīd* over 60 stadia. The determination of such measures of *barīd*, parasang, stadion, differed between ancient nations and dynastic epochs. It is worth noting the ambiguity in the manuscripts regarding the full vocalization of the term that figures in the text as بريدات. This Arabic term can be rendered as *barīdāt*, as I have presented it here, or it can be vocalized as *buraydāt* (as the plural of *burayda*), consequently referring by this to 'papyri' in an indication of the paperwork that establishes quantity surveys in connection with measuring magnitudes of distance.

42 The cord of *al-ashl* (pl., *al-ashūl*) refers to threads taken from a type of a *Juncus* plant (*al-ashl*) that is used to make cables; it is also referred to as *al-ḥabl*, namely, as 'the rope', and it is equivalent to the Greek *skhoinion*, being almost twice longer than the Greek *amma* (the latter being an Attic Greek unit that measures lengths). The arm, as *al-dhirāʿ*, is akin to the cubit (*cubitus*), though, unlike the cubit, it is the full stretched arm as it extends from the fingertip to the shoulder instead of stretching from the fingertip to the elbow. As for the fist, *al-qabḍa*, it is akin to the palm (*palmus*, or hand), though closed rather than open. The finger, *al-iṣbaʿ*, is the *digitus* as the finger-width at the midpoint of the finger or at the knuckles

Know that the finger is equal to six straw-stalks [*shaʿīrāt*; *shuʿayrāt*] that are contiguous and stacked tightly on top of one another. The fist is equal to four fingers. The arm is equal to eight fists [palms], which is thirty-two fingers. The doorway has a length of six [stretched] arms, which is forty-eight fists, or 192 fingers. The cord is a rope [*ḥabl*] that is of the length of ten doorways, [equal to] sixty arms, [or to] 480 fists [palms], or to 1,920 fingers.[43]

Know, O brother, that if these [five] magnitudes [finger, fist, arm, doorway, and cord] are multiplied by one another [respectively] the

---

(this gives a range in width). These measurements of lengths, and the ones that are used in determining the surface areas of plots of land, as mentioned below in the body of the text, are all also noted with their definitions in the *Mafātīḥ al-ʿulūm* (Keys to the Sciences) of Muḥammad ibn Aḥmad ibn Yūsuf al-Kātib al-Khwārizmī, who was a contemporary of the Brethren of Purity. His treatise was composed around 977CE, and the various magnitudes of measurement that were specifically used in Iraq in his time were defined in Book (*maqāla*) I, Section (*bāb*) 4, Chapter (*faṣl*) 6. This author should not be confused with the celebrated mathematician and algebraist Muḥammad ibn Mūsā al-Khwārizmī (d. ca. 850CE). For the Arabic critical edition and annotated Latin rendition of the *Mafātīḥ al-ʿulūm*, refer to *Liber Mafâtîh al-Olûm explicans vocabula technica scientiarum tam arabum quam peregrinorum*, ed. G. van Vloten (Leiden: Lugduni-Batavorum–E. J. Brill, 1895). The revised Arabic text figures also in a Cairo edition published in 1930, as follows: al-Khwārizmī, *Mafātīḥ al-ʿulūm*, ed. Muḥammad Kamāl al-Dīn al-Adhamī (Cairo: ʿUthmān Khalīl, 1930). Even though the genre of the *Mafātīḥ al-ʿulūm* treatise is different from that of the *Epistles of the Brethren of Purity*, the definitions it includes on the magnitudes of measure and also on arithmetic and geometry (in Sections 4–5 of Book II) are almost the same as those noted in the Brethren's epistles on arithmetic and geometry. Whilst the definitions and technical mathematical vocabularies of the *Rasāʾil* and the *Mafātīḥ* have resonances with those of al-Fārābī's *Enumeration of the Sciences* (*Iḥṣāʾ al-ʿulūm*), the style and overall contents of each of these proto-encyclopaedic compendia are different from one another. See C. E. Bosworth, 'A Pioneer Arabic Encyclopedia of the Sciences: al Khwārizmī's Keys of the Sciences', *Isis* 54, 1 (1963), pp. 97–111.

43 If the width of one finger = 2cm (at the knuckles), then a fist (palm-width) = 4 fingers = 8cm; and the length of a stretched arm = 8 fists = 32 fingers = 64cm; and a doorway = 6 arms = 48 fists = 192 fingers = 384cm; and the cord = 10 doorways = 60 arms = 480 fists = 1,920 fingers = 3,840cm. The measuring units of finger, fist, arm, doorway, and cord are respectively quantified in approximate values in centimetres as: 2, 8, 64, 384, and 3,840. Let a finger be x = 2cm, then: a finger multiplied by itself = x^2 = 4cm^2; a fist multiplied by itself = $16x^2$ = 64cm^2; an arm multiplied by itself = $1{,}024x^2$ = 4,096cm^2; a doorway multiplied by itself = $36{,}864x^2$ = 147,456cm^2; a cord multiplied by itself = $3{,}686{,}400x^2$ =14,745,600cm^2.

products will be fractional [*taksīr*].[44] If these are added, they result in the [magnitudes] *jarībāt*, *qafīzāt*, and *ʿashīrāt*.[45]

The calculation of these [magnitudes] is as such: a fist multiplied by itself is equal to sixteen [square] fingers.[46] An arm [multiplied] by itself is sixty four [square] fists, or 1,024 [square] fingers; and this is equal to a ninth of the quarter of the tenth of a tenth of a *jarīb*. A doorway [multiplied] by itself is thirty-six [square] arms [p. 98]; and this is its form: 36, which is 2,304 [square] fists, equal to 36,864 [square] fingers, and which is a tenth of a tenth of a *jarīb*.[47]

A *cord* [multiplied] by itself equals a *jarīb*,[48] which equates ten *aqfiza* [pl. of *qafīz*, which is also referred to as *qafīzāt*], and it is equal to one hundred *ʿashīr*; and this its form: 3,600 [square] arms, 230,400 [square] fists, (fol. 19b) or 3,686,400 [square] fingers.

As for the *qafīz*,[49] it equates ten *ʿashīrāt* [pl. of *ʿashīr*], and is equal to ten [square] doorways. It [also] amounts to multiplying a [magnitude

---

44 In this context of surveying, the terms *taksīr*, *mukassar*, and *mukassara* all designate the squaring of surface area magnitudes, as in square feet, square inches, etc.

45 These refer respectively to the plots of land that are decreasing in the scale of their surface areas. The *jarīb* (pl. *jarībāt*) is of an optimal minimal size for arable land which is ready to be ploughed and potentially will yield crops. The *qafīz* (pl. *qafīzāt*; Latinized in the term *cafiz*) is a tenth of the size of the *jarīb*, whilst the *ʿashīr* (pl. *ʿashīrāt*) is smaller than both.

46 The Arabic term that is used here is *taksīr*, which is derived from the root word *kasr*, as 'fraction'. As noted above, the term *taksīr*, in this context, designates the multiplication of a given magnitude by another that is equal to it in such a way that they form a square (i.e., what is fractured/fractioned, is turned into two magnitudes of length that are perpendicular to each other, and defines by this the two sides of a given square; the term used by the Brethren for this is *mukassar* or *mukassara*). This is similar to saying 'square feet' or, in the case of the measurements that are mentioned by the Brethren in this epistle, to saying 'square fingers', 'square fists', 'square arms', 'square doorways', and 'square cords' (whereby 'square' is designated by the appellations *mukassar* or *mukassara* instead of *murabbaʿ* or *murabbaʿa*).

47 To calculate these magnitudes, we have 36,864 fingers squared, with each finger being 2cm in width = 147,456cm^2 = 14.7456m^2.

48 By rough approximation, this ancient Iraqi surface area calculation is close to a quarter of an ancient Egyptian *faddān* (5,929m^2), a bit less than one and a half of a Levantine *dūnum* (1,000m^2), and a bit more than a third of an acre (4,047m^2).

49 The *qafīz* is accordingly equal to: 10 (3.84m x 3.84m) = 147.456m^2.

that is a] bit less than nineteen arms by itself, which yields [about] 360 [square] arms.[50]

Regarding the *ʿashīr*, it is equal to a doorway multiplied by itself; or such as: 36 [square] arms, 2,304 [square] fists, or 36,864 [square] fingers.[51]

With the cords [*al-ashūl*] multiplied by cords, one [square] cord amongst these [magnitudes] equals a *jarīb*;[52] its tenth equals the tenth of the *jarīb*.[53]

If cords are multiplied by doorways, then one amongst these [magnitudes] yields a *qafīz*, and ten times its magnitude makes the *jarīb*. Whilst with cords multiplied by arms, one amongst these [magnitudes] equals an *ʿashīr* and two-thirds of an *ʿashīr*,[54] and six of its magnitude equals a *qafīz*.

If cords are multiplied by fists, then one amongst these [magnitudes] equals a sixth of an *ʿashīr* and a quarter of a sixth of an *ʿashīr*;[55] and each three fifths of it equals one *ʿashīr*;[56] and every thirty-six of it are equal to a *qafīz*.[57]

If cords are multiplied by fingers, then one amongst these [magnitudes] equals a quarter of a sixth of an *ʿashīr* and a quarter of a quarter of a sixth of a *ʿashīr*;[58] and each ten of these equals a quarter of an *ʿashīr* and a sixth of an eighth of an *ʿashīr*.[59]

If doorways are multiplied by doorways, then one amongst these [magnitudes] equals an *ʿashīr*;[60] and ten of these yield a *qafīz*.[61]

---

50 In other words, equating 18.973666 multiplied by itself with $\sqrt{360}$.

51 The *ʿashīr* is accordingly equal to: 3.84m x 3.84m = 14.7456m^2.

52 As noted earlier a cord multiplied by a cord (i.e., 'one square cord') is equal to a *jarīb*, that is equal to 38.4m x 38.4m = 1,474.56m^2.

53 The remainder of this section on the magnitudes of surveying measures is computed in a convoluted way via sequences of fractions that result from the multiplication of cords, doorways, arms, fists, and fingers with one another, resulting in fractions of the surface area magnitudes of the *ʿashīr*. The notation of these magnitudes and their fractions will be indicated based on the following symbols: cord: *c*; doorway: *d*; arm: *ar*; fist: *fis*; finger: *fin*; *jarīb*: *J*; *qafīz*: Q; and *ʿashīr*: *A*.

54 I.e., $c \,.\, ar = A + \frac{2}{3}A$.

55 I.e., $c \,.\, fis = \frac{1}{6}A + \frac{1}{4}(\frac{1}{6}A)$.

56 I.e., $\frac{3}{5}c \,.\, fis = A$.

57 I.e., $36(c \,.\, fis) = Q$.

58 I.e., $c \,.\, fin = \frac{1}{4}(\frac{1}{6}A) + \frac{1}{4}(\frac{1}{4}[\frac{1}{6}A])$.

59 I.e., $10(c \,.\, fin) = (\frac{1}{4}A) + \frac{1}{6}(\frac{1}{8}A)$.

60 I.e., $d^2 = A$.

61 I.e., $10d^2 = Q$.

If doorways are multiplied by arms, then one amongst these [magnitudes] equals a sixth of an *ʿashīr*;[62] and six of these equal an *ʿashīr*.[63]

If doorways are multiplied by fists, then one amongst these [magnitudes] equals three-quarters of a quarter of a ninth of an *ʿashīr*.[64]

If doorways are multiplied by fingers, then each eighty-five amongst these [magnitudes] equals approximately [*taqrīban*] a third of an *ʿashīr* and a quarter of a sixth of an *ʿashīr* and a ninth of an *ʿashīr*;[65] and every four of these [magnitudes] equal a three-quarter and a ninth of an *ʿashīr*;[66] and every 128 of these [magnitudes] equal two-thirds of a third of an *ʿashīr*.[67]

If arms are multiplied by arms, then one amongst these [magnitudes] equals a quarter of a ninth of an *ʿashīr*;[68] and each four of these [magnitudes] equals a ninth of an *ʿashīr*;[69] and each 100 of these [magnitudes] equals a couple of *ʿashīr* and two-thirds of an *ʿashīr*, and a ninth of an *ʿashīr*.[70]

This is our explicative commentary on surveying widths and lengths [i.e., as they form surface areas]. As for the measure of the magnitude of depth [*misāḥat al-ʿumq*],[71] it obtains from multiplying the length by the width, and then multiplying the product by the depth; what results is the magnitude of the body [*taksīr al-mujassam*; literally, the fractioning of the body in its three dimensions, hence yielding the magnitude of its volume]. Such [surveying] art is needed when excavating water-streams, water-wells, or [when] digging holes [p. 99] or setting the measure-markers for roadways [*al-barīdāt*], erecting water-barriers [*al-masniyyāt*],[72] and laying down of the foundations of houses and of buildings, and the likes of these.

62 I.e., $d \,.\, ar = \frac{1}{6}A$.

63 I.e., $6(d \,.\, ar) = A$.

64 I.e., $d \,.\, fis = \frac{3}{4}(\frac{1}{4}[\frac{1}{9}A])$.

65 I.e., $85(d \,.\, fin) \approx (\frac{1}{3}A + \frac{1}{4}[\frac{1}{6}A] + \frac{1}{9}A)$.

66 I.e., $4(d \,.\, fin) = (\frac{3}{4}A + \frac{1}{9}A)$.

67 I.e., $128(d \,.\, fin) = (\frac{2}{3}[\frac{1}{3}A])$.

68 I.e., $ar \,.\, ar = \frac{1}{4}(\frac{1}{9}A)$.

69 I.e., $4(ar \,.\, ar) = \frac{1}{9}A$.

70 I.e., $100(ar.\, ar) = (2A + \frac{2}{3}A + \frac{1}{9}A)$. All of these calculations above are mixed up in the various manuscripts.

71 The expression ‘*misāḥat al-ʿumq*’ refers in this context to the volumetric magnitude.

72 The *masniyyāt* are types of water-barriers that are less grand than dams (*sudūd*).

Chapter 16

Know, O brother, may God aid you and us with a spirit from Him, that errors may occur in any of the arts [*al-ṣanāʾiʿ*] if those who apply them are not specialists in them, or if they be deficient or distracted in [applying] them. For example, it is mentioned that a man bought a lot of land from another for 1,000 dirhams, on the basis that it was 100 feet in length, (fol. 20a) and 100 feet in width. The other then said to him, 'In compensation, take instead two lots of land each of which is fifty feet in length and fifty feet in width'. The man was under the illusion that this fulfilled his rights. When they appealed before a judge [*qāḍī*] who was not [trained as] a geometer [*muhandis*], he approved this exchange, which was an erroneous [judgement]. The two men appealed after that to an arbiter who was versed in the art of geometry, and he ruled that the [exchange] constituted half of what was owed. Similarly, it is mentioned that a man offered eight dirhams to a workman in order for him to dig a pond that would be four arms in length, four arms in width, and four arms in depth. The workman dug for him two arms in length, in width, and in depth. He asked in return for four and a half dirhams as remuneration. Both got into a dispute with one another over this. They appealed before a jurist [*faqīh*] who was not a geometer, and he ruled that this was the workman's right. They then sought the judgement of specialists in the art of geometry who judged that the workman should be paid one dirham [only]. It was also said to someone who dealt with the art of reckoning, and who was not a specialist in this field, 'What is the ratio of one-thousand thousand [i.e., 10^6] to one-thousand thousand thousand [i.e., 10^9]?' He replied, 'Two-thirds' [which is evidently wrong, since it is 10^{-3}]. The specialists in the art [*ahl al-ṣināʿa*] then said that 'It is a tenth of a tenth of a tenth [i.e., 10^{-3}]'. Similar to these examples, errors occur with those who apply arts that they are not versed in. It is due to this that it is said, 'Seek assistance with any art from its experts!'

Chapter 17
On the Need of Humans to Collaborate

Know, O brother, may God aid you and us with a spirit from Him, that a single human being cannot live alone in but a miserable way, since his needs for well-being depend on the workings [p. 100] of various arts, which cannot be satisfied by a single individual alone, given that life is short, and the arts are manifold. Consequently, in every town and village many people are gathered to assist each other. The divine wisdom and the grace of His Lordship necessitated that a group amongst them works at mastering the arts, another on trade, and yet another group on construction, another on agriculture, and a group on regulating policies, and a group on the sciences and teaching them, and a group that serves all the other ones and seeks to respond to their needs. Their example is that of brothers from a single father who live within the same household and who co-operate in meeting their livelihood, each of them dealing with an aspect of it. What they determine by convention in terms of dry measure [*al-kayl*], weight [*al-wazn*], price [*al-thaman*], and remuneration [*al-ujra*; *al-ajr*], pertains to wisdom and policy-making, in order that these might motivate the exercising of effort in labours, the arts, and in collaborations, all in such a way that each individual is remunerated according to the effort he engages in labouring and in the workmanship of the arts.

Know, O brother, may God aid you and us with a spirit from Him, that you ought to be assured that you cannot be saved by yourself alone from what has befallen you in this world by way of crisis [*miḥna*] and its ills due to the offense committed by our father Adam, peace be upon him! This is the case since, to secure your success and salvation from this world, which is the realm of generation and corruption, and from the sufferings of hell and the company of demons and Iblīs' soldiers, and by way of ascending to the domain of the celestial spheres and the vastness of the heavens, to the abode of the lofty ones [*al-ʿaliyyīn*], and by way of neighbouring (fol. 20b) the angels of the Compassionate One who abide in His proximity, you need the help of those who are brothers to you, who are counsellors to you and virtuous friends, and who are knowledgeable about the articles of faith and are knowers of the truths of things. [This is the case] since they would guide you on the pathway of the afterlife and the way to reach it, in order to be saved from what has entrapped us all because of the offense of our father Adam, peace be upon him!

O brother, take heed of the tale of the ring-dove [*al-ḥamāma al-muṭawwaqa*] that is mentioned in the book of *Kalīla wa-Dimna*,[73] and how it escaped from the net, in order that you grasp the truth of what we have said.

Know that when the philosophers [*al-ḥukamāʾ*] give an example from worldly affairs, they allude by this to the matters of the afterlife, by way of hinting at them through examples, and based on what the minds of people are capable of grasping across space and time.

[p. 101]

Chapter 18
On Intellective Geometry

Since we have noted some of the aspects of sensible geometry [*al-handasa al-ḥissiyya*], by way of an introduction and prolegomena, we also want to mention something about intellective geometry [*al-handasa al-ʿaqliyya*]. [This is the case] since it is one of the aims of the scholars who are well grounded in the theological sciences and who are well trained in the philosophical propaedeutic [sciences], given that the aim behind their presentation of geometry after arithmetic is to elevate the apprentices from the realm of the sensible entities [*al-maḥsūsāt*] to that of the intelligible entities [*al-maʿqūlāt*], as well as elevating their students and progenies from corporeal matters to the spiritual ones.

Know, O brother, may God aid you and us with a spirit from Him, that studying sensible geometry yields dexterity in all the practical arts [*al-ṣanāʾiʿ al-ʿamaliyya*], and that studying intellective geometry results in skilfulness in the theoretical arts [*al-ṣanāʾiʿ al-ʿilmiyya*]. This is the case, given that this art [of geometry] constitutes one of the gateways

73 A famous animal fable that carries the names of two jackals Kalīla and Dimna was translated from the original Indian sources into the Arabic language by the litterateur Abū Muḥammad ʿAbd-Allāh ibn al-Muqaffaʿ around 750 CE.

to reach knowledge about the substance of the soul [*maʿrifat jawhar al-nafs*], which is at the root of the sciences and is the element [*ʿunṣur*] of wisdom, and [which is also] the origin of all the theoretical and practical arts, meaning by this the knowledge about the substance of the soul — so, grasp all that we have said!

Chapter 19
On Imagining Distances[74]

The conceptual line [*al-khaṭṭ al-ʿaqlī*][75] cannot be seen but in-between two surfaces. It is like differentiae [*al-faṣl al-mustarak*][76] common to sun [light] and shadows. If there was no sun [light] or shade, then you would not see a line with two imagined [*wahmiyyatayn*] points. If you imagine that one point is moving and the other is at rest, in such a way that [the mobile] returned to where it began its movement, a surface is imagined in your mind [*fikrak*; literally, 'your thought'].[77]

Likewise, a conceptual surface [*al-saṭḥ al-ʿaqlī*] cannot be seen but in-between two solids. It is like differentiae shared between water and grease.

The conceptual point [*al-nuqṭa al-ʿaqliyya*] cannot be seen but where a line is imagined as being divided into two halves. Any position that is pointed out would be where a point ends.

[p. 102] Know, O brother, that if you imagine the motion of a point rectilinearly, then an imagined straight line occurs in your mind. If you imagine the motion of this line in a direction other than that taken by the point, then an imagined surface occurs in your mind. If you imagine the motion of this surface to be in a direction other than

74 The title of this chapter is '*Fī tawahhum al-abʿād*'; *wahm* is used herein in the mathematical sense of *takhayyul*, *qua* imagining or postulating.

75 I have used the term 'conceptual' to render the Arabic designator *al-ʿaqlī*, in avoiding the use of the word 'mental' or the more literal expression 'intellective'; although the term 'intellective' has been used in the body of the text earlier to refer to a branch of geometry (*al-handasa al-ʿaqliyya*), the text potentially reads in a smoother way if *al-ʿaqlī* is translated as 'conceptual' when mentioning points, lines, surfaces, and solids.

76 Meaning by this here, 'a shared distinguishing factor'.

77 An image of a surface delimited by the motion of the mobile point around the point at rest is formed in the mind as it traces a circle.

that of the surface or the point, then an imagined solid body occurs in your mind, and it would have six square surfaces that meet at right angles, and this would be a cube.

<The thread of the spider is a sensible line, but, according to the wise philosophers, it is a body such as no matter how much smaller it is cut (fol. 21a) what results from it would [still] be a body. [However], following the opinion of the dialectical theologians [*al-mutakallimūn*; the exponents of *kalām*], if its parts were made smaller and [repeatedly] divided, the last of its [indivisible] parts would not be a body but what they call a 'substance' [*jawhar*].[78] This is the case, since what admits division is a body, according to them, whilst what does not admit it is a substance. The philosophers oppose them on this>.[79]

If the distance of the motion of the surface were lesser than the distance of the motion of the line, then a *libnī* solid [i.e., a parallelepiped shaped like a brick] would result from this; however, if it were greater, then the product would be a *bi'rī* solid [i.e., a parallelepiped shaped like a water-well]; moreover, if they were equal, then this would yield a cube.[80]

Know, O brother, that every straight line that is hypothetically postulated in the imagination ought to have two extremities, which are

---

78 The term 'substance' (*al-jawhar*) is used in this context differently from its Aristotelian sense. Its significance is to be grasped through the atomist physical theory of *kalām*, which takes substance to be the indivisible smallest part, which is the atom (it is usually referred to as the part [*juz'*] that cannot be partitioned or divided: '*al-juz' al-ladhī lā yatajazza' wa lā yanqasim*'). Anything that admits division and partitioning would not yet be an atom, and is therefore a body, no matter how small it might be.

79 The passage that is included here between <…> figures in the Atif Efendi manuscript, and whilst it is a digression from the main theme being discussed in this chapter, its content is significant and noteworthy, since it signals an engaging familiarity with the physical theory of the exponents of *kalām*, which is presented here without explicit objections to its conceptual entailments, even though it conflicts with the Aristotelian outlook of the philosophers, and, more importantly, that it disaccords with what is more commonly seen as being of the order of the belief of the Brethren of Purity themselves.

80 Let distance ds be that of the motion of the surface, and dl be the distance of the motion of the line, then: If $ds < dl$, then, the result is a *libnī* solid; if $ds > dl$, then, the result is a *bi'rī* solid; and if $ds = dl$, then, the result is a cube. These solids are mentioned also in Chapter 13 above, with geometrical illustrations of their forms. They are also noted in Chapter 21 of the Epistle 1: 'On Arithmetic' in connection with the properties of numbers.

its two heads [*ra'sayh*], and these are called the 'two imagined points' [*al-nuqṭatayn al-wahmiyyatayn*]. If you imagine that one of these points is set in motion whilst the other is at rest in such a way that the mobile one returns to where it started in movement, then, from this, a circular surface would occur in your mind, with the point at rest being its centre, and the mobile point, through its motion, tracing the circumference of the circle in your mind. Then know that the first surface to result from its [i.e., the mobile point's] motion would be a third of a circle, then a quarter of a circle, then a semicircle, and then the [whole] circle. If you were then to imagine that the two extremities of the arched line of a semicircle were at rest, then, if that line were set in motion in such a way that it would return to where it started its mobility, then a spherical body would result in your mind from its motion.

Chapter 20

It is evident to you from what we have stated that intellective geometry consists of studying the three dimensions of length, width, and breadth as abstracted from natural bodies. This is the case, given that if those who study sensible geometry, which we mentioned earlier, are well practised in this [art] and have strengthened their thought in studying it, then they derive these three dimensions in the line, the surface, and the solid. They visualize these in their own souls [*nufūs*] and they reflect upon them as abstracted from matter [*hayūla*]. They consequently arrive at the substance [*jawhar*] of their own souls through these images that are visualized in them. These [appear] like matter, and yet they are in it as the form [*ṣūra*],[81] which they call 'the surveying magnitudes'

81 If the form of a given object is abstracted from the matter of the body of this object by way of being imagined in the mind (or the soul), then the image of the form in the mind, as abstracted from matter, is itself indicative of the essence of the mind as that which can imagine things in their formal shape as abstracted from their own material constitution. The imagined forms have the semblance of being corporeal when visualized in the mind, even though they are immaterial and simply mental. They are rather like the form entangled with matter in the case of the existents in the terrestrial sphere and the sublunary realm as composite entities from matter and form.

[*maqādīr masāḥiyya*]. They consequently discard the examination of sensory magnitudes [*maqādīr ḥissiyya*]. They then comment on these [p. 103] and report on their genus, kinds, and attributes, and on all the properties that result when they are added to one another.

They accordingly say, 'The line is a magnitude with one dimension, and the surface is a magnitude with two dimensions, and the solid is a magnitude with three dimensions'. [And] 'The straight line is the shortest line that connects two points, and the point is the extremity of the line'. [And] 'The arched line is that which cannot have three points on a rectilinear stretch [*samat*]'. [And] 'Angles occur as the contact between two lines that are extended in (fol. 21b) a non-rectilinear stretch'. [And] 'The shape is what has been delimited by a single line or by multiple lines'. 'The circle is a shape delimited by a single line, which is called the "circumference", and within it a point is posited in such a way that all straight lines that are extended from it would be equal'. And 'The triangle is a shape delimited by three lines, while having three angles'. [And] 'The square is a shape delimited by four lines, and it has four right angles'. The like of what they say follows this pattern and example in talking about the geometric shapes, without making reference to any given body of the natural bodies.

Chapter 21
On the Reality of Dimensions in Intellective Geometry

Know that numerous geometers and scientists believe that dimensions, like length, width, and breadth, have an existence in themselves and by their own constitution. They do not realize, therefore, that such existence is in the substance of the body or in the substance of the mind, and that these [dimensions] are for it like matter [*al-hayūlī*], or they are in it like in the form that has been derived by the thinking faculty from sensible entities.[82]

82 Dimensions, like length, width, and depth, do not have an existence in themselves or due to their own existential constitution. The existence of these dimensions is either integral to the substance of the body, and they exist therefore as part of a material constitution in natural corporeal entities, or the existence of these dimensions is secured at the level of imagination by our faculty of thought, and they belong as such to the substance of the mind as they get derived in abstractive

If they were to be informed that the ultimate aim of studying the propaedeutic sciences is training the apprentices' minds to receive the forms of the sensible entities through the sensory faculties [*al-qiwā al-ḥassāsa*], and to imagine their forms as they are in themselves through the faculty of ratiocination [*al-quwwa al-mufakkira*]. If the sensible entities were to be concealed from being perceivable through the senses, then the images [*al-rusūm*] formed of them, which have been transmitted from the sensory faculties to the faculty of imagination [*al-quwwa al-mutakhayyila*], and from the faculty of imagination to the faculty of ratiocination,[83] would persist. The faculty of ratiocination then transmits [these forms] to the faculty of memory [*al-quwwa al-ḥāfiẓa*] in such a way that they are visualized [*muṣawwara*][84] in the essence of the soul. In consequence, the soul abandons the use [p. 104] of the sensory faculties in grasping data when being self-reflective, since it finds all the forms of the sense data in its own essence. It [the soul] is therefore in no need of the body [*al-jasad*], and is ascetic while dwelling in the world with it. It is alert to the slumber of inattentiveness [*nawm al-ghafla*] and wakeful from the slothfulness of ignorance [*raqdat al-jahāla*]. [The soul] rises up by its own potency and is autonomous by itself. It departs from the body and exits the ocean of matter, and is saved from being captivated by nature. It is freed from slavishness to its bodily desires, and it is delivered from the pain of missing its corporeal pleasures. It is elevated higher as He said: *Good words rise up to Him and He lifts up the righteous deed.*[85] He meant by it the 'chastened soul'

---

reflections from the sensible corporeal entities. This reflects the binary dualism of mind and body; dimensions are either bodily or they are mental abstractions derived by way of imaginative postulations from natural bodies.

83 It is worth mentioning that the transmission of forms from the senses to imagination, and then from this faculty to that of ratiocination was itself omitted in the body of the text within the Atif Efendi manuscript, and it was corrected afterwards within the margins of that same manuscript by being stated in the form that is translated herein. This process of revision and correction shows a more nuanced grasp of the cognitive character of the interactions between the various faculties of the mind.

84 The Atif Efendi manuscript uses the poetic term *maqṣūra* (instead of *muṣawwara*), whereby it offers the alternative and pertinent reading: '... that they take a dwelling in the essence of the soul', rather than simply being 'visualized in the essence of the soul'.

85 Qur'an 35:10. This Qur'anic verse was omitted from the Atif Efendi manuscript

[*al-nafs al-zakiyya*] that is recompensed with the finest of rewards. This is the ultimate aim behind studying the propaedeutic sciences that the sons of the philosophers and the pupils of the ancients graduated from. This is the credo of our righteous and esteemed brethren. May God grant you and us success on the pathway of guidance; He is most kind towards His servants! Our righteous and merciful brethren direct unto Him their invocations. May God aid them and us wherever we may be, for He is most bountiful and most munificent![86]

* * * * * * *

Chapter 22
On the Properties of Geometrical Figures

Know, O brother, may God aid you and us with a spirit from Him, that the geometrical figures have properties, like numbers have properties, and that combinations of them have characteristics too, and so is the case with their individual constituents. We have shown in the epistle on arithmetic some of the properties of numbers, and we seek to mention in this present epistle some aspects of the properties of geometrical figures, in order that those who study these two sciences might be alerted to the aims that lie behind them, (fol. 22a) and so that it might constitute a guidance to those who seek to know the properties of things and the manner of their comportment.

We begin firstly with the triangles [*al-muthallathāt*], since they are the first amongst the geometrical figures;[87] we therefore say: [p. 105]

---

but it figured in the Ṣādir edition.

86 The Atif Efendi copyist indicates at this stage (fifth line from the bottom of fol. 21b) that 'the epistle is completed [*tammat al-risāla*]', as if signalling by this that this treatise is concluded, and that what follows is either an appendix or an afterthought addition, even though the thematic content is still connected with the art of geometry.

87 The Atif Efendi manuscript and also the text of the Ṣādir edition both mention in this context that 'this has been demonstrated in the epistle on geometry [*bayyannā fī risālat al-jūmāṭrīyyā*]'. This corresponds with what the copyist of the Atif Efendi indicated at the end of Chapter 21 (in the fifth line of fol. 21b) that 'the epistle is completed'. This is the case even though what is being noted in the remainder of the text is still integral to the art of geometry, and, at the end of this whole treatise, it is noted that the epistle is completed, and a transitional

The triangle [*al-muthallath*] is that which has three sides and three angles, and it is of seven kinds:

1. The first is the acute-angled equilateral [*al-mutasāwī al-aḍlāʿ al-ḥādd al-zawāyā*], like this:

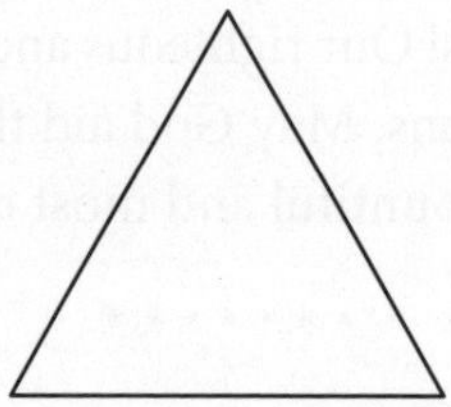

2. The second is the acute-angled isosceles [*al-ḥādd al-zawāyā al-mutasāwī al-ḍilʿayn*], like this:

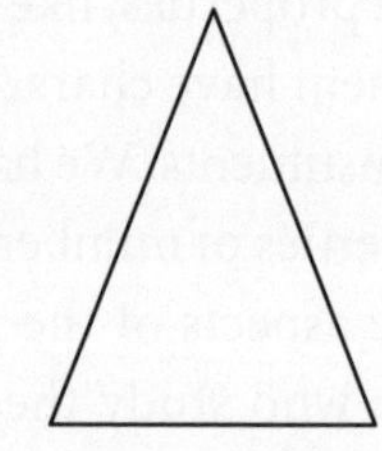

3. The third is the acute-angled scalene triangle [*al-ḥādd al-zawāyā al-mukhtalif al-aḍlāʿ*], like this:

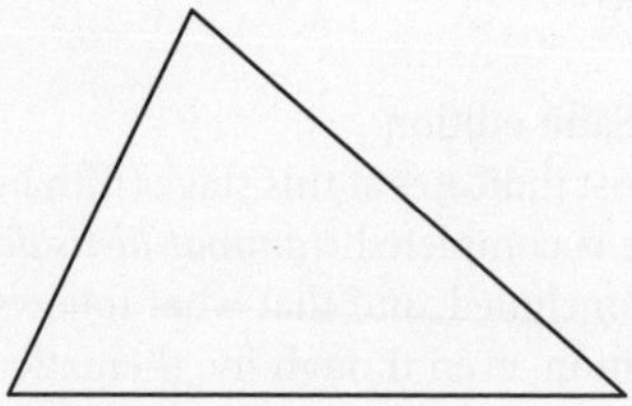

4. The fourth is the right-angled isosceles triangle [*al-mutasāwī al-ḍilʿayn al-qāʾim al-zāwiya*], like this:

phrase is included by way of introducing the third epistle of the compendium, which deals with astronomy.

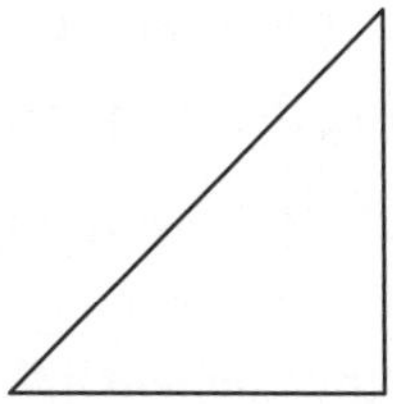

5. The fifth is the right-angled scalene triangle [*al-qāʾim al-zāwiya al-mukhtalif al-aḍlāʿ*], like this:

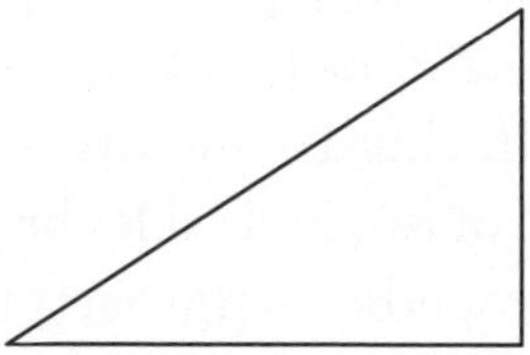

6. The sixth is the obtuse-angled isosceles triangle [*al-munfarij al-zāwiya al-mutasāwī al-ḍilʿayn*], like this:

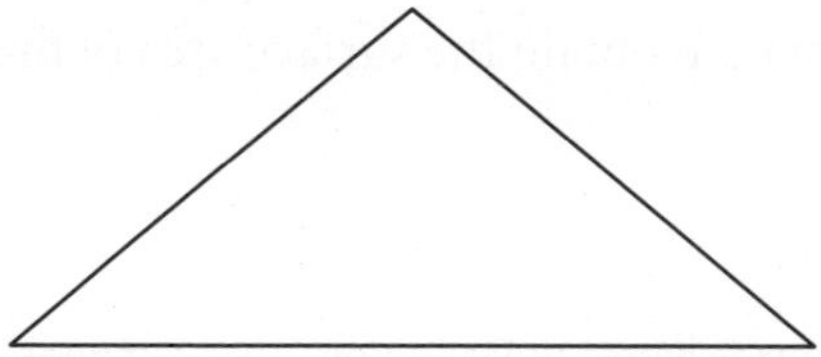

7. [p. 106] The seventh is the obtuse-angled scalene triangle [*al-munfarij al-zāwiya al-mukhtalif al-aḍlāʿ*], like this:

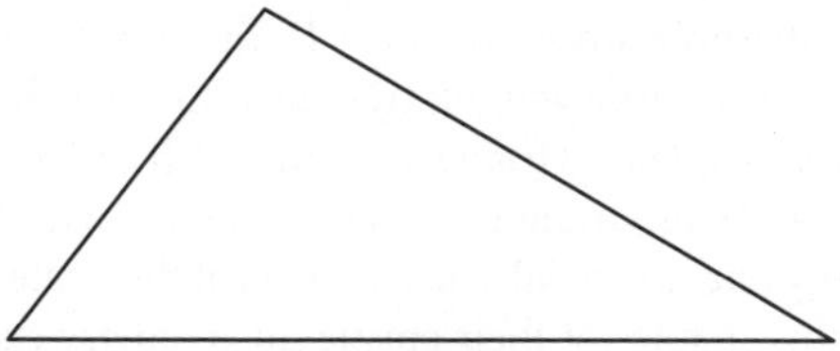

Chapter 23
On Demonstrating the Properties [of Triangles]

Know, O brother, that each of these triangles has a property that is not shared with another. This has been demonstrated with proofs in Book I of Euclid's treatise [the *Elements*].[88] Nonetheless, we shall note the property that is common to all the seven [kinds], for the property of any trilateral figure [*shakl muthallath*], no matter what triangle it constitutes, is that it ought to have two acute angles, whilst the third angle can be acute, a right angle, or obtuse. One of its properties also is that the sum of the three angles of every given triangle is equal to two right angles. Another of its properties is that the longer side of every given triangle is the chord of the largest of its angles. It also has the property of the sum of two of its sides being longer than its third side. Also of its other properties is [the fact] that if any side amongst its sides is extended rectilinearly outside the [triangle's] perimeter, it generates an external angle to the triangle, which would be larger than the [internal] angle that is adjacent to it, whilst being equal to the sum of the other two [internal] angles that are opposite to it. Moreover, another of its properties is that if you multiply the height of a triangle by half of its base, you then obtain the surface area of that given triangle.[89]

---

88 The reference is made herein to Definitions 10–12 of Book I of Euclid's *Elements* on the types of angles (respectively: right [*orthê*], obtuse [*ambleia*], acute [*oxeia*]), and then in Definitions 20–21 in the same text on the types of triangles. The latter read as follows: Definition 20: 'Of trilateral figures, an equilateral [*isopleuron*] triangle is that which has its three sides equal, and an isosceles [*isosceles*] triangle that which has two of its sides alone equal, and a scalene [*skalênon*] triangle that which has its three sides unequal'; Definition 21: 'Further of trilateral figures, a right-angled [*orthogônion*] triangle is that which has a right angle, an obtuse-angled [*amblugônion*] triangle is that which has an obtuse angle, and an acute-angled [*oxugônion*] triangle is that which has an acute angle'. Whilst the Brethren of Purity refer to Euclid's *Elements*, what they state about the properties of triangles in this chapter of their epistle on geometry exceeds what is noted in the definitions of Book I of the *Elements* in this regard.

89 To find the area of a given triangle, multiply its height by its base then divide the product by two. This can be geometrically determined by having a parallelogram that is divided into two equal triangles when its surface is divided in half across one of its diagonals.

Chapter 24

The property of a right-angled triangle is that the squaring of the chord that is subtended by its right angle is equal to the squaring of its two other sides.[90]

One of the properties of the acute-angled triangle is that the square on the chord [*al-watar*] is lesser than the sum of the squares on its remaining two sides [p. 107] by twice the magnitude of the segment that extends between the [acute] angle and the point where the perpendicular falls [i.e., twice the square of the height or altitude of this acute-angled triangle].[91]

Of the properties of the obtuse-angled triangle is [the fact] that the square on the chord is greater than the sum of the squares on the other sides by twice the magnitude of the square of the segment that extends between the [obtuse] angle and the point where the perpendicular falls [i.e., twice the square of the height or altitude of this obtuse-angled triangle], such as in this [form]:[92]

---

90 This describes the Pythagorean Theorem, whereby: in any right triangle, the area of the square whose side is the hypotenuse (as the chord subtended by the right angle; or as the side opposite the right angle) is equal to the sum of the areas of the two squares whose respective sides are the two other sides of the triangle. Let a and b be the two sides of a given right-angled triangle, and let c be its hypotenuse, then $a^2 + b^2 = c^2$. This is stated in Proposition 47 of Book I of Euclid's *Elements* as follows: 'In right-angled triangles the square on the side subtending the right angle is equal to the squares on the sides containing the right angle'. This theorem was commented on by various scholars, such as Proclus, Plutarch, Pappus, Heron, and Thābit ibn Qurra.

91 In the figure shown below, let a, b, c be the sides of this acute-angled triangle, such that c is the side subtending the acute angle, and also let h be the height of the triangle, then: $a^2 + b^2 = c^2 + h^2$

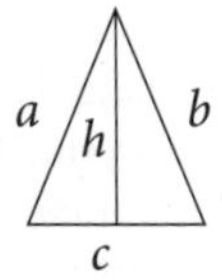

92 In the figure shown in the body of the text, let a, b, c be the sides of this obtuse-angled triangle, such that c is the side subtending the obtuse angle, and also let h be the height of the triangle, then: $a^2 + b^2 = c^2 - h^2$. The more generalized equation would be: $c^2 = a^2 + b^2 - h^2$

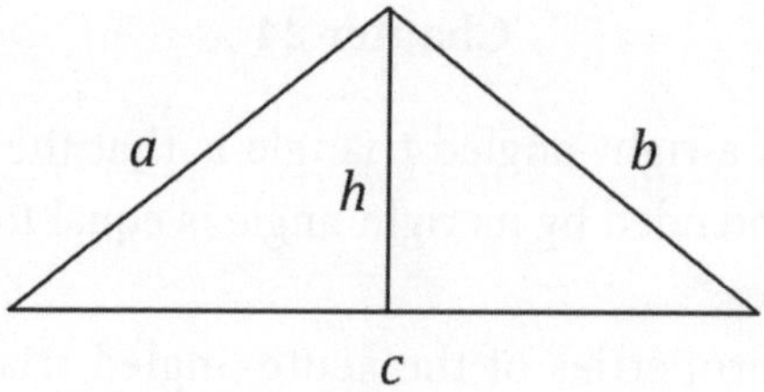

Chapter 25[93]

The quadrilateral figure [*al-shakl al-murabbaʿ*] is that which has four sides and four angles, and it is of five kinds:[94]

1. The first is equilateral and right-angled [*al-mutasāwī al-aḍlāʿ al-qāʾim al-zawāyā*; i.e., a square], like this:

93 The copyist of the Atif Efendi manuscript includes a new untitled chapter heading in this context, which seems pertinent and acts as a valid thematic transition within the text.

94 These five kinds of quadrilaterals are noted in Definition 22 of Book I of Euclid's *Elements* as follows: 'Of quadrilateral [*tetrapleurôn*] figures, a square [*tetragônon*] is that which is both equilateral [*isopleuron*] and right-angled [*orthogônion*]; an oblong [*heteromêkes*] that which is right-angled but not equilateral; a rhombus [*rombos*] that which is equilateral but not right-angled; and a rhomboid [*romboiedes*] that which has its opposite sides and angles equal to one another but is neither equilateral nor right-angled. And let equilaterals other than these be called trapezia'. Having noted that, there are in fact seven species of quadrilateral figures that are determined in two main sub-classes: parallelograms and non-parallelograms. The parallelograms can be rectangular and non-rectangular; with the rectangular encompassing the square and the oblong, and the non-rectangular covering the rhombus and rhomboid. Of the non-parallelograms are classed the trapezium and the trapezoid, the former having two of its sides parallel, whilst the latter does not have any parallel sides. Of the trapezia (as non-parallelograms with two parallel sides), there is the isosceles trapezium and the scalene trapezium. Refer to Thomas L. Heath's commentary on this sevenfold classification in *The Thirteen Books of the Elements*, vol. 1, p. 189.

2. The second is oblong and right-angled [*al-mustaṭīl al-qā'im al-zawāyā*; i.e., a rectangle], which has each (fol. 22b) of its opposite [and parallel] sides equal [*al-mutasāwī fī kull ḍilʿayan mutaqābilayn*]:

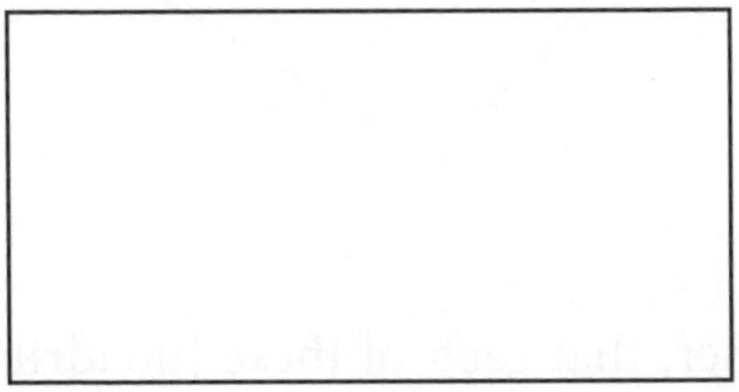

3. The third is the rhombus [*al-muʿayyan*], which has equal sides with unequal angles [*al-mutasāwī al-aḍlāʿ, al-mukhtalif al-zawāyā*], like this:

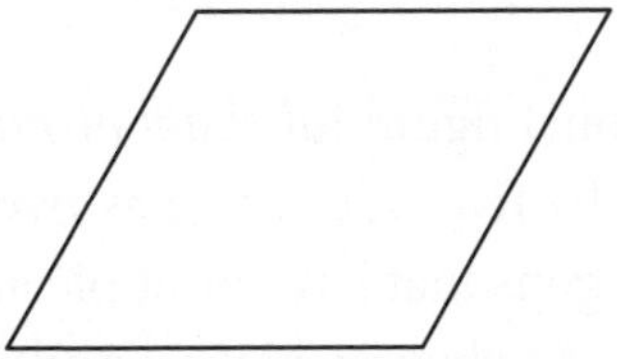

4. The fourth is a rhomboid [*al-shabīh bi'l-muʿayyan*] that has each of its two opposite sides equal [*al-mutasāwī fī kull ḍilʿayn mutaqābilayn*], like this:

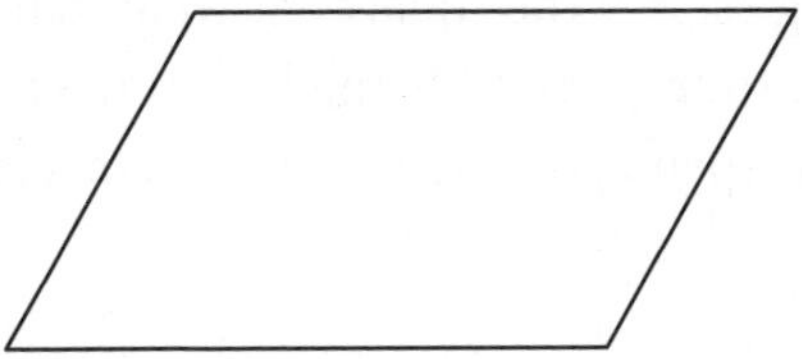

5. [p. 108] The fifth is the trapezoid that has unequal sides and unequal angles [*al-mukhtalif al-aḍlāʿ wa'l-zawāyā*],[95] like this:

95 As indicated in the footnote above, the trapezoid is different from the trapezium, even though they are both non-parallelograms. The *trapezoid* does not have any parallel sides (as shown in the body of the text by the Brethren of Purity),

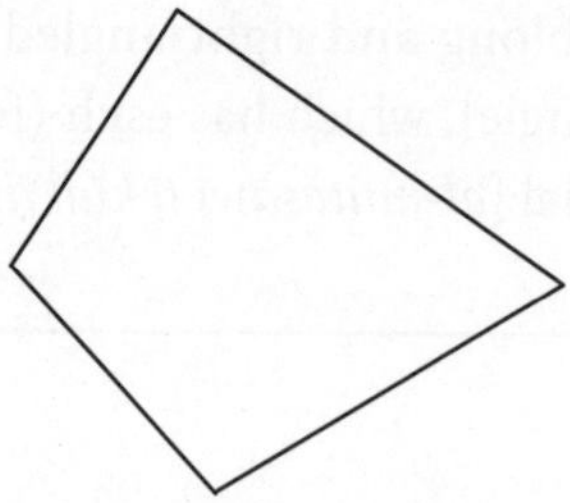

Know, O brother, that each of these [quadrilateral] figures has properties that require extensive commentaries to explicate them. Nonetheless, we note here properties that are common to them all, namely, that each quadrilateral, whatever [kind] it may be, will have the sum total of its four angles equal to four right angles. Moreover, every quadrilateral [figure] can possibly be divided into two triangles, and if another triangle be added to it, the product would be a pentagonal solid.[96]

As for the pentagonal figure [*al-shakl al-mukhammas*], it is that which is surrounded by five sides and has five angles. It is also the first amongst the polygons that have multiple angles and equal sides.

Of the properties of polygons having multiple angles and equal sides, is [the fact] that a circle might envelop each one of them, and that each amongst them might also surround a circle. Each figure of these [i.e., the pentagons], which has more angles, is greater and larger in surface area than a surface that is smaller than it [i.e., that also has fewer angles, like the square], if what delimits them both is of the same magnitude.[97] Multiply the height of each of the triangles that constitute this [pentagon] with half the [correlative] base in each

whilst a *trapezium* has two of its sides parallel, and it could be either an isosceles trapezium or a scalene trapezium.

96 The Brethren of Purity do not add in this context a crucial parameter, that in order for this shape, which consists of three triangles, to constitute a solid that has a hollowed pyramidal form, its trilateral elements should be spatially folded twice along the lines of their respective contacts with one another.

97 If a circle contains a pentagon, a square, and a triangle, then the pentagon will have a larger surface area than that of the square or the triangle, and, furthermore, the square will be greater in surface area than the triangle.

of these triangles, then [the sum of their areas] constitutes the area of this polygon. And this is its form:

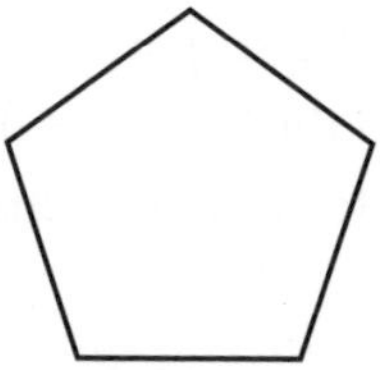

Concerning the hexagonal figure [*al-shakl al-musaddas*] that has equal sides, each of its sides is equal to half the diameter of the circle that surrounds it.[98]

In general, every figure has a property or multiple properties that are its own, and we have left off their mention here in view of eschewing verbosity.

As for the properties of the circular figure [*al-shakl al-mustadīr*], these were noted in Book III of Euclid's *Elements*, and we note some of its aspects here. We therefore say that the circular figure is a surface that is surrounded by a single [curved] line, which has its centre within its middle, and its diameters are equal,[99] and that it is the largest of all figures that are polygonal if they were all isoperimetric. This [figure] shares the properties of the circle, and its relation to other bodies is like the relation of the circle to all other surfaces. The properties of this figure have been demonstrated [p. 109] in the last Book of Euclid's [*Elements*][100] by way of commentaries and proofs.

---

98 These remarks on the pentagon and the hexagon are inspired from adaptations of Propositions 7–11 of Book XIII of Euclid's *Elements*.

99 These remarks on the circle repeat some of the earlier comments in Chapters 7–8, and they all correspond with Definitions 15–17 of Book I of Euclid's *Elements*: Definition 15 states that 'A circle [*kuklos*] is a plane figure contained by one line such that all the straight lines falling upon it from one point among those lying within the figure are equal to one another'; Definition 16 states that 'The point is called the centre [*kentron*] of the circle'; Definition 17 states that 'A diameter [*diametros*] of a circle is any straight line drawn through the centre and terminated in both directions by the circumference [*periphereias*] of the circle, and such a straight line also bisects the circle'.

100 It is unclear if the Brethren squarely meant by this 'Book XIII of Euclid's *Elements*'.

On the whole, if you reflect, O brother, on the aims of what is noted in Euclid's treatise from such demonstrations, and grasp what is set down in all the books of geometry,[101] you will then find that they all consist of an inquiry about the properties of magnitudes and knowing their truths as comprised in the lines, the surfaces, and the solids, and in what obtains to them in terms of dimensions, angles, and relations between one another.

Chapter 26

Since we have shown some of the properties of these figures in this present epistle, and, prior to that, having shown some of the properties of numbers in the epistle 'On Arithmetic', we want to (fol. 23a) mention furthermore some of the properties of their association. This is the case given that when some numbers are associated with some of the geometrical figures, other properties become manifest from them that would not appear in either of them by itself.[102] An example of this would be if the [set of the first] nine units [i.e., from 1– 9][103] are noted in a

---

101 It is unclear what is meant by this expression as it figures with variants in the manuscripts. Perhaps what the Brethren of Purity intended by this is the thirteen books of Euclid's *Elements*.

102 The examples that the Brethren of Purity offer in this regard focus on figurate numbers that are arranged within sets of cells that are shaped as squares in orthogonal horizontal rows and vertical columns, which correspond to the following numbers and their roots: 9 (3 [rows] multiplied by 3 [columns]), 16 (4 [rows] multiplied by 4 [columns]), 25 (5 [rows] multiplied by 5 [columns]), 36 (6 [rows] multiplied by 6 [columns]), 49 (7 [rows] multiplied by 7 [columns]), 64 (8 [rows] multiplied by 8 [columns]), and 81 (9 [rows] multiplied by 9 [columns]).

103 What the Brethren of Purity mean by this is that the units of ones are now taken as the *āḥād*, instead of the numbers per se as the *aʿdād*. This nuance is significant in order that what they indicate here corresponds with their classifications in the epistle 'On Arithmetic', given that the 1 is not a number per se but is an iteration unit. The first number is thus 2; and the first set of nine numbers should be from 2 to 10 instead of 1 to 9. However, if they adopted this view in this context within the epistle 'On Geometry', hence starting from 2 and ending in 10, the set of nine numbers that this encompasses as also set within the ninefold geometric figure would not have corresponded with the recurrent result of having 15 as the sum-total of the addition of each set of three of these numbers whether this is done horizontally, vertically, or diagonally.

ninefold figure that has the following form [as shown below], their property as embedded within this ninefold figure would be such that if you were to add them up in any way [i.e., horizontally, vertically, diagonally], the sum [*al-jumla*] would be fifteen, like this:

2	7	6
9	5	1
4	3	8

The same would apply to the case of [the set of the first] sixteen units [i.e., 1–16] if they were to be noted in a figure that consists of sixteen cells [*baytan*] like in this following form [as shown below]; its property is such that if you were to add them up in any way [i.e., horizontally, vertically, diagonally], the sum would be thirty-four, like this:[104]

4	14	15	1
9	7	6	12
5	11	10	8
16	2	3	13

[p. 110] The same would apply to the case of [the set of the first] twenty-five units [i.e., 1–25] if they were to be noted in a figure that consists of twenty-five cells, like in this following form [as shown below]; its property is such that if you were to add them up in any way

104 It is worth noting here that the copyist of the Atif Efendi manuscript included within the body of the text what appears to be a brief 'instruction note' written on a separate piece of paper. This would seem to be a note for the draftsman to assist in drawing the table (*al-jadwal*) containing the sixteen square cells. However, the copyist of the manuscript did not set this note as being separate from the text itself, or indicate it in the margins; rather, he (mistakenly?) thought of it as being integral to the content of the epistle. The instruction note literally reads as follows: 'use for the form of the table that which is indicated on the paper' (*yuʿtamad fī ṣūrat al-jadwal ʿalā mā fī al-waraqa* [lines 6–7 fol. 23a]).

[i.e., horizontally, vertically, diagonally], the sum would be sixty-five, like this:

21	3	4	12	25
15	17	6	19	8
10	24	13	2	16
18	7	20	9	11
1	14	22	23	5

<The same would apply to the case of [the set of the first] thirty-six units [i.e., 1–36] if they were to be noted in a figure that consists of thirty-six cells, like in this following form [as shown below]; its property is such that if you were to add them up in any way, the sum would be 111, like this:

11	22	32	5	23	18
25	16	7	30	13	20
27	6	35	36	4	3
10	31	1	2	33	34
14	19	8	29	26	15
24	17	28	9	12	21

The same would apply to the case of [the set of the first] forty-nine units [i.e., 1–49] if they were to be noted in a figure that consists of forty-nine [p. 111] cells, like in this following form [as shown below]; its property is that if you were to add them up in any way, the sum would be 175, like this:

47	11	8	9	6	45	49
4	37	20	17	16	35	46
2	18	26	21	28	32	48
43	19	27	25	23	31	7
38	36	22	29	24	14	12
40	15	30	33	34	13	10
1	39	42	41	44	5	3

The same would apply to the case of [the set of the first] sixty-four units [i.e., 1–64] if they were to be noted in a figure that consists of sixty-four cells, like in this following form [as shown below]; its property is such that if you were to add them up in any way, the sum would be 260, like this:

52	61	4	13	20	29	36	45
14	3	62	51	46	35	30	19
53	60	5	12	21	28	37	44
11	6	59	54	43	38	27	22
55	58	7	10	23	26	39	42
9	8	57	56	41	40	25	24
50	63	2	15	18	31	34	47
16	1	64	49	48	33	32	17

The same would apply to the case of [the set of the first] eighty-one units [i.e., 1–81] if they were to be noted in a figure that consists of eighty-one cells [p. 112], like in this following form [as shown below];

its property is such that if you were to add them up in any way [i.e., horizontally, vertically, diagonally], the sum would be 369, like this>:[105]

78	65	64	27	1	18	19	17	80
25	5	47	49	68	39	40	74	22
46	45	6	50	15	44	73	33	57
34	43	48	7	16	72	37	52	60
69	56	71	72	31	41	14	12	3
29	42	31	11	66	79	34	51	26
32	30	9	36	67	24	77	35	59
54	8	23	57	13	28	53	75	58
2	61	62	63	81	55	20	21	4

It is by this example that other numbers and forms get combined in such a way that other properties of theirs appear.

We have noted some of the aspects of the benefit and interest from this in the epistle 'On Magic' [*risālat al-ṭalismāt wa'l-'azā'im*; literally, 'the epistle on talismans and incantations'],[106] and we mention one example of these in this chapter so that it might point to the truthfulness of what we stated [there]. We therefore say that [in the case of] the ninefold figure, its benefit is in easing birth [*al-wilāda*] if it is drafted on two pieces of ceramics [*khazafayn*], which water has not touched, and these are hung on the woman in labour [*ḍarabahā al-ṭalaq*], then, if the moon happens to be in the ninth [house],[107] and is connected

105 This particular case seems to have problems with it in terms of yielding results that accord with what the Brethren of Purity indicate. The text that is set between <...>, which covers also the last four tables and their contents, were all omitted from the Atif Efendi manuscript, but they all figure in the Ṣādir edition.

106 This is Epistle 52: 'On Magic'. The significance of the figurate numbers is not only linked to the art of magic but it is also connected with astrological symbolism.

107 What is hinted at in this context would be the astrological phenomenon of the moon falling in Sagittarius, as the ninth astrological house, whilst also taking into account that Jupiter is the Lord of that house.

with the lord of the ninth, then birthing is eased; or [also if it] is in the lord of its house in the ninth; and similarly with [other] ninefold entities [*al-muttasi'āt*].[108]

د	ج	ح
ط	ه	ا
ب	ز	و

[p. 113] This is the pathway that has been followed by those who assemble the talismans, since none of the things amongst existents, be it those that are mathematical, natural, or divine, does not have a property that is its own alone and is not that of something else. Their sum has properties that are not theirs individually, whether in numbers, figures, forms, place, time, remedies [*'aqāqīr*], tastes, colours, scents, sounds, words, deeds, letters, or movements. If you add them to one another based on harmonic proportions [*al-nisab al-ta'līfiyya*], then their properties and actions become manifest. The evidence of the correctness of what we have said is in the effects of the antidotes [*tiryāqāt*; theriac], ointments [*marāhim*], potions [*mashrūbāt*], the melodies of music, and all their influences on the bodies and souls, which are not concealed from every intelligent and wise philosopher as we demonstrated in the epistle 'On Music'.[109]

Chapter 27
On the Fruits of this Art[110]

Know, O brother, that the study of sensible geometry aids skilfulness in the arts, whilst the study of intellective geometry and the knowledge of the properties of numbers and figures help in grasping the manner by which the (fol. 23b) heavenly bodies affect the lower natural entities, and also in understanding how the sounds of music affect the souls of the listeners. Studying the manner by which these two types influence

108 For example, the enneagon, or nine-sided polygon.
109 This is Epistle 5: 'On Music'.
110 The 'art' is, of course, geometry.

their effects is a prerequisite to knowing the way by virtue of which the separate [i.e., incorporeal] souls [*al-nufūs al-mufāriqa*] impact the embodied souls [*al-nufūs al-mutajassida*] in the realm of generation and corruption. Those who study intellective geometry have a way to reach knowledge of it through God's aid and guidance.

* * * * * * *

The epistle 'On Geometry' is concluded, as the second epistle of the first section. Praise be to God alone, and prayers upon Muhammad and his household after him! The epistle 'On Astronomy' follows after this, as the third from the first of the four sections [of the *Epistles of the Brethren of Purity*].

Appendix to Epistle 2

From MS 7437 [ط]: The Mahdavī Collection (Tehran)

The long passage that is included in this appendix figured in folios 27–28 of MS 7437 from the Mahdavī Collection in Tehran. It correlates with material that figured towards the middle of Chapter 14 in the epistle 'On Geometry', as also highlighted in footnote 40 on page 130 above. Even though the content of the text that is noted herein in this appendix has an epistemic and mathematical value, in terms of its connection with Book I of Euclid's *Elements*, it nonetheless constitutes a digression from the main topic of the epistle. I have not incorporated this long passage into the body of the Arabic edition or its apparatus, but, rather, noted it in its literal and unedited form in an appendix (*mulḥaq*), which is translated below almost verbatim. I did this to demarcate its distinction from the content of the epistle, given that it is uniquely incorporated in the Tehran manuscript and does not figure in any of the versions of the text as they appear in the other manuscripts or in the Dār Ṣādir Beirut edition. The passage in the Tehran manuscript is attributed to a certain 'scribe' (*al-nāsikh*), who is perhaps another person than the one who served as the copyist of the manuscript. The ending folio of Epistle 2 of the Tehran manuscript carries the copyist's signature in its left margin, which reads, in relatively unclear handwriting, as follows:

The weak servant,
The one with the proper refined honours
Son of Saʿd
Of the people of Shahrābād
Nicknamed
'The one famed for his staff'

Beginning in line 10 of the verso of folio 27, the long passage from the Tehran manuscript reads as follows:

> *The scribe said*: 'One of the properties of the hexagon, which is traceable from within a circle too, is that its side is equal to half the diameter of the circle. By opening the compass, from which can be traced the roundness of a circle, the circumference of the circle can be divided into six equal sides and equal angles. If these were to be grouped together, they would then become composed and gathered in such a manner that they fill in the surface in which they are contained without there being a gap in between them that is left out'; as he said. 'This property is not shared with any other shapes besides the hexagon, with the exception of the square and the triangle. The reason behind this is that the straight lines that intersect at a given single point will have the total sum of their angles equal to four right angles. This has been shown in the figure that was set in Book I of Euclid's treatise in such a way that the magnitude of the angles of its shape resulted in a magnitude that is that of their grouping together in four right angles. If the angles are gathered around a single point that is common to them all, then from it alone (fol. 28a) the surface can be filled without excess or deficiency. From it alone four right angles are obtained. As for the hexagon, the magnitude of [each of] its angles is equal to one and a third of a right angle since it is divisible into four triangles and angles, and the sum of the angles of each triangle is two right angles. So, it would be eight right angles. And each of the angles [of the hexagon] is one and a third of a right angle. Consequently, if three of the angles [of the hexagon] were to be grouped at one point, the result [of their sum] is four right angles. As for the square, the magnitude of its angles is one right angle, so that four of these result in four right angles. Whilst each of the angles of the triangle is two-thirds of a right angle, so that six of these result in four right angles. Regarding other forms, the gathering of their angles is either lesser than four right angles or greater than that. They would not ever result in four right angles. Consider a pentagon, its angle is one and a fifth of a right angle, and it is divisible into three triangles, with the sum of their angles being six right angles. Each of its angles is one and a fifth of a right angle, which [together] do not amount to four right angles, given that if they were added to one another, three of them would be lesser than four right angles, since they would be three right angles and four fifths of a right angle. And the angle of a heptagon is also similar in this regard to what we have shown, given that it is divisible into

five triangles, each of which is one and three sevenths of a right angle. And the magnitude of an angle of the octagon is one and half of a right angle. And the magnitude [of an angle] of a nine-sided figure is one and five ninths of a right angle. And the magnitude [of an angle] of a ten-sided figure is one and three fifths of a right angle. Nothing is perfected from the angles of these figures, or from those that come after them, in terms of resulting in four right angles. That is why you find all the entities made by nature or by the animals to be either circular, or in a[nother] perfect shape, which is either that of a triangle, or of a square, or of a hexagon, whether in circles or in segments of circles.' *The statement of this aforementioned scribe is concluded, and he returned to the original.*[1]

1 It is as if meant by this is that the aforementioned scribe finished with his statement, which has been inserted in the Tehran manuscript by way of an interjection, and that he then returned to copying from an 'original' version, which did not contain his interpolating declarations as recorded in the long passage above. The body of the text continues at this stage in Chapter 14 of the epistle 'On Geometry' by describing the spider's web.

Select Bibliography

Rasā'il Ikhwān al-Ṣafā'

Complete Editions

Ikhwān al-Ṣafā'. *Kitāb Ikhwān al-Ṣafā' wa-Khullān al-Wafā'*, ed. Wilāyat Ḥusayn. 4 vols. Bombay: Maṭba'at Nukhbat al-Akhbār, 1305–1306/ ca. 1888.

——. *Rasā'il Ikhwān al-Ṣafā' wa-Khillān al-Wafā'*, ed. Khayr al-Dīn al-Ziriklī, with two separate introductions by Ṭāhā Ḥusayn and Aḥmad Zakī Pasha. 4 vols. Cairo: al-Maktaba al-Tijāriyya al-Kubrā, 1347/1928.

——. *Rasā'il Ikhwān al-Ṣafā'*, ed. Buṭrus Bustānī. 4 vols. Beirut: Dār Ṣādir and Dār Bayrūt, 1377/1957. Reprint, Beirut: Dār Ṣādir, 2004 and 2006.

——. *Rasā'il Ikhwān al-Ṣafā' wa-Khullān al-Wafā'*, ed. 'Ārif Tāmir. 5 vols. Beirut and Paris: Manshūrāt 'Uwaydāt, 1415/1995.

——. *Al-Risāla al-jāmi'a*, ed. Jamīl Ṣalībā. 2 vols. Damascus: Maṭba'at al-Taraqqī, 1949–1951.

——. *Al-Risāla al-jāmi'a*, ed. Muṣṭafā Ghālib. Beirut: Dār Ṣādir, 1974.

——. *Al-Risāla al-jāmi'a*, ed. 'Ārif Tāmir (as the fifth volume of this edition of the *Rasā'il*, for which see above). Beirut and Paris: Manshūrāt 'Uwaydāt, 1415/1995.

——. *Risālat jāmi'at al-jāmi'a*, ed. 'Ārif Tāmir. Beirut: Dār Ṣādir, 1959.

Partial Editions/Translations

Ikhwān al-Ṣafā'. 'A Treatise on Number Theory from a Tenth-Century Arabic Source', ed. and tr. Bernard R. Goldstein. *Centaurus* 10 (1964), pp. 129–160.

——. *Epistles of the Brethren of Purity. The Case of the Animals Before the King of the Jinn: An Arabic Critical Edition and Annotated English Translation of Epistle 22*, ed. and tr. Lenn E. Goodman and Richard

McGregor. New York–London: Oxford University Press–Institute of Ismaili Studies, 2009.

——. *Epistles of the Brethren of Purity. On Logic: An Arabic Critical Edition and Annotated English Translation of Epistles 10–14*, ed. and tr. Carmela Baffioni. New York–London: Oxford University Press–Institute of Ismaili Studies, 2010.

——. *Epistles of the Brethren of Purity. On Music: An Arabic Critical Edition and Annotated English Translation of Epistle 5*, ed. and tr. Owen Wright. New York–London: Oxford University Press–Institute of Ismaili Studies, 2010.

——. *Epistles of the Brethren of Purity. On Astronomy and Geography: An Arabic Critical Edition and Annotated English Translation of Epistles 3 & 4*, ed. and tr. Jamil Ragep, Taro Mimura, and Ignacio Sánchez with James Montgomery. New York–London: Oxford University Press–Institute of Ismaili Studies, forthcoming.

Primary Sources

Apollonius of Perga. *Les coniques d'Apollonius de Perge*, tr. Paul ver Eecke. Bruges: Desclée De Brouwer, 1923.

Aristotle. *Metaphysics*, ed. W. David Ross. Oxford: Clarendon Press, 1997.

——. *Physics*, ed. W. David Ross. Oxford: Clarendon Press, 1998.

Boethius. *De Institutione Arithmetica, De Institutione Musica, Geometria*, ed. G. Friedlein. Leipzig: Teubner, 1867.

Diophantus. *Diophanti Alexandrini Opera omnia*, ed. Paul Tannery. 2 vols. Leipzig: Teubner, 1893–1895. Reprint, 1974.

——. *Books IV to VII of Diophantus' Arithmetica in the Arabic translation attributed to Qusṭā ibn Lūqā*, ed. and tr. Jaques Sesiano. New York: Springer, 1982.

Euclid. *The Thirteen Books of the Elements*, tr. Thomas L. Heath. 3 vols. Cambridge: Cambridge University Press, 1925. Reprint, New York: Dover, 1956.

——. *Euclides opera omnia*, ed. J. L. Heiberg and H. Menge. 8 vols. Leipzig: Teubner, 1883–1916.

Ibn al-Nadīm. *Kitab al-Fihrist*, ed. M. Riḍā Tajadudd. Tehran, 1971.

Ibn Qurra, Thābit. *Kitāb al-Madkhal ilā ʿilm al-ʿadad*, in *Arabische Übersetzung der Arithmêtikê Eisagôgê des Nikomachus von Gerasa*, ed. Wilhelm Kutsch. Beirut: Imprimerie Catholique, St. Joseph, 1959.

ʿImād al-Dīn, Idrīs. *ʿUyūn al-akhbār*, ed. Muṣṭafā Ghālib. Vol. 4. Beirut, 1973.

——. *ʿUyūn al-akhbār wa-funūn al-āthār,* ed. Ma'mūn al-Ṣāghirjī. Vol. 4. Damascus and London: Institut Français du Proche Orient and the Institute of Ismaili Studies, 2007.

al-Khwārizmī, Muḥammad ibn Aḥmad ibn Yūsuf al-Kātib. *Mafātīḥ al-ʿulūm*, ed. Gerlof van Vloten as *Liber Mafâtîh al-Olûm. Explicans vocabula technica scientiarum tam arabum quam peregrinorum.* Leiden: Brill, 1895.

——. *Mafātīḥ al-ʿulūm*, ed. Muḥammad Kamāl al-Dīn al-Adhamī. Cairo: ʿUthmān Khalīl, 1930.

al-Khwārizmī, Muḥammad ibn Mūsā. *Le commencement de l'algèbre*, ed. and tr. Roshdi Rashed. Paris: Albert Blanchard, 2007.

——. *Al-Khwārizmī: The Beginnings of Algebra*, ed. Roshdi Rashed, tr. Judith Field, with a revision of the annotated English translation by Nader El-Bizri. London: Saqi Books, 2009.

Nicomachus of Gerasa. *Theologumena arithmeticae* (The Theology of Numbers), ed. Fridericus Astius. Leipzig: Libraria Weidmannia, 1817.

——. *Nicomachi Geraseni Pythagorei Introductionis arithmeticae*, ed. Richard Hoche. Leipzig: Teubner, 1886.

——. *Theologumena arithmeticae: Theological Principles of Arithmetic*, ed. Vittorio de Faco. Leipzig: Teubner, 1922.

——. *Introduction to Arithmetic*, tr. M. L. D'Ooge, with studies by F. E. Robbins and L. C. Karpinski. New York, 1926.

——. *Introduction arithmétique*, tr. Janine Bertier. Paris: J. Vrin, 1978.

——. *The Theology of Arithmetic: On the Mystical, Mathematical, and Cosmological Symbolism of the First Ten Numbers*, tr. Robin Waterfield. Grand Rapids, MI: Phanes Press, 1988.

Plato. *Timaeus*, ed. and tr. R. G. Bury. Loeb Classical Library, Vol. IX. Cambridge, MA.: Harvard University Press, 1929. 8th rep., 1999.

Thābit ibn Qurra. *Kitāb al-Madkhal ilā ʿilm al-ʿadad*, in *Arabische Übersetzung der Arithmêtikê Eisagôgê des Nikomachus von Gerasa*, ed. Wilhelm Kutsch. Beirut: Imprimerie Catholique, St. Joseph, 1959.

Secondary Sources

Artmann, Benno. *Euclid, the Creation of Mathematics*. New York: Springer-Verlag, 1999.

Baffioni, Carmela. 'Euclides in the *Rasāʾil* by Ikhwān al-Ṣafāʾ'. *Études Orientales*, 5–6 (1990), pp. 58–68.

——. 'Traces of "secret sects" in the *Rasā'il* of the Ikhwān al-Ṣafā'', in *Shī'a Islām, Sects and Sufism: Historical dimensions, religious practice and methodological considerations*, ed. Frederick De Jong. Utrecht: M. Th. Houtsma, 1992, pp. 10–25.

——. 'The Concept of Science and its Legitimation in the Ikhwān al-Ṣafā'', in *Religion versus Science in Islam: A Medieval and Modern Debate*, ed. C. Baffioni. *Oriente Moderno* 19 (2000), pp. 427–441.

——. 'The "General Policy" of the Ikhwān al-Ṣafā': Plato and Aristotle Restated', in *Words, Texts and Concepts Cruising the Mediterranean Sea: Studies on the Sources, Contents and Influences of Islamic Civilization and Arabic Philosophy and Science Dedicated to Gerhard Endress on his Sixty-fifth Birthday*, ed. A. Arnzen and J. Thielmann. Leuven, Paris, and Dudley, MA: Peeters, 2004, pp. 575–592.

Beaujouan, Guy. 'L'enseignement du *quadrivium*', in *La scuola nell'Occidente latino dell'alto medioevo*. Settimane 19. Spoleto: Centro italiano di studi sull'alto medioevo, 1972, vol. 2, pp. 639–723.

——. 'The Transformation of the Quadrivium', in *Renaissance and Renewal in the Twelfth Century*, ed. Robert Louis Benson and Giles Constable. Cambridge, MA: Harvard University Press, 1982, pp. 463–487.

Blumenthal, David. 'A Comparative Table of the Bombay, Cairo, and Beirut Editions of the *Rasā'il Iḫwān al-Ṣafā*'. *Arabica* 21, 2 (1974), pp. 186–203.

Bosworth, C. E. 'A Pioneer Arabic Encyclopedia of the Sciences: al Khwārizmī's Keys of the Sciences'. *Isis* 54, 1 (1963).

Brentjes, Sonja. 'Die erste *Risāla* der *Rasā'il Ikhwān al-Ṣafa'* über elementare Zahlentheorie'. *Janus* 71 (1984), pp. 181–274.

Burnyeat, Myles F. 'Plato on Why Mathematics is Good for the Soul', in *Mathematics and Necessity: Essays in the History of Philosophy*, ed. Timothy Smiley. Proceedings of the British Academy, vol. 103. Oxford: Oxford University Press, 2000, pp. 1–81.

Butterworth, Charles E. 'Paris est et sagesse ouest. Du *Trivium* et *Quadrivium* dans le monde arabe médiéval' in *L'enseignement des disciplines à la Faculté des arts (Paris et Oxford, XIIIe–XVe siècles)*. Actes du colloque international, Studia Artistarum, 4, ed. Olga Weijers and Louis Holtz. Turnhout: Brepols, 1997, pp. 477–493.

Casanova, Paul. 'Alphabets magiques arabes'. *Journal Asiatique* 18 (1921), pp. 37–55; *Journal Asiatique* 19 (1922), pp. 250–262.

Conway, J. H., and Richard K. Guy. *The Book of Numbers*. New York: Springer-Verlag, 1996.

Daftary, Farhad. *The Ismāʿīlīs: Their History and Doctrines*. Cambridge: Cambridge University Press, 1990; 2nd ed., Cambridge: Cambridge University Press, 2007.

de Boer, T. J. *Geschichte der Philosophie im Islam*. Stuttgart, 1901. Tr. into English by Edward Jones as *The History of Philosophy in Islam*. London: Luzac and Co., 1903. Reprint, 1965.

de Callataÿ, Godefroid. *Ikhwan al-Safa': A Brotherhood of Idealists on the Fringe of Orthodox Islam*. Oxford: Oneworld, 2005.

De Smet, Daniel. *La quiétude de l'intellect. Néoplatonisme et gnose ismaélienne dans l'œuvre de Ḥamīd al-Dīn al-Kirmānī*. Leuven: Peeters, 1995.

de Young, Gregg. 'The Arabic Textual Traditions of Euclid's *Elements*'. *Historia Mathematica* 11 (1984), pp. 147–160.

——. 'New Traces of the Lost al-Ḥajjāj Arabic Translations of Euclid's *Elements*', *Physis* 23 (1991), pp. 647–666.

——. 'Isḥāq ibn Ḥunayn, Ḥunayn ibn Isḥāq, and the Third Arabic Translation of Euclid's Elements', *Historia Mathematica* 19 (1992), pp. 188–199.

Dieterici, Friedrich. *Die Philosophie der Araber im X. Jahrhundert n. Chr. aus den Schriften der Lauteren Brüder*. 8 vols. Leipzig, 1858–1872.

Diwald, Susanne, ed. and tr. *Arabische Philosophie und Wissenschaft in der Enzyklopädie Kitāb Iḫwān aṣ-ṣafāʾ (III): Die Lehre von Seele und Intellekt*. Wiesbaden: Otto Harrassowitz, 1975.

El-Bizri, Nader. 'La perception de la profondeur: Alhazen, Berkeley et Merleau-Ponty'. *Oriens-Occidens: Sciences, mathématiques et philosophie de l'antiquité à l'âge classique (Cahiers du Centre d'Histoire des Sciences et des Philosophies Arabes et Médiévales, CNRS)* 5 (2004), pp. 171–184.

——. 'Variations autour de la notion d'expérience dans la pensée arabe', in *L'expérience, collection les mots du monde*, ed. N. Tazi. Paris: Editions la Découverte, 2004, pp. 39–58.

——. 'The Conceptions of Nature in Arabic Thought', in *Keywords: Nature*, ed. N. Tazi. New York: Other Press, 2005, pp. 63–92.

——. 'Ibn al-Haytham', in *Medieval Science, Technology, and Medicine: An Encyclopedia*, ed. Thomas F. Glick, Steven J. Livesey, and Faith Wallis. New York–London: Routledge, 2005, pp. 237–240.

——. 'A Philosophical Perspective on Alhazen's *Optics*'. *Arabic Sciences and Philosophy* 15, 2 (2005), pp. 189–218.

——. 'Ikhwān al-Ṣafā' (Brethren of Sincerity)', in *Encyclopedia of Philosophy*. 2nd ed., ed. D. M. Borchert. Detroit: Macmillan Reference USA, 2006, pp. 575–577.

——. 'The Microcosm/Macrocosm Analogy: A Tentative Encounter between Graeco-Arabic Philosophy and Phenomenology', in *Islamic Philosophy and Occidental Phenomenology on the Perennial Issue of Microcosm and Macrocosm*, ed. Anna-Teresa Tymieniecka. Dordrecht: Kluwer Academic Publishers, 2006, pp. 3–23.

——. 'In Defence of the Sovereignty of Philosophy: al-Baghdādī's Critique of Ibn al-Haytham's Geometrisation of Place'. *Arabic Sciences and Philosophy* 17, 1 (2007), pp. 57–80.

——, ed. *The Ikhwān al-Ṣafā' and their 'Rasā'il': An Introduction*. New York–London: Oxford University Press–Institute of Ismaili Studies, 2008.

Encyclopaedia of Islam, ed. H. A. R. Gibb et al. 2nd ed. 12 vols. Leiden and London: Brill, 1960–2004.

Euclid. *Stoikheia* (*Elements*), ed. J. L. Heiberg and H. Menge. Teubner Classical Library. 8 vols., with a supplement entitled: *Euclid opera omnia*. Leipzig: Teubner, 1883–1916.

——. *The Thirteen Books of the Elements*, ed. Thomas L. Heath. 3 vols. Cambridge, 1905; repr. 1925. Repr., New York: Dover, 1956.

Fakhry, Majid. 'The Liberal Arts in the Mediaeval Arabic Tradition from the Seventh to the Twelfth Centuries', in *Arts Libéraux et Philosophie au Moyen Âge*. Actes du quatrième congrès international de philosophie médiéval, Montreal, 1967. Montreal and Paris: Institut d'études médiévales, 1969, pp. 91–97.

——. *A History of Islamic Philosophy*. New York: Columbia University Press, 1983.

Farhān, Muḥammad Jalūb. 'Philosophy of Mathematics of Ikhwan al-Safa''. *Journal of Islamic Science* 15 (1999), pp. 25–53.

Farrūkh, 'Umar. 'Ikhwān al-Ṣafā'', in *History of Muslim Philosophy*, ed. M. M. Sharif. Vol. 1. 4th reprint, Delhi: Low Price Publications, 1999, pp. 289–310.

Fermat, Pierre de. *Oeuvres de Fermat*, ed. Paul Tannery and Charles Henry. Vol. 2. Paris: Gauthier–Villars et fils, 1894.

Gardies, Jean-Louis. 'Sur l'axiomatique de l'arithmétique Euclidienne'. *Oriens-Occidens* 2 (1998), pp. 125–140.

Gödel, Kurt. 'Über formal unentscheidbare Sätze der "Principia Mathematica" und verwandter Systeme'. *Monatshefte für Mathematik und Physik* 38 (1931), pp. 173–198.

Gutas, Dimitri. 'Paul the Persian on the Classification of the Parts of Aristotle's Philosophy: A Milestone Between Alexandria and Baghdad'. *Der Islam* 60 (1983), pp. 231–267.

——. *Greek Thought, Arabic Culture.* London: Routledge, 1998. Reprint, 2002.

Heath, Thomas L. *Diophantus of Alexandria: A Study in the History of Greek Algebra.* Cambridge: Cambridge University Press, 1910. 2nd ed., New York: Dover, 1964.

Hughes, Barnabas B. 'Gerard of Cremona's Translation of al-Khwārizmī's *al-Jabr*: A Critical Edition'. *Mediaeval Studies* 48 (1986), pp. 211–263.

——. *Robert of Chester's Latin Translation of al-Khwārizmī's al-Jabr. A New Critical Edition.* Collection Boethius XIV. Stuttgart: F. Steiner Verlag Wiesbaden, 1989.

Hugonnard-Roche, Henri. 'La classification des sciences de Gundissalinus et l'influence d'Avicenne', in *Études sur Avicenne*, ed. Jean Jolivet and Roshdi Rashed. Paris: Belles Lettres, 1984, pp. 41–75.

Hume, David. *An Enquiry Concerning Human Understanding*, ed. L. A. Shelby-Bigge. Oxford: Oxford University Press, 1972.

Irani, Rida A. K. 'Arabic Numeral Forms'. *Centaurus* 4 (1955), pp. 1–12.

Ivry, Alfred. 'Al-Kindī on *First Philosophy* and Aristotle's *Metaphysics*', in *Essays on Islamic Philosophy and Science*, ed. George F. Hourani. Albany, NY: State University of New York Press, 1975, pp. 15–24.

Jolivet, Jean. 'Classifications des sciences', in *Histoire des sciences arabes*, ed. Roshdi Rashed. Vol. 3. Paris: Seuil, 1997.

Khalidi, Tarif. *Arabic Historical Thought in the Classical Period.* Cambridge: Cambridge University Press, 1996.

Lloyd, A. C. 'The Later Neoplatonists', in *The Cambridge History of Later Greek and Early Medieval Philosophy*, ed. A. H. Armstrong. Cambridge: Cambridge University Press, 1970.

Mahdi, Muhsin. 'Science, Philosophy, and Religion in Alfarabi's *Enumeration of the Sciences*', in *The Cultural Context of Medieval Learning*. Proceedings of the First Colloquium on Philosophy, Science, and Technology in the Middle Ages, ed. John E. Murdoch and Edith D. Sylla. Dordrecht–Boston: D. Reidel Publishing Company, 1973, pp. 113–147.

Marquet, Yves. 'Ikhwān al-Ṣafāʾ'. *EI2*, ed. H. A. R. Gibb et al. Leiden and London: Brill, 1960–2004, vol. 3, pp. 1071–1076.

——. *La philosophie des Ikhwān al-Ṣafā de Dieu à l'homme*. Lille: Service de reproduction des thèses, 1973.

——. *La philosophie des Ikhwān al-Ṣafā': Thèse présentée devant l'Université de Paris IV, 1971*. Algiers: Société nationale d'édition et de diffusion, 1973.Reprint, Paris and Milan: Archè, 1999.

——. '910 en Ifrīqiya: Une épître des Ikhwān al-Ṣafā''. *Bulletin d'Études Orientales* 30 (1978), pp. 61–73.

——. 'Les Ikhwān al-Ṣafā' et l'ismaélisme', in *Convegno sugli Ikhwân al-Ṣafā', Roma, 1979*. Rome: Accademia Nazionale dei Lincei, 1981, pp. 69–96.

——. 'Les Épîtres des Ikhwān al-Ṣafā', œuvre ismaïlienne'. *Studia Islamica* 61 (1985), pp. 57–79.

——. 'Les références à Aristote dans les Épîtres des *Ikhwān aṣ-Ṣafā*", in *Individu et société: L'influence d'Aristote dans le monde méditerranéen*, ed. T. Zarcone. Istanbul, Paris, Rome, and Trieste: Isis, 1988, pp. 159–164.

——. 'Ibn al-Rūmī et les Ikhwān al-Ṣafā''. *Arabica* 47 (2000), pp. 121–123.

——. *Les 'Frères de la Pureté', pythagoriciens de l'Islam: La marque du pythagorisme dans la rédaction des Épîtres des Ikhwān aṣ-Ṣafā'*. Paris: EDIDIT, 2006.

Mueller, Ian. *Philosophy of Mathematics and Deductive Structure in Euclid's Elements*. Cambridge, MA: MIT Press, 1981.

Nasr, Seyyed Hossein. *An Introduction to Islamic Cosmological Doctrines: Conceptions of Nature and Methods Used for Its Study by the Ikhwān al-Ṣafā', al-Bīrūnī and Ibn Sīnā*. Revised ed. London: Thames and Hudson, 1978. Originally published by Cambridge, MA: Harvard University Press, 1964.

Necipoglu, Gülru. *The Topkapi Scroll: Geometry and Ornament in Islamic Architecture*. Santa Monica, CA: The Getty Center for the History of Art and the Humanities, 1995.

Netton, Ian Richard. *Allah Transcendent: Studies in the Structure and Semiotics of Islamic Philosophy, Theology and Cosmology*. Richmond: Curzon Press, 1989. Reprint, 1994.

——. *Seek Knowledge: Thought and Travel in the House of Islam*. Richmond: Curzon, 1996.

——. *Muslim Neoplatonists. An Introduction to the Thought of the Brethren of Purity (Ikhwān al-Ṣafā')*. Reprint, London: Routledge Curzon, 2000. First published by George Allen and Unwin, 1982.

——. 'Private Caves and Public Islands: Islam, Plato and the Ikhwān al-Ṣafā''. *Sacred Web* 15 (2005).
O'Meara, Dominic J. *Pythagoras Revived: Mathematics and Philosophy in Late Antiquity*. 2nd ed. Oxford: Clarendon Press, 1990.
——. *Plotinus: An Introduction to the Enneads*. Oxford: Clarendon Press, 1993.
Pellat, Charles. 'Les encyclopédies dans le monde arabe'. *Cahiers d'histoire mondiale* 9 (1966) pp. 631–658.
Peters, Francis E. *Aristoteles Arabus: The Oriental translations and commentaries on the Aristotelian corpus*. Leiden: Brill, 1968.
Pinès, Shlomo. 'Some Problems of Islamic Philosophy'. *Islamic Culture* 11, 1 (1937), pp. 66–80.
——. 'Une encyclopédie arabe du 10e siècle. Les Épîtres des Frères de la Pureté, *Rasā'il Ikhwān al-Ṣafā*''. *Rivista di storia della filosofia* 40 (1985), pp. 131–136.
Poonawala, Ismail K. *Biobibliography of Ismā'īlī Literature*. Malibu, CA: Undena, 1977.
Rashed, Roshdi. *Les mathématiques infinitésimales du IXe au XIe siècle*. Vol. 1: Fondateurs et commentateurs. London: al-Furqān Islamic Heritage Foundation, 1996.
——. 'Analyse combinatoire, analyse numérique, analyse diophantienne et théorie des nombres', in *Histoire des sciences arabes,* ed. Roshdi Rashed with Régis Morélon. Vol. 2. Paris: Editions du Seuil, 1997.
——. *Les mathématiques infinitésimales du IXe au XIe siècle*. Vol. 4. London: al-Furqān Islamic Heritage Foundation, 2002.
——. *Geometry and Dioptrics in Classical Islam*. London: al-Furqān Islamic Heritage Foundation, 2005.
—— and Christian Houzel. 'Thābit ibn Qurra et la théorie des parallèles'. *Arabic Sciences and Philosophy* 15 (2005), pp. 9–55.
——, ed. and tr. into French. *Al-Khwārizmī, Le commencement de l'algèbre*. Paris: Editions Albert Blanchard, 2007.
——, ed. *Al-Khwārizmī: The Beginnings of Algebra*, tr. Judith Field, with a revision of the annotated English translation by Nader El-Bizri. London: Saqi Books, 2009.
——. *Founding Figures and Commentators in Arabic Mathematics: A History of Arabic Sciences and Mathematics*, ed. Nader El-Bizri, tr. Roger Wareham, with Chris Allen and Michael Barany. London: Routledge, 2011.
Rosenthal, Franz. *Das Fortleben der Antike in Islam*. Zurich and Stuttgart: Artemis Verlag, 1965.

——. *Knowledge Triumphant: The Concept of Knowledge in Medieval Islam*. Leiden: E. J. Brill, 1970.

Rudolph, U. *Islamische Philosophie. Von den Anfängen bis zur Gegenwart*. Munich: Beck, 2004.

Sabra, A. I. ''Ilm al-ḥisāb'. *EI2*, ed. H. A. R. Gibb et al. Leiden and London: Brill, 1960–2004, vol. 3, pp. 1138–1141.

——. 'The Appropriation and Subsequent Naturalization of Greek Science in Medieval Islam: A Preliminary Statement'. *History of Science* 25 (1987), pp. 223–243.

——. 'Situating Arabic Science: Locality versus Essence'. *Isis* 87 (1996), pp. 654–670.

Saidan, Ahmad S. 'The Earliest Extant Arabic Arithmetic: *Kitāb al-Fuṣūl fī al-Ḥisāb al-Hindī* of Abū al-Ḥasan Aḥmad ibn Ibrāhīm al-Uqlīdisī'. *Isis* 57 (1966), pp. 475–490.

——. *The Arithmetic of al-Uqlīdisī*. Dordrecht: D. Reidel, 1978.

——. 'Numeration and Arithmetic', in *Encyclopedia of the History of Arabic Science*, ed. Roshdi Rashed with Régis Morélon. Vol. 2. London: Routledge, 1996, pp. 331–348.

Straface, Antonella. 'Testimonianze pitagoriche alla luce di una filosofia profetica: la numerologia pitagorica negli Ikhwān al-Ṣafā''. *Annali dell'Istituto Universitario Orientale* 47 (1987), pp. 225–241.

Tarán, Leonardo. 'Nicomachus of Gerasa', in *Dictionary of Scientific Biography*, ed. Charles Coulston Gillispie et al. Vol. 10. New York: Charles Scribner's Sons, 1974, p. 112–114.

al-Tawḥīdī, Abū Ḥayyān. *Kitāb al-Imtāʿ wa'l-muʾānasa*, ed. Aḥmad Amīn and Aḥmad al-Zayn. 2 vols. Beirut: Manshūrāt Dār Maktabat al-Ḥayāt, 1939–1944. Reprint, Cairo, 1953; 2nd ed. Beirut, 1965.

——. *Risālat al-saqīfa*, ed. Ibrāhīm al-Kaylānī in *Trois Épîtres d'Abū Ḥayyān al-Tawḥīdī*. Damascus, 1951.

Ṭībāwī, 'Abd al-Laṭīf (or, Abdul Latif Tibawi). 'Ikhwān aṣ-Ṣafāʾ and their *Rasāʾil*: A Critical Review of a Century and a Half of Research'. *Islamic Quarterly* 2 (1955), pp. 28–46.

——. 'Further Studies on Ikhwān aṣ-Ṣafā''. *Islamic Quarterly* 20–22 (1978), pp. 57–67.

Vandoulakis, Ioannis M. 'Was Euclid's Approach to Arithmetic Axiomatic?' *Oriens-Occidens* 2 (1998), pp. 141–181.

Vilchez, José M. Puerta. *Historia del pensamiento estético árabe*. Madrid: Ediciones Akal, 1997.

Vuillemin, Jules. *Mathématiques pythagoriciennes et platoniciennes*. Paris: Albert Blanchard, 2001.

Wallis, R. T. *Neoplatonism*. London: Duckworth, 1972.

Weber, Edouard. 'La classification des sciences selon Avicenne à Paris vers 1250', in *Études sur Avicenne*, ed. Jean Jolivet and Roshdi Rashed. Paris: Belles Lettres, 1984, pp. 77–101.

Wensinck, A. J. et al. *Concordance et Indices de la Tradition Musulmane*. Leiden: E. J. Brill, 1967.

Whitehead, Alfred North, and Bertrand Russell. *Principia Mathematica*. 3 vols. 2nd ed. Cambridge: Cambridge University Press, 1925–1927.

Widengren, Geo. 'Macrocosmos–Microcosmos Speculation in the *Rasā'il Ikhwān al-Ṣafā'* and Some Ḥurūfī Texts'. *Archivio di Filosofia* 1 (1980), pp. 297–312.

Subject Index

Index Locorum

رَسَائِلُ إِخْوَانِ الصَّفَاءِ وخُلاَّنِ الوَفَاءِ

رَسَائِلُ إِخْوَانِ الصَّفَاءِ وخُلَّانِ الوَفَاءِ

(١ - ٢)

الرِسالَةُ الأولى والرِسالَةُ الثانيةُ

في العَدَدِ والهَنْدَسَةِ

مِن القِسمِ الأَوَّلِ في العُلومِ الرياضيَّةِ التَعْليميَّةِ

حَقَّقَها وتَرجَمَها وقَدَّمَها وعَلَّقَ عَلَيها

نادِر البِزري

دار نَشْر جامِعة أُكسْفُورْد

بالتَعاونِ مع مَعْهَد الدِّراساتِ الإسْماعيليَّة في لَنْدَن

٢٠١٢

﴿يَٰٓأَيَّتُهَا ٱلنَّفۡسُ ٱلۡمُطۡمَئِنَّةُ (٢٧) ٱرۡجِعِيٓ إِلَىٰ رَبِّكِ رَاضِيَةٗ مَّرۡضِيَّةٗ (٢٨)
فَٱدۡخُلِي فِي عِبَٰدِي (٢٩) وَٱدۡخُلِي جَنَّتِي (٣٠)﴾
(سورة الفَجْر)

لِذكرى والدي الحَبيب

مُحمَّد مُهيب البِزري

(١٩٢٧ - ٢٠١١)

|ع ٥ و| [٤٨](*) الرِّسالَةُ الأولى مِنَ القِسْمِ الأَوَّلِ في العُلومِ الرِّياضيَّةِ التَّعْليميَّة

(*) تمّ الاستناد إلى مخطوط عاطف أفندي، المشار إليه هنا بالرمز [ع]، في التحقيق النقديّ لهذه الرسالة وهو الأقدم من مجمل مخطوطات رسائل إخوان الصفاء. وقد اعتُمِد على ما ورد في هذا المخطوط كما لو أنه المصدر الأساسي في وضع متن النص في هذا المجلّد من دون الأخذ به بشكل مطلق على أنه المخطوط الأمّ. وقد قوبل هذا المخطوط على طبعة دار صادر، التي أشير إليها هنا بالرمز اللاتينيّ [Ṣ]، كما أن تحقيقه استند إلى مقارنته مع بقيّة المخطوطات الأخرى. وقد وردت الإشارة إلى رموزها المعتمدة في التحقيق في توطئة هذا الكتاب المكتوبة باللّغة الإنكليزيّة (Foreword, pp. xx-xxi). وللدلالة على أوراق مخطوط عاطف أفندي تمّ الاعتماد على هذا الشكل الرمزي في الرجوع إلى محتواها: |ع ٥ و|. الحرف «ع» يدلّ هنا على مخطوط عاطف أفندي، ثمّ الرقم «٥» يشير إلى الورقة (folio) الخامسة مثلاً من المخطوط، وإذا كانت الكتابة على وجه (recto) مثل هذه الورقة أُشير إلى ذلك عندها بالحرف «و»، أما إذا وردت الكتابة على ظهر (verso) الورقة فيشار إليها بالحرف «ظ». أما بقيّة المخطوطات فإن الاختلاف في محتوياتها عمّا تضمّنه مخطوط عاطف أفندي قد تمّت الإشارة إليه في الحواشي، مع الدلالة عليه من خلال الرموز المعتمدة في التحقيق، حسب ما ذكرنا أعلاه، وتبعاً لما ورد في توطئة هذا الكتاب. أما الإشارة إلى صفحات طبعة دار صادر في متن النصّ فترد بين [. . .] على هذه الصورة: [٤٨]؛ أي أن الرقم الوارد هنا يشير إلى الصفحة ٤٨ على سبيل المثال. ولقد أشرنا أيضاً إلى بعض التنّبيهات والتّعليقات المقتضبة حول النصّ العربيّ في الحاشية، ووضعناها هنا بين < . . . > للدلالة عليها بوضوح، ولإبراز تباين محتواها عن الرموز المعتمدة في التحقيق النقدي للمخطوطات في الحواشي. أما الشروح المفصّلة فقد أوردناها في مقدمة (Introduction) الكتاب باللّغة الإنكليزيّة، حيث ذكرنا أيضاً حيثيات المنهج المتّبع في التحقيق النقديّ والترجمة في المقدمات التقنيّة (Technical Introduction). نتقدم أيضاً هنا بالشكر للأستاذ صالح الأشمر على تدقيقه النص العربي وضبط شكله.

في العَدَدِ[١]

< المَوسومة[٢] بالأرِثماطيقي > [٣]

|ع ٥ ظ| بسمِ اللهِ الرَّحمٰنِ الرَّحيمِ[٤]

< فصل ١ >

إعلَمْ أيّها الأخ البارّ الرحيم، أيّدنا الله وإيّاك[٥] بروح منه، أنه لما كان من مذهب إخواننا الكرام، أيّدهم الله، النظرُ في

(١) «في العدد»: في العدد وماهيّته وكميّته [ن]، في العدد وخواصّه [ط، ك، ل].

(٢) «الموسومة»: المسمّاة [ط، ك، ل].

(٣) «من جملة إحدى وخمسين رسالة لإخوان الصفاء من القسم الأول في الرياضيّات في تهذيب النفس وإصلاح الأخلاق من كلام الصّوفيّة» زيادة في [ع]؛ ورد العنوان على مثل هذا الشكل في [ط]، ولكن لم تنسب فيه الرسالة إلى كلام الصّوفيّة. < وقد ناقشنا الشكوك حول نَسب هذه الرسالة إلى الصّوفيّة في مقدّمة هذا الكتاب > .

(٤) ورد في مخطوط [ع]: «ربّ سهّل. الحمد لله الذي لا تحسن الأشياء إلا أن يكون بدؤها حمده وكل ناطقٍ وساكتٍ فهو عبده الذي ناهت الألباب في عظمته ودلّت له عقول أهل المعرفة عندما شاهدت من عزّ جبروته وصلواته على خير خلقه وبزينة محمّد النبيّ وآله (؟ < الكلمة مطموسة >) والمنتَجين من أصحابه وعشيرته والصالحين من عباده وأمّته» // ورد في [S]: «الحمد لله وسلام على عباده الذين إصطفى». < وقد تباينت هذه المقدّمات عن بعضها البعض فيما إشتملته المخطوطات من الأدعية وألوان صياغاتها. ولم نذكرها كلّها هنا، على سبيل المثال، كونها خارجة عن صلب الموضوع التعليميّ الرياضيّ المتطرّق إليه في هذه الرسالة، وكونه كذلك من باب التنوّع اللفظيّ اللغويّ، وليس له صلة مباشرة ههنا بالقياس المعرفيّ المراد من هذه الرسالة > .

(٥) «أيّدنا الله وإيّاك»: أيّدك الله وإيّانا [ط، ك].

جميع علوم الموجودات التي هي في العالم، من الجواهر والأعراض والبسائط[١] والمجُرّدات والمُفردات[٢] والمُركّبات، والبحث عن مبادئها وعن كميّة أجناسها وأنواعها وخواصّها، وعن ترتيبها ونظامها، على ما هي عليه الآن، وعن كيفيّة حدوثها ونشوئها عن عِلَّةٍ واحدة، ومبدإٍ واحد، من مُبدع واحد، جلَّ جلالُه[٣]، ويُستَشهَدُ على تِبيانِها[٤] بمثالاتٍ عدديّةٍ، وبراهينَ هندسيّةٍ، مثل ما كان يفعل[٥] الحكماء الفيثاغوريون، احتجنا أن نُقدِّم هذه الرسالة قبل رسائلنا كلّها، ونذكر فيها طرفاً من علم[٦] العدد وخواصّه، التي تُسمَّى الأرِثماطيقي[٧]، شِبْهَ المدخلِ والمقدِّمات، كيما يسهُل الطريقُ على المتعلّمين إلى طلب العلم الذي يُسمّى[٨] الفلسفة[٩]، ويقرُب تناولُها للمبتدئين بالنظر في العلوم الرياضيّة الفلسفيّة، فنقول[١٠]:

---

(١) «البسائط» في [ك، Ṣ]: البسايط [ع،ط، ن، ف، ل]. <الكلمات المشابهة لهذا اللّفظ وردت بمجملها في المخطوطات على هذا الشكل، وذلك من خلال استخدام «ي» (الياء في: «. . .يع») بدلاً عن «ئ» (الهمزة المتوسّطة في: «. . .ئع»). ولقد استخدمنا هنا في سياق تحقيق هذه الرسالة الهمزة المتوسّطة بدلاً من الياء>.

(٢) «والمجردات والمفردات»: سقطت من [ع، ن، ف، ك، ل]، وردت في [Ṣ].

(٣) «من مبدع واحد، جلّ جلاله»: سقطت من [ع، ف، ن]، وردت في [Ṣ].

(٤) «ويُستَشهَدُ على تِبيانِها»: ويستشهدون على بيانها [ن، ط، ف، Ṣ].

(٥) «كان يفعل» :كانوا يعملون [ل].

(٦) «علم»: علوم [ل].

(٧) «التي تسمى الأرِثماطيقي»: سقطت من [ك].

(٨) «العلم الذي يسمى»: الحكمة التي تسمى [ل، ك، Ṣ].

(٩) «الفلسفة»: مقدّمات الفلسفة [ك].

(١٠) «ويقرب تناولها للمبتدئين بالنظر في العلوم الرياضيّة الفلسفيّة، فنقول»: سقطت من [ل].

الفلسفةُ أوّلُها محبَّةُ العلومِ، وأوسطُها معرِفةُ حقائقِ الموجودات، بحسَبِ الطاقةِ الإنسانية، وآخرُها القولُ والعمل بما يوافق العلمَ[(١)].

< فصل ٢ >

العلومُ الفلسفيّةُ[(٢)] [٤٩] أربعةُ أنواعٍ: أوّلُها الرياضيّاتُ، والثاني المنطقيّاتُ، والثالثُ الطبيعيّاتُ[(٣)]، والرابعُ الإلهيّاتُ[(٤)].

والرِّياضيّاتُ أربعةُ أنواع: أوّلُها الأرِثماطيقي، والثاني الجومَطريا، والثالثُ الأسطُرُنومِيا، والرابعُ الموسيقى[(٥)]. فالموسيقى هو معرفةُ تأليفِ الأصواتِ وبه استخراجُ أصولِ[(٦)] الألحان. والأسطُرُنومِيا[(٧)] هو معرفةُ[(٨)] النجومِ بالبراهين التي ذكِرت في كتاب المَجِسْطيّ[(٩)]. والجومَطْرِيا، وهو علمُ الهندسةِ

---

(١) «العلم»: الله العلم والعلوم [ل].

(٢) «العلوم الفلسفيّة»: الفلسفة [ك].

(٣) «الطبيعيّات»: العلوم الطبيعيّات [ف].

(٤) «الإلهيّات»: العلوم الإلهيّات [ك، ل].

(٥) < كلمة «الموسيقى» وردت في المخطوطات [ع ، ف، ل] على شاكلة: «الموسيقي»، وهي بذلك أقرب إلى أصل الكلمة اليونانيّة من المصطلح العربيّ المتداول في شكله الحديث، ولكن لأجل المحافظة على قراءة سلسة للنص أورَدّتُ هذه الكلمة بشكلها المعاصر تماشياً بذلك مع مفردات اللغة العربيّة ومع ما ذكر أيضاً في [ط، ن، ك] >.

(٦) «أصول»: سقطت من [ع]، وردت في [Ṣ].

(٧) < وقع خطأ في متن نصّ المخطوط [ل] بنَسب علم الأسطُرُنوميا إلى أوقليدِس بدلاً من بَطلاميوس، ولكنّ الناسخ عاد فصحّحه في الهامش >.

(٨) «معرفة»: علم [Ṣ].

(٩) < المَجِسْطيّ (*Almagest*): هو كتاب في علم الفلك لبَطلاميوس العالم

بالبراهين التي ذكِرت في كتاب أوقليدِس[١] والأرِثماطيقي[٢] وهو معرفةُ خواصِّ العددِ وما يطابقها من معاني الموجودات التي ذكرها فيثاغورِس ونيقوماخِس. فأوَّلُ ما يُبتدأ بالنظر به في هذه العلوم الفلسفيّة الرياضيّاتُ. وأوّلُ[٣] الرياضيّاتِ معرفةُ خواصِّ العدد لأنه أقربُ العلوم تناولاً، ثمّ الهندسةُ، ثمّ التنجيمُ، ثمّ التأليفُ[٤]، ثمّ المَنْطِقِيّاتُ، ثمّ الطبيعيّاتُ، ثمّ الإلهيّاتُ.

وهذا أوَّلُ ما نقولُ في علمِ العدد، شِبْهَ المَدخَلِ |ع ٦ و| والمقدِّمات:

الألفاظُ[٥] تدلُّ على المعاني، والمعاني هي المسمّياتُ، والألفاظُ هي الأسماءُ، وأعمُّ الألفاظِ والأسماءِ قولُنا «الشيء». والشيء إمّا أن يكون واحداً أو أكثرَ من واحد. فالواحدُ يُقال على وجهَين: إمّا بالحقيقة وإمّا بالمجاز. فالواحدُ بالحقيقةِ هو

اليوناني، نقله الحجاج بن يوسف بن مطر من اليونانيّة إلى العربيّة في العصر العباسي الأول>.

(١) <هو كتاب أوقليدِس في الأصول أو الأسطُقسات أو الأركان (*Stoikheia; The Elements*)، الذي وضع في ثلاث عشرة مقالة. وموضوعه علم الهندسة أو الجومَطْريا. نقله أيضاً الحجاج بن يوسف بن مطر من اليونانيّة إلى العربيّة في العصر العباسي الأول، ومن ثم وضعت نسخة جديدة منقّحة منه على يد حنين ابن إسحٰق وراجعها من بعده ثابت ابن قرّة الحرّانيّ>.

(٢) <ويعنى به كتاب الأرثماطيقي أو المدخل في علم خواص العدد الذي وضعه نيقوماخِس الجَرَشيّ في اللّغة اليونانيّة>.

(٣) «وأوّل»: وأوّل معرفة [ع، ن، ف]، سقطت من [S، ك].

(٤) «ثم التنجيم، ثم التأليف» في [ع، ن، ف، ك، ل]: ثم التأليف، ثم التنجيم [S].

(٥) «الألفاظ»: إعلم أن الألفاظ [ك].

الشيءُ الذي لا جُزْءَ له البتَّةَ، ولا ينقسمُ، وكلُّ ما لا ينقسم فهو واحدٌ من تلك الجهة التي بها لا ينقسم، وإن شئت قلت: الواحد ما ليس فيه غيرُه، بما هو واحد[1]. وأمّا الواحدُ بالمجاز[2] فهو كلُّ جملةٍ يُقالُ لها واحِدٌ كما يُقالُ عشرةٌ واحدةٌ، ومائةٌ واحدةٌ، وألْفٌ واحد. فالواحدُ واحدٌ بالوَحْدَةِ كما أنَّ الأسْودَ أسْودُ بالسَّوادِ، والوَحْدَةُ صِفةٌ للواحِد، كما أنَّ السَّوادَ صِفةٌ للأسود.

< فصل ٣ >

وأمّا الكَثْرَةُ فهي جُملةُ الآحادِ، وأوَّل الكثرةِ اثنانِ، ثمّ الثلاثةُ، ثمّ الأربعةُ، ثمّ الخمسةُ، وما زاد على ذلك بالغاً ما بلغ[3]. والكثرةُ نوعانِ إمّا عددٌ وإما مَعْدودٌ، والفرقُ بينهما أنَّ العددَ إنّما هو كميّةُ صُوَرِ الأشياءِ في نفسِ العادِّ [٥٠]، وأمّا المعدوداتُ فهي الأشياءُ نفسُها، وأما الحسابُ فهو جَمْعُ العددِ وتفريقُه.

والعددُ نوعانِ: صحيحٌ وكُسورٌ، والواحدُ الذي قبل الاثنين هو أصلُ العددِ ومبدأُه، ومنه ينشأ العددُ كلُّه، صحيحُه وكسورُه، وإليه يَنْحَلُّ راجعاً. أمّا نشوءُ الصحيح فبالتزايُدِ، وأمّا الكسورّ

(١) «ولا ينقسم، وكل ما لا ينقسم فهو واحد من تلك الجهة التي بها لا ينقسم، وإن شئت قلت: الواحد ما ليس فيه غيره، بما هو واحد»: سقطت كلّها من [ع، ن، ف، ك، ل]، ووردت في [S].

(٢) «فالواحد بالحقيقة هو الشيء الذي لا جزء له البتّة، ولا ينقسم، وكل ما لا ينقسم فهو واحد من تلك الجهة التي بها لا ينقسم ... وأما الواحد بالمجاز»: سقطت كلّها من [ل].

(٣) «بالغاً ما بلغ»: سقطت من [ن، ل].

فبالتجزُّؤ. والمِثالُ في ذلك ما أقولُ في نُشوءِ العدد الصحيح، إنّه إذا أُضيف إلى الواحد واحدٌ آخرُ يقالُ عند ذلك إنهما اثنان، وإذا أُضيف إليهما واحدٌ آخرُ يُقالُ لتلك الجملة ثلاثةٌ، وإذا أُضيف إليها واحدٌ آخرُ يُقالُ لها أربعةٌ، وإذا أُضيف إليها واحدٌ < آخرُ > يقالُ لها خمسةٌ.

وعلى هذا القياس نُشوءُ العددِ الصحيحِ بالتزايُدِ واحِداً واحِداً، بالِغاً ما بلغ. وهذه صورتُها:

١ ٢ ٣ ٤ ٥ ٦ ٧ ٨ ٩

وأمّا تحليلُ العددِ إلى الواحد فعلى هذا المثال الذي أقول إنّه إذا أُخِذَ مِنَ العشَرةِ واحدٌ تبقَى تِسْعة، وإذا أُلقيَ من التسعةِ واحدٌ تبقَى ثمانية، وإذا أُسقِطَ من الثمانيةِ واحدٌ تَبقَى سبعة، وعلى هذا المِثال والقِياس، أنْ يُلقَى واحِدٌ واحِدٌ حتى يبقَى واحد. فالواحدُ لا يمكنُ أنْ يُلقَى منه شيءٌ لأنّه لا جُزءَ له البتَّة. فقد تبيَّن كيف ينشأ[١] العددُ الصحيحُ من الواحد وكيف ينحلُّ إليه.

< فصل ٤ >

وأمّا نشوءُ العددِ الكُسورِ من الواحد فعلى هذا المِثال الذي أقولُ: إنّه إذا رُتِّبَ العددُ الصحيحُ على النَّظْم[٢] الطبيعيّ، الذي هو: واحدٌ، اثنانِ، ثلاثةٌ، أربعةٌ، خمسةٌ، سِتّةٌ، سبعةٌ، ثمانيةٌ،

---

(١) «ينشاء»: نشؤ [ن].

(٢) «النظم»: نظمه [ن، ل].

تِسعةٌ، عشَرةٌ، ثم أُشيرَ إلى الواحد من كلِّ جُملةٍ، فإنّه يتبيّن[١] كيف يكون نُشوؤه من الواحد، وذلك أنّه إذا أُشيرَ إلى الواحد من الاثنين، يقال للواحد عند ذلك نِصفٌ، وإذا أُشيرَ إلى |ع ٦ ظ| الواحد من جُملة الثلاثةِ فيقال له الثُّلثُ، وإذا أُشيرَ إليه من جُملة[٢] الأربعةِ، يقال له الرُّبعُ، وإذا أُشيرَ إليه من جُملةِ الخمسةِ، يقال له الخُمسُ، وكذلك إذا أُشيرَ له من جمُلةِ السِّتّةِ < يقال له > السُّدسُ[٣]، ثمّ[٤] السُّبعُ والثمنُ والتُّسعُ والعُشرُ. وأيضاً إذا أُشيرَ إلى الواحد من جُملةِ الأحدَ عشرَ فيقال له جُزءٌ من أحدَ عشرَ جُزءاً، ومن اثني عشر نصف السّدس، ومن ثلاثةَ عشرَ جُزءٌ من ثلاثةَ عشرَ، ومن [٥١] أربعةَ عشرَ نِصفُ السُّبعِ، ومن خمسةَ عشرَ ثُلثُ الخمُسِ[٥]، وعلى هذا المثال يُعتبرُ سائرُ الكسُور. فقد بان على هذا القياس كيف يكون[٦] نشوءُ العدد من الواحد الصحيح منها، والكسورِ جميعاً، وكيف هو أصلٌ لهما جميعاً، وهذه صورتُها:

ب	ج	د	هـ	و	ز	ح	ط	ي
نِصف	ثُلث	رُبع	خُمس	سُدس	سُبع	ثُمن	تُسع	عُشر

(١) «يتبيّن»: سيتبيّن [ل].

(٢) «جملة»: سقطت من [ل].

(٣) «وكذلك إذا أشير له من جملة السّتّة < يقال له > السّدس»: وكذلك السدس [ف، ل]؛ وسقطت كلّها من [ن].

(٤) «ثمّ»: وكذا [ن].

(٥) «ثلث الخمس» [ن، ف، S]: خمس الثلث [ع].

(٦) «فقد بان على هذا القياس كيف يكون»: فقدّ تبيّن كيف [ل].

يه	يد	يج	يب	يا
ثُلث الخُمس	نِصفُ السُّبع	جزءٌ من ١٣	نِصفُ السُّدس	جزءٌ من ١١

واعلَمْ يا أخي[١] [٢] بأنَّ العددَ الصحيحَ رُتِّب أربعَ مراتِبَ: آحادٌ وعشراتٌ ومِئاتٌ، وأُلوفٌ. فالآحادُ من واحدٍ إلى تسعة، والعشراتُ من عشرةٍ إلى تسعين، والمئاتُ من مئةٍ[٣] إلى تسعِ مائة، والأُلوفُ من ألفٍ إلى تسعةِ آلاف.

فأسماؤها كلُّها < تَشْتَمِلُها >[٤] اثنتا عشرةَ لفظةً بسيطةً، وذلك أنّ من واحدٍ إلى عشرةٍ، عشرةَ ألفاظٍ، ولفظةَ مئةٍ، ولفظةَ ألفٍ، فصار الجميعُ اثنتي عشرةَ لفظةً بسيطة[٥].

وأمّا سائرُ الألفاظِ فمشتقّةٌ منها أو مركَّبةٌ أو مكرَّرةٌ، فالمكرّرةُ كالعشرينَ من العشرة، والثلاثين من الثلاثة، والأربعين من الأربعة، والخمسين من الخمسة، والستّين من الستّة[٦] [٧]، وأمثال[٨] ذلك.

وأمّا المركَّبةُ[٩] كالمئتين[١٠] وثلثمائةٍ وأربعمائةٍ وخمسمائةٍ،

---

(١) «يا أخي»: سقطت من [ع]، وردت في [Ṣ].

(٢) زيادة في [ف]: «أيّدك الله وإيّانا بروح منه».

(٣) «مئة»: مائة [ع].

(٤) «فأسماؤها كلها < تشتملها >»: وتشتملها كلّها [ن]، فأسماؤها كلّها [ف]، ويشتملها كلها [Ṣ].

(٥) «بسيطة»: سقطت من [ع]، وردت في [ن، ف، Ṣ].

(٦) «والخمسين من الخمسة، والستّين من الستّة»: سقطت من [Ṣ].

(٧) «والستّين من الستّة»: سقطت من [ن، ف، ك، ل].

(٨) «أمثال»: مثال [ل].

(٩) «وأما المركّبة»: ومثال المركّب [ن].

(١٠) «كالمئتين»: سقطت من [ع].

فإنّها مركّبةٌ من لفظة المئة مع سائر لفظة الآحاد. وكذلك ألفان وثلاثةُ آلافٍ وأربعةُ آلافٍ، فإنّها مركّبةٌ من لفظة الألف مع سائر الألفاظ من الآحاد والعشرات والمئات، كما يقال خمسةُ آلافٍ، وستّةُ آلافٍ[1]، وسبعةُ آلافٍ، وعشرون ألفاً ومئةُ ألفٍ، وسائرُ ذلك، وهذه صورتُها[2] :

أ	ب	ج	د	هـ	و	ز	ح	ط	ي	ك	ل	م
١	٢	٣	٤	٥	٦	٧	٨	٩	١٠	٢٠	٣٠	٤٠
ن	س	ع	ف	ص	ق	ر	ش	ت	ث	خ	ذ	
٥٠	٦٠	٧٠	٨٠	٩٠	١٠٠	٢٠٠	٣٠٠	٤٠٠	٥٠٠	٦٠٠	٧٠٠	

[٥٢]

ض	ظ	غ	بغ	جغ	دغ	هغ	وغ	زغ	حغ
٨٠٠	٩٠٠	١٠٠٠	٢٠٠٠	٣٠٠٠	٤٠٠٠	٥٠٠٠	٦٠٠٠	٧٠٠٠	٨٠٠٠
طغ	يغ	كغ	لغ	مغ	نغ	سغ	عغ		
٩٠٠٠	١٠٠٠٠	٢٠٠٠٠	٣٠٠٠٠	٤٠٠٠٠	٥٠٠٠٠	٦٠٠٠٠	٧٠٠٠٠		
فغ	صغ	قغ	رغ	شغ	تغ	ثغ			
٨٠٠٠٠	٩٠٠٠٠	١٠٠٠٠٠	٢٠٠٠٠٠	٣٠٠٠٠٠	٤٠٠٠٠٠	٥٠٠٠٠٠			
خغ	ذغ	ضغ	ظغ						
٦٠٠٠٠٠	٧٠٠٠٠٠	٨٠٠٠٠٠	٩٠٠٠٠٠						

(١) «وستّة آلاف»: سقطت من [S].

(٢) «وهذه صورتها»: كما ترى في هذا المثال [ل].

أمّا الآحادُ فهي: «أ ب ج د ه و ز ح ط». وأمّا العشراتُ فهي: «ي ك ل م ن س ع ف ص». وأمّا المئاتُ فهي: «ق ر ش ت ث خ ذ ض ظ». وأمّا الألوفُ فهي: «غ، بغ، جغ، دغ، هغ، وغ، زغ، حغ، طغ»[١] [٢].

< فصل ٥ >

واعلَمْ يا أخي[٣]، أيّدك الله وإيّانا بروح منه، بأنّ كونَ العددِ على أربعِ مراتِبَ التي هي الآحادُ والعشراتُ والمئاتُ والألوفُ، ليس هو أمراً ضرورياً لازماً |ع ٧ و| لطبيعة[٤] العدد مثلَ كونِه أزواجاً وأفراداً صحيحاً وكُسوراً، بعضُها يتلو[٥] بعضاً، لكنّه أمرٌ وضعيٌّ رتّبته[٦] الحكماءُ باختيارٍ منهم. وإنّما فعلوا ذلك لتكونَ الأمورُ العدديةُ مطابقةً لمراتبِ الأمورِ الطبيعيّة، وذلك أنّ الأمورَ الطبيعيّةَ أكثرُها جعلَها البارئ[٧]، جلَّ ثناؤه[٨]، مربّعاتٍ مِثل الطبائعِ الأربعِ، التي هي الحرارةُ والبرودةُ والرطوبةُ

---

(١) < وقد وردت هذه الأحرف وألفاظ الأعداد في مخطوط [ع]. على شاكلة جدّول مكوّن من مربّعات. وسقطت كلّها من [ن] وكانت الأوضح في [ل]> .

(٢) زيادة في [ل]: «فإنهم فيما بيّناه من المثال وقس على ذلك ما يقرب إليه إن شاء الله».

(٣) «يا أخي»: أيّها الأخ الصالح [ف].

(٤) «لطبيعة»: سقطت من [ن].

(٥) «يتلو»: تحت [ن، Ṣ].

(٦) «رتّبته»: وضعته [ف].

(٧) «البارئ »: وردت في هامش [ل].

(٨) «جعلها البارئ، جلّ ثناؤه»: سقطت من [ع، ن، ف، ك، ل]، وردت في [Ṣ].

واليبوسةُ(١)؛(٢) ومِثلَ الأركانِ الأربعةِ، التي هي(٣) النارُ والهواءُ والماءُ والأرضُ(٤)؛ ومِثلَ الأخلاطِ الأربعةِ، التي هي الدمُّ والبلغمُ والمِرَّتانِ المِرَّةُ الصفراءُ، والمِرّةُ السوداءُ؛ ومِثلَ الأزمانِ الأربعةِ، التي هي الربيعُ والصيفُ والخريفُ والشتاءُ(٥)؛ ومِثلَ الرياحِ الأربعِ، التي هي الصَّبا والدَّبُورُ(٦) [٥٣] والجَنوبُ والشَّمالُ(٧)؛ ومِثلَ الجهاتِ الأربعِ، التي هي فوقٌ وأسفلٌ ويمينٌ وشِمالٌ(٨)؛ والأوتادِ الأربعِ(٩)، التي هي الطالعُ والغارِبُ

---

(١) «الحرارة والبرودة والرطوبة واليبوسة»: أضيفت في هامش المخطوط [ع]، ووردت في متن النصّ في [ن، ف، Ṣ].

(٢) «الرطوبة واليبوسة»: اليبوسة والرطوبة [ف].

(٣) «ومثل الأركان الأربعة، التي هي»: سقطت من [ع]، وردت في [Ṣ].

(٤) زيادة في [ف]: «ومثل الجهات الأربع التي هي الشرق والغرب والشمال والجنوب».

(٥) زيادة في هامش [ل]: «ومثل الجهات الأربع التي هي الشرق والغرب والجنوب والشمال».

(٦) < وكما ورد في طبعة دار صادر [Ṣ] ، فإن الصبا والدبور يعنى بها تباعاً الريح الشرقيّة والريح الغربيّة > .

(٧) «والجنوب والشمال»: والخرباء واليمن [ط، ل]. < وهي شبه مطموسة بين أسطر كلا المخطوطين > .

(٨) «ومثل الجهات الأربع، التي هي فوق وأسفل ويمين وشمال»: الشرق والغرب والجنوب والشمال [ط]. سقطت من [Ṣ].

(٩) < وكما ورد في طبعة دار صادر [Ṣ]، فإن الأوتاد الأربعة هي المنازل الأربع الرئيسة، بين الإثنتي عشرة منزلة من منطقة البروج، سمّيت أوتاداً لأنها أقوى منازل منطقة البروج، وهي التي تقرر المصير في التنجيم، ولهذا سمّي كل منها برج السعادة وأصل الكائن، وقولهم هنا الأوتاد الأربع لأنها بمعنى المنازل > .

ووتَدُ(١) السماءِ ووتَدُ الأرضِ(٢)؛(٣) والمكوِّناتِ الأربعةِ، التي هي المعادنُ والنباتُ والحيوانُ والإنسانُ(٤).

وعلى هذا المثال(٥) وُجِدَ أكثرُ الأمورِ الطبيعيّةِ مُربّعاتٍ.

< فصل ٦ >

واعلَمْ يا أخي(٦) بأنّ هذهِ الأمورَ الطبيعيّةَ إنّما صارت أكثرُها مربّعاتٍ بعناية البارئ، جلّ اسمُه(٧)، واقتضاءِ حِكمته، لتكونَ مراتِبُ الأمورِ الطبيعيّةِ مطابقةً للأمورِ الروحانيّة التي هي فوقَ الأمورِ الطبيعيّة(٨)، وهي التي ليست بأجسامٍ(٩)، وذلك أنّ الأشياءَ التي ما بعد(١٠) الطبيعةِ(١١) على أربع مراتبَ، أوَّلُها

---

(١) «وتد»: وسط [ط].

(٢) «الطالع والغارب ووتد السماء ووتد الأرض»: أضيفت على مثل هذه الصورة في هامش مخطوط [ع]، ووردت كذلك على تلك الشاكلة في [Ṣ]. وقد أورد ناسخ مخطوط [ع] في متن النص بدلاً منها: «الزير والمثنى والمثلث واليمّ». ولم تذكر في [ن].

(٣) سقط من [ن]: «ومثل الرياح الأربع التي هي الصَبا والدَبور . . . ومثل الجهات الأربع، التي هي فوق وأسفل ويمين وشمال . . . الطالع والغارب ووتد السماء ووتد الأرض».

(٤) «الإنسان» في [ع، ن]: الإنس [Ṣ].

(٥) «المثال»: القياس والمثال [ف].

(٦) «يا أخي»: أيّها الأخ البارّ، أيّدك الله وإيّانا بروح منه [ف، ل].

(٧) «إسمه»: ثناؤه [ن، ل، Ṣ]، جلاله [ف].

(٨) «الأمور الطبيعيّة»: الطبيعة [ن، ف].

(٩) «التي ليست بأجسام»: سقطت من [ع، ن، ف، ل]، وردت في [Ṣ].

(١٠) «ما بعد»: ما فوق [Ṣ].

(١١) «وذلك أن الأشياء ما بعد الطبيعة»: سقطت من [ف].

الباريء جلّ ثناؤه[١]، ثمّ دونه العقلُ الكلّيُ الفعّال، ثمّ دونَه النفسُ الكلّيةُ الفلكيّة[٢]، ثمّ دونَها الهَيُولى الأولى. وكلٌّ هذه[٣] ليست بأجسام[٤].

واعلَمْ يا أخي، أيّدك الله وإيّانا بروح منه[٥]، بأنّ نِسبةَ الباريء، جلّ ثناؤه[٦]، من الموجودات، كنسبة الواحد من العدد، ونِسبةَ العقل منها، كنسبة الاثنين من العدد، ونِسبةَ النفس من الموجودات، كنسبة الثلاثة من العدد[٧]، ونِسبةَ الهَيُولى الأولى كنسبة الأربعة من العدد[٨].

< فصل ٧ >

واعلَمْ يا أخي، أيّدك الله وإيّانا بروح منه[٩]، بأنّ العددَ كلَّه، آحاده وعشراته ومئاته وألوفه، أو ما زاد بالغاً ما بلغ، فأصلُها كلُّها من الواحد إلى الأربعة؛ وهي هذه[١٠]:

١ ٢ ٣ ٤

---

(١) «ثناؤه»: سبحانه وتعالى [ن]، وعزّ [ف]، سبحانه [ل]، جلاله [Ṣ].

(٢) «الفلكيّة»: زيادة في [ط، ن، ل].

(٣) «هذه»: هذه الأربعة [ف].

(٤) «وكل هذه ليست بأجسام»: سقطت من [ع، ن]، وردت في [ف، Ṣ].

(٥) «أيّدك الله وإيانا بروح منه»: سقطت من [ع، ن، ل]، وردت في [ف، Ṣ].

(٦) «جلّ ثناؤه»: سبحانه وتعالى [ن]، سبحانه [ل].

(٧) «من العدد»: سقطت من [ع]، وردت في [ن، ف، Ṣ].

(٨) زيادة في [ف]: «الأربعة».

(٩) «أيّدك الله وإيّانا بروح منه»: سقطت من [ع، ن، ف]، وردت في [Ṣ].

(١٠) «وهي هذه»: على هذا المثال [ف].

وذلك أنّ سائرَ الأعداد كلِّها من هذه تَركَّبت، ومنها تنشأ[1] [2]. وهي أصلٌ فيها كلِّها[3].

بيانُ ذلك أنّه إذا أُضيفَ واحدٌ إلى أربعةٍ، كانت خمسةً، وإنْ أُضيف اثنان إلى أربعةٍ، كانت ستّةً، وإنْ أُضيفَ ثلاثةٌ إلى أربعةٍ، صارت سبعةً، وإنْ أُضيفَ واحدٌ وثلاثةٌ إلى أربعةٍ، كانت ثمانيةً، وإنْ أُضيفَ اثنان وثلاثةٌ إلى أربعةٍ[4]، كانت تسعةً، وإنْ أُضيفَ واحِدٌ واثنانِ وثلاثةٌ إلى أربعةٍ، صارت عشرةً.

وعلى هذا المِثال حُكمُ سائرِ [٥٤] الأعداد من الآحاد[5] والعشرات والمئات والألوف، وما زاد بالغاً ما بلغ. وكذلك أصولُ الخطِّ أربعةٌ، وسائرُ الحروفِ منها يتركّب. والكلامُ من الحروف يتركّب كما بيَّنا فيما بعد[6]. فاعتَبِرْها، فإنّك تجدُ ما قلنا حقّاً صحيحاً. ومَنْ يريدُ أنْ يعرفَ كيفيّةَ اختراع[7] البارئ، عزّ اسمه[8]، الأشياءَ في العقل الكُلّيِّ[9]، وكيف أوجدها في

(١) «تركّبت، ومنها تنشأ»: يتركّب وينشأ [S]، مركّب ومنها ينشؤ [ف].

(٢) « ومنها تنشأ»: سقطت من [ن].

(٣) «كلّها»: سقطت من [ف].

(٤) «وإن أضيف واحد وثلاثة إلى أربعة، كانت ثمانية، وإن أضيف اثنان وثلاثة إلى أربعة»: شطبها ناسخ [ن] في متن النصّ ثمّ عاد فصححها في الهامش.

(٥) «الآحاد»: أضافها ناسخ [ل] في الهامش.

(٦) «وكذلك أصول الخط أربعة، وسائر الحروف منها يتركّب. والكلام من الحروف يتركّب كما بيّنا فيما بعد»: سقطت من [ع، ن، ط، ف، ل]، ووردت في [S].

(٧) «كيفيّة اختراع»: كيف اخترع [ن، ف، S].

(٨) «عزّ اسمه»: جلّ ثناؤه [S]، سبحانه [ن، ل]، سبحانه تعالى [ف].

(٩) «الكليّ»: زيادة في [ف].

النفس، وكيف صوّرها في الهَيُولى، فليعتَبِرْ بما ذكرنا في هذا الفصل[١] .[٢]

< فصل ٨ >

واعلَمْ يا أخي[٣]، أيّدك |ع ٧ ظ| الله وإيّانا بروح منه[٤]، أنّ البارئ، عزّ اسمه[٥]، أوّلُ شيءٍ أبدعه واخترعه[٦] من نور وحدانيّته جوهرٌ بسيطٌ يقال له: العقلُ الفعّالُ، كما أنشأ الاثنين من الواحد بالتكرار. ثم أنشأ النفسَ الكليّةَ الفلكيّةَ من نور العقل، كما أنشأ الثلاثةَ بزيادة الواحد على الاثنين. ثم أنشأ الهيولى الأولى[٧] من حركة النفس، كما أنشأ الأربعة بزيادة الواحد على الثلاثة. ثم أنشأ سائر الخلائق من الهيولى ورتّبها بتوسّط العقل والنفس، كما أنشأ سائر العدد من الأربعة، بإضافة ما قبلها إليها، كما مثّلنا <من> قبل.

---

(١) «الفصل»: المقام [ف].

(٢) «ومن يريد أن يعرف كيفيّة اختراع البارئ، عزّ اسمه، الأشياء في العقل الكليّ، وكيف أوجدها في النفس، وكيف صوّرها في الهيولى، فليعتبر بما ذكرنا في هذا الفصل»: وردت في [ط] بعد الفصل ٩ وليس في نهاية الفصل ٧.

(٣) «يا أخي»: أيّها الأخ البارّ الرحيم [ف].

(٤) «أيّدك الله وإيانا بروح منه» في [ع، ن، ل]: سقطت من [Ṣ].

(٥) «عزّ اسمه»: جلّ ثناؤه [ن، ل، Ṣ]، جلّت قدرته [ف].

(٦) «واخترعه»: واخترعه وأبدعه [ن، ل]، سقطت من [ف].

(٧) «الأولى»: سقطت من [ع، ف]، وردت في [ن، Ṣ]. <لقد وجبت الإشارة إليها على أنّها «الأولى» كون ذلك التوصيف يتوافق بدقّة أكثر مع المعنى الفلسفيّ المراد من ذكرها على ما ذهب إليه أرِسطوطاليس> .

< فصل ٩ >

واعلَمْ يا أخي، أيّدك الله بروح منه[١]، [٢] بأنّك إذا تأمّلتَ ما ذكرنا من تركيب العدد من الواحد الذي قبل الاثنين، ونشوئه منه، وجدتَه من أدلِّ الدلائل[٣] على وحدانيّةِ الله تعالى[٤]، وكيفيةِ اختراعِهِ الأشياءَ وإبداعِه لها.

وذلك أنّ الواحد الذي قبل الاثنين، وإن كان منه يُتصوّر وجودُ العددِ وتركيبُه، كما بيَّنا قبلُ، فهو لم يتغيّر عمّا كان عليه، ولم يتجزّأ، كذلك الله، عزّ وجلّ، وإنْ كان هو الذي اخترع الأشياءَ من نور وحدانيّتِه، وأبدعها وأنشأها، وبه قِوامُها وبقاؤها وتمامُها وكمالُها، فهو لم يتغيّر عمّا كان عليه من الوحدانية قبل اختراعه وإبداعه لها، كما بيَّنا[٥] في رسالة المبادئ العقليّة[٦]. فقد[٧] أنبأناك بما ذكرنا بأنّ نسبة الباري، جلّ

(١) «أيّدك الله بروح منه»: سقطت من [ع]، وردت في [ن، ف، Ṣ].

(٢) < كرر ناسخ [ف] مجدداً ما ذكره في الفصل السابق بزيادة سِمَتها التَرداد الذي لا نفع فيه: «ومن يريد أن يعرف كيفيّة اختراع الباري، عزّ اسمه، الأشياء في العقل، وكيف أوجدها في النفس، وكيف صوّرها في الهيولى، فليعتبر بما ذكرنا في هذا الفصل» >.

(٣) «الدلائل»: الدليل [ن، ط، Ṣ]، الأدِلّاء [ل].

(٤) «الله تعالى»: الباري جلّ ثناؤه [Ṣ]، الباري جلّ اسمه [ن، ل]، الباري سبحانه [ف].

(٥) «بيّنا»: وردت في متن نص [ل]، ثمّ غيّرت في هامش نفس ذلك المخطوط إلى «سنبيّن».

(٦) < وهي الرسالة الثانية من القسم الثالث في النفسانيّات والعقليّات، أيّ الرسالة الثالثة والثلاثون من مجمل رسائل إخوان الصفاء >.

(٧) زيادة في [ع]: بان.

اسمه[١]، من الموجودات كنسبة الواحد من العدد، وكما أنّ الواحدَ أصلُ العددِ ومنشأُه وأوّلُه وآخِرُه، كذلك اللهُ، عزّ وجلّ[٢]، هو عِلّةُ الأشياءِ، وخالِقُها[٣] [٥٥] وأوّلُها وآخِرُها، وكما أنّ الواحدَ لا جزءَ له ولا مِثل له في العدد، فكذلك اللهُ، جلّ سبّحانه[٤]، لا مِثلَ له في خَلقِه، ولا شِبهَ، فكما أنّ الواحدَ محيطٌ بالعدد كُلِّه ويَعُدُّه[٥]، كذلك اللهُ، عزّ وجلّ[٦]، عالِمٌ بالأشياءِ وماهيّاتِها، تعالى اللهُ عمّا يقول الظالمون عُلُوّاً كبيراً[٧].

< فصل ١٠ >

واعلَمْ يا أخي[٨] بأنّ مراتِبَ العددِ عند أكثرِ الأُمَم على أربعِ مراتبَ، كما تقدّم ذِكرُها، وأمّا عند الفيثاغوريين فعلى ستَّ عشرةَ مرتبةً، وهذه صورتُها[٩]:

---

(١) «اسمه»: ثناؤه [ن، ل، Ṣ]، سبحانه وتعالى [ف].

(٢) «عزّ وجلّ»: جلّ إسمه [ن].

(٣) زيادة في [ف]: «وباريها ومصوّرها».

(٤) «سبّحانه»: ثناؤه [ك، Ṣ]، جلّت عظمته [ن]، وعزّ وجلّ [ف].

(٥) «ويعدّه»: وتعدّها كلّها أزواجاً وأفراداً [ف].

(٦) «عزّ وجلّ»: جلّ جلاله [Ṣ]، جلّ ثناؤه [ن، ك]، جلّ وعزّ [ف].

(٧) «تعالى الله عما يقول الظالمون علواً كبيراً»: سقطت من [ع، ن، ك]، وردت في [Ṣ]، وردت في [ف] مع سقوط «علواً كبيراً».

(٨) «يا أخي»: أيّها الأخ البارّ الرحيم، أيّدك الله وإيّانا بروح منه [ف].

(٩) زيادة في [ف]: «كما بيّنا في الجدول وهذا مثاله. اعتبِر بذلك من هذه الصورة والشكل إن شاء الله تعالى».

آحاد: ١

عشرات: ١٠

مِئات: ١٠٠

أُلوف: ١٠٠٠

رِبْوات عشرات أُلوف: ١٠٠٠٠

نَوعات مِئات أُلوف: ١٠٠٠٠٠

غايات أُلوف أُلوف: ١٠٠٠٠٠٠

سُورات عشرات أُلوف أُلوف: ١٠٠٠٠٠٠٠

حَلبات مِئات أُلوف أُلوف: ١٠٠٠٠٠٠٠٠

البطات أُلوف أُلوف أُلوف: ١٠٠٠٠٠٠٠٠٠

هنيّات عشرات أُلوف أُلوف أُلوف: ١٠٠٠٠٠٠٠٠٠٠

دَعورات مِئات أُلوف أُلوف أُلوف: ١٠٠٠٠٠٠٠٠٠٠٠

وَهوات أُلوف أُلوف أُلوف أُلوف: ١٠٠٠٠٠٠٠٠٠٠٠٠

مَجوات عشرات أُلوف أُلوف أُلوف أُلوف: ١٠٠٠٠٠٠٠٠٠٠٠٠٠

ومور مئات أُلوف أُلوف أُلوف أُلوف: ١٠٠٠٠٠٠٠٠٠٠٠٠٠٠

مارو أُلوف أُلوف أُلوف أُلوف أُلوف: ١٠٠٠٠٠٠٠٠٠٠٠٠٠٠٠[١]

---

(١) <أُشيرَ إلى المراد من هذا الجدول في مخطوط [ع] بكتابة أفقيّة دلّت على مراتب العدد من خلال جملة من الحروف المفردة والمركّبة على هذه الصورة: ق غ يغ قغ غغ . . . يغغغ قغغغ قغغغغ . . . وقد وردت مركّبات هذه الأحرف كلّها في هامش [ن]، ولكن من دون الأرقام. وسقطت جميعها من [ك] ومن [ل]>.

واعلَمْ يا أخي[١]، أيّدك الله وإيّانا بروح منه[٢]، بأنّ للعددِ الكُسورِ مراتبَ[٣] [٥٦] كثيرةً لأنّه ما من عدد صحيح إلاّ وله جزءٌ أو جزآن أو عِدّة أجزاء، كالاثني عشر فإنّ له نِصفاً |ع ٨ و| وثُلثاً ورُبعاً وسُدساً ونِصفَ سُدس، وكذلك الثمانيةُ والعشرون وغيرُهما من الأعداد. إلاّ أنّ العددَ الكسورَ وإنْ كثُرت مراتِبهُ وأجزاؤه، فهي مرَتّبةٌ بعضُها تحت[٤] بعضٍ، ويشمَلُها كلَّها عشرةُ ألفاظٍ: لفظةٌ منها عامّة[٥] مبهمةٌ، وتِسعة[٦] مخصوصةٌ مفهومةٌ، ومن التسعةِ الألفاظِ[٧] لفظةٌ موضوعةٌ، وهي النِصفُ، وثمانيةٌ مشتقّةٌ وهي: الثُّلثُ من الثلاثةِ، والرُّبعُ من الأربعةِ، والخُمسُ من الخمسةِ، والسُّدسُ من السّتةِ، والسُّبعُ من السّبعةِ، والثُمنُ من الثمانيةِ، والتُّسعُ من التِّسعةِ، والعُشرُ من العشرةِ. وأمّا اللفظةُ العامّةُ المُبهمةُ[٨] فهي الجزءُ، لأنّ الواحدَ من أحدَ عشَر يقال له جزءٌ من أحدَ عشَر، وكذلك من ثلاثةَ عشَر ومن سبعةَ عشرَ وما شاكلَ ذلك. وأمّا باقي الألفاظ الكُسورِ فمضافةٌ إلى هذه العشرةِ ألفاظٍ، كما يقال للواحد من اثني عشرَ نِصفُ السُّدسِ، ولواحدٍ من خمسةَ عشرَ خمُسُ الثُّلث، ولواحدٍ من عشرينَ نِصفُ العُشر،

---

(١) «اعلم يا أخي»: سقطت من [ف].

(٢) «أيّدك الله وإيّانا بروح منه»: سقطت من [ن، ف].

(٣) «للعدد الكسور مراتب»: العدد الكسور مراتبه [Ṣ].

(٤) «تحت»: فوق [ك، ل].

(٥) «عامّة» في [ن، ك، ل، Ṣ]: عاميّة [ع، ف].

(٦) «تسعة»: سقطت من [ع]، وردت في [ن، Ṣ].

(٧) «الألفاظ»: سقطت من [ع، ن]، وردت في [Ṣ].

(٨) «المبهمة»: سقطت من [ع، ف]، وردت في [ن، Ṣ].

وعلى هذا المِثال، وهكذا يتبيّن[١] سائرُ معاني[٢] الكُسورِ بإضافة بعضها لبعض.

واعلَمْ يا أخي[٣] بأنّ نوعَي العددِ يذهبان في الكثرة بلا نهاية، غير أنّ العدد الصحيح يبتدئ من أقلّ الكمّية، وهو الاثنان، ويذهب في التزايد بلا نهاية. وأمّا الكسور فيبتدئ من أكثر الكمّية، وهو النصفُ، ويمرّ في التجزّؤ بلا نهاية. فكلاهما من حيث الابتداء ذو نهاية، ومن حيث الانتهاء غيرُ ذي نهاية[٤].

فصل < ١١ > في خواصِّ العدد[٥]

ثمّ اعلَمْ يا أخي أنّ ما مِن عددٍ إلاّ وله خاصّيَّةٌ أو عِدّةُ خواصّ[٦]، ومعنى الخاصّيّةِ أنّها الصِّفةُ المخصوصةُ للموصوفِ التي لا يشارِكُه فيها غيرُه.

فخاصّيَّةُ الواحدِ هي أنّه أصلُ العددِ ومنشأُه كما بيّنا قبلُ، وهو أنّه يَعُدُّ العددَ كلَّه: الأزواجَ والأفرادَ جميعاً.

ومِن خاصِّيَّةِ الاثنينِ أنّه أوّلُ العددِ مُطلقاً، وهو يَعُدُّ نِصفَ

---

(١) «يتبيّن»: سقطت من [ع]، وردت في [ن، S].

(٢) «معاني»: سقطت من [ك].

(٣) «يا أخي»: أيّها الأخ، أيّدك الله وإيّانا بروح منه [ف].

(٤) «وأما الكسور فيبتدئ من أكثر الكميّة، وهو النصف، ويمرّ في التجزؤ بلا نهاية. فكلاهما من حيث الابتداء ذو نهاية، ومن حيث الانتهاء غير ذي نهاية»: سقطت من [ن].

(٥) «في خواصّ العدد»: سقطت من [ع، ن، ك]، وردت في [ف، S].

(٦) «ثمّ اعلم يا أخي أنّ ما من عدد إلا وله خاصّية أو عدة خواص»: سقطت من [ن، ط، ك، ل].

العدد: [٥٧] الأزواجَ دُونَ الأفرادِ.

ومِن خاصِّيَّةِ الثلاثةِ أنّها أوّلُ[١] عددٍ[٢] الأفرادِ، وهي تَعُدُّ ثُلُثَ الأعدادِ، تارةً الأفرادَ وتارةً الأزواجَ.

ومِن خاصِّيَّةِ الأربعةِ أنّها أوّلُ عددٍ مَجذورٍ.

ومِن خاصِّيَّةِ الخمسةِ أنّها أوّلُ عددٍ دائرٍ، ويقال <له> كُرِيٌّ.

ومِن خاصِّيَّةِ السِّتَّةِ أنّها أوّلُ عددٍ تامّ.

ومِن خاصِّيَّةِ السَّبعةِ أنّها أوّلُ عددٍ كاملٍ.

ومِن خاصِّيَّةِ الثمانيةِ أنّها أوّلُ عددٍ مكعَّبٍ.

ومِن خاصِّيَّةِ التسّعةِ أنّها أوّلُ عددٍ فردٍ مَجذورٍ، وأنّها آخرُ مرتبةِ الآحاد.

ومِن خاصِّيَّةِ العشرةِ أنّها أوّلُ مرتبةِ العشرات.

ومِن خاصِّيَّةِ الأحدَّ عشرَ أنّها أوّلُ عددٍ أصمّ.

ومِن خاصِّيَّةِ الاثني عشرَ أنّها أوّلُ عددٍ زائد.

وبالجمُلةِ فإنَّ مِن خاصِّيَّةِ كلِّ عددٍ أنّه نصفُ حاشيتيهِ مجموعتَينِ، وإذا جُمِعت حاشيتاه تكونان مِثلَه |ع ٨ ظ| مرّتينِ، ومِثالُ ذلك خمسةٌ، فإنَّ إحدى حاشيتَيها أربعةٌ والأخرى سِتَّةٌ، ومجموعُهما عشرةٌ، وخمسةٌ نِصفُها[٣]؛ وعلى هذا القياس يوجد

---

(١) «أوّل»: أقلّ [ف].

(٢) «عدد»: سقطت من [ع، ن، ف]، وردت في [S].

(٣) زيادة في [ك]: «وإن إحدى حاشيتيها ثلاثة ومن الجانب الآخر سبعة، كانت مجموعهما عشرة وإن أحد حاشيتيها ثمانية ثمّ الجانب الآخر اثنتان، كان مجموعهما عشرة والخمسة نصفها». <كرّرت هذه الجملة في هامش المخطوط

سائر الأعداد، إذا اعتُبِر[1]، وهذه صورتُها:

١ ٢ ٣ ٤ - ٥ - ٦ ٧ ٨ ٩

وأمّا الواحدُ فليس له إلا حاشِيةٌ واحدةٌ وهي الاثنان، والواحدُ نِصفُها، وهي مثلُه مرّتين. وأمّا قولنا: إنّ الواحدَ أصلُ العددِ ومنشأُه فهو أنّ الواحدَ إذا رفعته من الوجود ارتفع العددُ بارتفاعه، وإذا رفعت العدد من الوجود، لم يرتفع الواحد. وأمّا قولُنا: إنّ الاثنين أوّلُ العددِ مُطلقاً فهو أنّ العدَد كثرةُ الآحاد، وأوّلُ الكثرةِ اثنان. وأمّا قولُنا: إنّ الثلاثةَ أوّلُ الأفرادِ فهي كذلك، لأنّ الاثنين أوّلُ العددِ وهو الزوج، ويليه ثلاثةٌ وهي فرد.

وأمّا قولُنا: إنها تَعُدُّ ثُلثَ العددِ، تارةً الأفرادَ وتارةً الأزواجَ، فلأنّها تتخطّى العددين، وتَعُدُّ الثالثَ منهما، وذلك الثالثُ يكون تارةً زوجاً وتارةً فرداً[2].

[ل]. ولعل ذلك ورد نقلاً عن [ك]. وقد خطّ في هامش [ل] بيدٍ اختلفت في شكل كتابتها عن الخطّ الذي ظهر في متن نصّ هذا المخطوط. وربما كان في ذلك دلالة على وجود ناسخ للنصّ ومن ثمّ تدخّل شخص آخر لاحقاً لتنقيحه > .

(١) «اعتُبِر» في [ن، ك]: اعتبرت [ع].

(٢) «وأما الواحد فليس له إلا حاشية واحدة وهي الاثنان، والواحد نصفها، وهي مثله مرّتين. وأما قولنا: إن الواحد أصل العدد ومنشأه فهو أن الواحد إذا رفعته من الوجود ارتفع العدد بارتفاعه، وإذا رفعت العدد من الوجود، لم يرتفع الواحد. وأما قولنا: إن الاثنين أوّل العدد مطلقاً فهو أن العدد كثرة الآحاد، وأوّل الكثرة اثنان. وأما قولنا: إن الثلاثة أوّل الأفراد فهي كذلك، لأن الاثنين أوّل العدد وهو الزوج، ويليه ثلاثة وهي فرد. وأما قولنا: أنها تعد ثلث العدد، تارةً الأفراد وتارةً الأزواج، فلأنها تتخطى العددين، وتعد الثالث منهما، وذلك الثالث يكون تارةً زوجاً وتارةً فرداً»: سقط هذا المقطع الطويل

وأمّا قولُنا: إنّ الأربعةَ أوّلُ عددٍ مجذورٍ، فلأنّها من ضربِ الاثنين في نفسه، وكلُّ عددٍ إذا ضُرب في نفسه يُسمّى[1] جذراً، والمجتمعُ من ذلك مجذوراً.

وأمّا ما قيل من أنّ الخمسةَ أوّلُ عددٍ دائرٍ فمعناه أنّها إذا ضُربت في نفسها [٥٨] رجعت إلى ذاتها، وإنْ ضُرب ذلك العددُ المجتمعُ من ضربِها في نفسه[2]، [3] رجع إلى ذاته[4] أيضاً، وهكذا حُكمُ الخمسةِ[5] دائماً. مِثالُ ذلك خمسةٌ في خمسةٍ: خمسةٌ وعِشرون، وإذا ضُرب خمسةٌ وعِشرون في مثلِه[6]، صار ستمّائةً وخمسةً وعِشرين، وإذا ضُرب هذا العددُ أيضاً[7] في نفسه خرج ثلثمائةُ ألفٍ وتسعون ألفاً وستمائةٌ وخمسةٌ وعِشرون، وإنْ ضُرب هذا العددُ في نفسه خرج عددٌ آخرُ وخمسةٌ وعشرون. ألا ترى أنّ

من [ع، ن، ف، ل]، وقد ورد في [ك، S̤]، وذكر جزء منه في [ط] على هذا الشكل: «وأما الواحد فليس له إلا حاشية واحدة وهي الاثنان، والواحد نصفها، وهي مثله مرّتين».

(١) «يسمّى»: يصير [S̤].

(٢) «نفسه»: ، نفسها [S̤].

(٣) «وكل عدد إذا ضُرب في نفسه يسمّى جذراً، والمجتمع من ذلك مجذوراً. وأما ما قيل من أن الخمسة أوّل عدد دائر فمعناه أنها إذا ضُربت في نفسها رجعت إلى ذاتها، وإن ضُرب ذلك العدد المجتمع من ضربها في نفسه»: سقطت من [ن].

(٤) «ذاته»: ذاتها [ع].

(٥) «حكم الخمسة»: زيادة في [ع]، سقطت من [S̤].

(٦) «في مثله»: في خمسة وعشرين [ع].

(٧) «أيضاً»: سقطت من [ع].

الخمسةَ كيف تحفظ نفسها دائماً[1] وما يتولّد منها أبداً[2] دائماً، بالغاً ما بلغ[3]، وهذه صورتُها:

٥ ٢٥ ٦٢٥ ٣٩٠٦٢٥

وأما الستّة فإنّ فيها مشابهةً للخمسة في هذا المعنى، لكنّها ليست مُلازمة كلُزومِ الخمسة ودوامها، وهذه صورتُها[4]:

٦ ٣٦ ١٢٩٦

ستّةٌ في ستّةٍ: ستّةٌ وثلاثون، فالستّةُ راجعةٌ إلى ذاتها، وظهر ثلاثون، وإذا ضُربت سِتّةٌ وثلاثون في نفسها، خرج ألفٌ ومئتان وستَةٌ وتسعون، فظهرت الستّةُ، ولم يظهر الثلاثون. فقد بان أنّ الستّةَ تحفظ نفسها، ولا تحفظ ما يتولّد منها[5]، وأمّا الخمسةُ فإنّها تحفظ نفسها، وما يتولّد منها دائماً أبداً[6].

وأمّا ما قيل من خاصِّيَّة السِتّة، أنّها أوّل عددٍ تامٍّ، فمعناه أنّ كلَّ عددٍ إذا جُمعت أجزاؤه فكانت مِثلَه سواءً سُمّي ذلك العددُ عدداً تامّاً. فالسِتّة أوّلُها، وذلك أنّ لها نِصفاً وهو ثلاثةٌ، وثلثاً وهو اثنان، وسُدساً وهو واحدٌ، فإذا جمعت هذه الأجزاءُ كانت سِتّةً سواءً. وليست هذه الخاصِّيَّةُ لعددٍ قبلها، ولكن لما بعدها

(١) «دائمًا»: زيادة في [ع]، سقطت من [S].
(٢) «أبداً»: زيادة في [ع]، سقطت من [S].
(٣) «بالغاً ما بلغَ»: سقطت من [ن، ل].
(٤) «وهذه صورتها»: زيادة في [ع]، سقطت من [ف، ل، S].
(٥) زيادة في [ف]: «أبداً دايماً».
(٦) «أبداً»: سقطت من [ك، ل].

كثمانيةٍ وعشرين، وكأربعِ مائةٍ وستّةٍ وتسعين، وثمانيةِ آلافٍ ومائةٍ وثمانيةٍ وعشرين، وهذه صورتها:

٨   ٢٨   ٤٩٦   ٨١٢٨

وأمّا ما قيل إنّ السّبعةَ هي أوّلُ عددٍ كاملٍ، فمعناه أنّ السّبعة[١] قد جَمَعَت معانيَ العددِ كلِّه[٢]، وذلك أنّ العددَ كلَّه أزواجٌ وأفرادٌ، والأزواجُ منها أوّلُ [٥٩] وثانٍ، فالاثنان أوّلُ الأزواجِ، والأربعةُ زوجٌ ثانٍ، والأفرادُ منها أوّلٌ وثانٍ، والثلاثةُ أوّلُ الأفراد، والخمسةُ فردٌ ثانٍ. فإذا جُمِعَ فردٌ |ع ٩ و| أوّلٌ إلى زوجٍ ثانٍ، أو زوجاً أوّلاً إلى فردٍ ثانٍ، كانت منها السّبعة. مِثالُ ذلك أنّك إذا جمعت الاثنين، الذي هو أوّلُ الأزواجِ، إلى الخمسةِ، الذي هو فردٌ ثانٍ، كانت منهما سبعةٌ. وكذلك إذا جمعت الثلاثةَ، التي هي أوّلُ فردٍ[٣]، إلى الأربعة، التي هي زوجٌ ثانٍ، كانت منهما سبعةٌ. وكذلك إذا أُخِذَ الواحدُ، الذي هو أصلُ العددِ، مع السّتةِ، التي هي عددٌ تامٌّ، يكون منهما السّبعةُ، التي هي عددٌ كاملٌ، وهذه صورتُها:

١ ٢ ٣ ٤ ٥ ٦ ٧

وهذه الخاصِّيَّةُ لا توجد لعددٍ قبل السّبعة، ولها خواصُّ أُخرُ سنذكرها عند ذكرنا أنّ المعدودات[٤] بحسب طبيعة العدد.

---

(١) «هي أوّل عدد كامل، فمعناه أن السّبعة»: سقطت من [ن].

(٢) «كلّه»: كلّها [Ṣ].

(٣) «أوّل فرد»: فرد أوّل [Ṣ].

(٤) «المعدودات» في [ع، ن، ط، ف، ك، ل]: الموجودات [Ṣ]. <إن ذكر

وأمّا ما قيل إنّ الثمانيةَ أوّلُ عددٍ مُكعّبٍ، فمعناه أنّ كلّ عددٍ إذا ضُرب في نفسه سُمّي جذراً، والمجتمعُ منهما مجذوراً، كما بيّنا من قبلُ[(١)]. وإذا ضُرب المجذورُ في جَذْرِه ثانياً[(٢)] سُمّي المجتمعُ من ذلك مُكعّباً، وذلك أنّ الاثنين، الذي هو[(٣)] أوّلُ العددِ، إذا ضُرب في نفسه كان المجتمعُ منه أربعةً، وهي أوّلُ عددٍ مجذورٍ، ثمّ ضُرب المجذورُ في جذرِه الذي هو اثنان، فخرج من ذلك ثمانيةٌ، فالثمانيةُ أوّلُ عددٍ مُكعّب[(٤)].

وأمّا ما قيل إنّها أوّلُ عددٍ مجسَّمٍ، فلأنّ الجسمَ لا يكون إلاّ من سطوحٍ مُتراكمةٍ، والسطحُ لا يكونُ إلاّ من خطوطٍ متجاورةٍ، والخطُّ لا يكون إلاّ من نُقطٍ منتظمةٍ كما بيَّنا[(٥)] في رسالة الهندسة[(٦)]. فأقلُّ خطّ من جُزأين، وأضيقُ سطحٍ من خطَّين، وأصغرُ جسمٍ من سطحين، فينتج من هذه المقدّماتِ أنَّ أصغرَ جسمٍ من ثمانيةِ أجزاءٍ، أحدُها الخطُّ، وهو جُزآن، فإذا ضُرب الخطُّ في نفسه كان منه السطحُ، وهو أربعةُ أجزاءٍ، وإذا ضُرب السطحُ في أحدِ طوليه، كان منه العمقُ، فيصير جُملةُ ذلك ثمانيةَ

«الموجودات» في هذا الصدد صائب عندما ينظر إلى الأعداد من المنطلق الميتافيزيقيّ على رأي إخوان الصفاء والفيثاغوريّين > .

(١) «كما بيّنا من قبل»: سقطت من [ك].

(٢) «ثانياً»: زيادة في [ع]، سقطت من [ن، ك، S].

(٣) «الذي هو»: زيادة في [ع]، سقطت من [ن، S].

(٤) «ثم ضرب المجذور في جذره الذي هو إثنان، فخرج من ذلك ثمانية، فالثمانية أوّل عدد مكعّب»: وإذا ضرب الأربعة في إثنين خرج ثمانية، فهي أوّل عدد مكعّب وهذه صورتها ٢ ٤ ٨ [ك].

(٥) «بيّنا» في [ع، ن، S]: سنبيّن [ف، ك، ل].

(٦) < وهي الرسالة الثانية من القسم ألأوّل في العلوم الرياضيّة التعليميّة > .

أجزاءٍ، طولَ اثنين في عرضِ اثنينِ في عمق اثنينِ، وهذه صورتُها[١]:

٢ ٤ ٨

وأمّا ما قيل إنّ التسعةَ أوّلُ فردٍ مجذورٍ، فلأنّ الثلاثةَ في الثلاثةِ تسعةٌ، وليس من السبعة[٢] والخمسة والثلاثة شيءٌ منها مجذوراً.

[٦٠] وأمّا ما قيل إنّ العشرةَ أوّلُ مرتبةِ العشراتِ فهو بيّنٌ، كما أنّ الواحدَ أوّلُ مرتبةِ الآحاد، وهذا بيّنٌ ليس يحتاج إلى الشرح. ولها خاصِّيَّةٌ أُخرى وهي تشبه خاصِّيَّةَ الواحد، وذلك أنّه ليس لها من جنسها[٣] إلاّ طرفٌ واحدٌ[٤] وهو العشرون، وهي نِصفُها كما بيّنا للواحد أنّه نِصفُ الاثنين.

وأمّا ما قيل إنّ الأحدَ عشرَ أوّلُ عددٍ أصمَّ، فلأنّه ليس لهُ جزءٌ يُنطَقُ به ولكن يقال واحِدٌ من أحدَ عشرَ واثنان منه. وكلُّ عددٍ هذه صورتُه وصِفتُه[٥] يُسمّى أصمَّ، مِثلُ ثلاثةَ عشرَ وسبعةَ عشرَ وما شاكل ذلك، وهذه صورتُها[٦]:

(١) «وهذه صورتها»: زيادة في [ع، ن]، سقطت من [Ṣ]. <وقد وردت على هيئة نقط في [ك]>.

(٢) «السبعة»: التسعة [ل]. <ما ورد في [ل] خطاء حسابيّ عدديّ>.

(٣) «جنسها»: حاشيتها [ع، ك]. <ورد اللّفظ «جنسها» في متن نصّ [ط] ثمّ غيّرت إلى «حاشيتها» تحت السطر في ذلك الموضع عينه>.

(٤) «إلا طرف واحد»: سقطت من [ن].

(٥) «صورته وصفته»: وصفه [ن، Ṣ].

(٦) <الصورة الواردة هنا من خلال المقاربة بين جدول مركّبات أحرف الأبجديّة

يا	يج	يز	يط	كج	كط	لا	لز	ما	مج	مز	نج	نط
١١	١٣	١٧	١٩	٢٣	٢٩	٣١	٣٧	٤١	٤٣	٤٧	٥٣	٥٩

سا	سز	عا	عج	عط	فج	فط	صا
٦١	٦٧	٧١	٧٣	٧٩	٨٣	٨٩	٩١

وأمّا ما قيل إنّ الاثني عشرَ أوّلُ عددٍ |ع ٩ ظ| زائدٍ، فلأنّ كلَّ عددٍ إذا جمُعت أجزاؤه، وكانت أكثرَ منه سُمّي عدداً زائداً، والاثنا عشرَ أوّلُها، وذلك أنّ لها نِصفٌ، وهو ستّةٌ، ولها ثُلثٌ وهو أربعةٌ، ورُبعٌ وهو ثلاثةٌ، وسُدسٌ وهو اثنان، ونِصفُ سُدسٍ وهو واحد. وإذا جُمعت هذه الأجزاءُ كانت ستّةَ عشرَ، وهي أكثرُ من الاثني عشرَ بزيادةِ أربعةٍ[١]، وهذه صورتُها:

١٢

نِصف ٦ / ثُلث ٤ / رُبع سُدس ٢ / نِصف السُّدس ١

وبالجملة[٢]، ما من عدد صحيح[٣]، إلاّ وله خاصِّيّةٌ تختصُّ به دون غيره، ونحن تركنا ذِكرها كراهيةَ[٤] التطويل.

وما يطابقها من الأعداد رُسمت في مخطوط [ع] بشكل غير واضح ومسّه بعض من الالتباس. وكذلك، لم تُرسم هذه الصورة في [ط]> .

(١) «بزيادة أربعة»: سقطت من [ع، ن، ف، ك، ل]، وقد وردت في [Ṣ].

(٢) زيادة في [ك]: «على هذا القياس».

(٣) «صحيح»: سقطت من [ع، ن، ف، ك]، وردت في [Ṣ].

(٤) «كراهية» في [ع، ط، ف، ك، Ṣ]: كراهة [ن، ل].

< فصل ١٢ >

واعلَمْ يا أخي[١]، أيّدك الله وإيّانا بروح منه[٢]، أنّ العددَ ينقسم قِسمين: صحيحٌ وكسورُ كما بيّنا قبلُ. فالصحيحُ ينقسم قِسمين: أزواجاً وأفراداً. فالزوجُ هو كلُّ عددٍ ينقسم بنصفين[٣] صحيحين، والفردُ هو كلُّ عددٍ[٤] [٦١] يزيد على الزوج واحداً، أو ينقص عن الزوج بواحد[٥].

فأمّا نشوءُ عددِ الزوجِ[٦]، فيبتدئ من الاثنين بالتكرار دائماً على ما يُرى في صورتها هذه[٧]:

٢	٤	٦	٨	١٠	١٢	١٤	١٦	١٨	٢٠
ب	د	و	ح	ي	يب	يد	يو	يح	ك

فأمّا نشوءُ الأفرادِ فيبتدئ من الواحد، إذا أُضيف إليه اثنان، وأُضيف إلى ذلك اثنان دائماً، بالغاً ما بلغ، مثل هذا[٨]:

(١) «يا أخي»: أيّها الأخ [ف].

(٢) «أيّدك الله وإيّانا بروح منه»: سقطت من [ع، ن، ط، ف، ل]، وردت في [ك، Ṣ].

(٣) «بنصفين»: قسمين [ع، ن، ف]. < وردت في مخطوط [ع] «قسمين»، ثمّ أضاف الناسخ في الهامش: «متساويين» / أيّ يعنى بها: «نصفين» كما ورد في [Ṣ] >.

(٤) «ينقسم قسمين: صحيح وكسور كما بيّنا قبل. فالصحيح ينقسم قسمين: أزواجاً وأفراداً. فالزوج هو كل عدد ينقسم بنصفين صحيحين، والفرد هو كل عدد»: سقطت من [ك].

(٥) «عن الزوج بواحد»: عنه واحداً [ع]، واحداً [ف].

(٦) «الزوج»: سقطت من [ك].

(٧) «في صورتها هذه»: زيادة في [ع]، سقطت من [Ṣ]، وهذه صورتها [ف]، مثل [ك، ل].

(٨) «مثل هذا»: زيادة في [ع]، سقطت من [Ṣ]، مثال ذلك [ف، ك].

٣	٥	٧	٩	١١	١٣	١٥	١٧	١٩
ج	ه	ز	ط	يا	يج	يه	يز	يط

والزوج ينقسم على ثلاثة أنواع: زوج الزوجِ، وزوج الفردِ، وزوج الزوجِ والفردِ. فزوجُ الزوجِ هو كلُّ عددٍ ينقسم بنصفين صحيحين متساويين، ونصفُه بنصفين دائماً، إلى أن تنتهي القسمة إلى الواحد. مِثالُ ذلك أربعةٌ وسِتّون، فإنّه زوجُ الزوجِ، وذلك أنّ نصفَه اثنان وثلاثون، ونِصفُه ستةَ عشرَ، ونِصفُه ثمانيةٌ، ونِصفُه أربعةٌ، ونِصفُه اثنان، ونِصفُه واحِدٌ. ونشوء هذا العدد يبتدئ من الاثنين، إذا[١] ضُرب[٢] في الاثنين ثم ضُرب المجموع في الاثنين، وما يجتمع من ذلك في الاثنين، ثم ضُرب المجموع في الاثنين دائماً بلا نهاية.

ومن أراد أن يتبيّن هذا العددَ، أعني زوجَ الزوجِ[٣]،[٤] مُستقصى، فليضعِّفْ بيوتَ الشَطَرَنْجِ، فإنّه لا يخرج إلاّ من هذا العدد، أعني زوجَ الزوجِ، ولهذا العددِ خواصُّ أُخَرُ ذكرها نيقوماخِس في كتابه[٥] بشرحٍ طويل ونحن نذكر منها طرفاً، قال[٦]:

(١) «إذا»: سقطت من [ع]، وردت في [S].

(٢) زيادة في [ك]: «المجتمع».

(٣) «العدد، أعني زوج الزوج»: زيادة في [ع]، سقطت من [S].

(٤) «ومن أراد أن يتبيّن هذا العدد، أعني زوج الزوج»: ومن يرد أن يتبيّن هذا [ف، ك].

(٥) < المدخل إلى العدد >.

(٦) < نيقوماخِس >.

إنّ هذا العدد إذا رُتِّبَ على نظمِه الطبيعيّ، وهو واحد، اثنان، أربعة، ثمانية، ستة عشر، اثنان وثلاثون، أربعة وسِتّون، وعلى هذا القياس بالغاً ما بلغ، فإنّ من خاصِّيَّتِه أنّ ضربَ الطرفين أحدِهما في الآخرِ يكون مساوياً لضربِ الواسطة في نفسها، إن كان له واسطة واحدة، وإن كانت له واسطتان [٦٢] فمثلُ ضربِ إحداهما في الأخرى، مِثالُ ذلك أربعةٌ وسِتّون فإنّه الطرفُ الآخرُ والواحِدُ الطرفُ الأوّلُ، وله واسطةٌ واحدةٌ، وهي ثمانيةٌ، فأقول: إنّ ضربَ الواحدِ في أربعة وستّين، أو الاثنين في اثنين وثلاثين، أو الأربعة في سِتّةَ عشرَ، مُساوٍ لضربِ ثمانيةٍ في نفسها، وهذه صورتُها:

أ	ب	د	ح	يو	لب	سب
١	٢	٤	٨	١٦	٣٢	٦٤

|ع ١٠ و| وإن تَزِد فيه مرتبة أخرى حتى تصير[١] له واسطتان فأقول: إنّ ضربَ الطرفين أحدهما في الآخر، يكون مُساوياً لضربِ الواسطتين إحداهما في الأخرى، مِثالُ ذلك مئةٌ وثمانيةٌ وعشرون إذا ضُرب في واحد، وأربعةٌ وستّون في اثنين، أو اثنان وثلاثون في أربعةٍ يكون مُساوياً لضرب ستّةَ عشرَ في ثمانية وهذه صورتُها:

(١) «وإن تَزِد فيه مرتبة أخرى حتى تصير»: وإن زيدَت فيه رتبة أخرى حتى يصير [S]، وإن زيدَت فيه مرتبة أخرى تصير [ن]، فإن زيدَ فيه مرتبة أخرى تصير [ف، ك، ل]، وإن زيدَ فيه مرتبة أخرى حتى تصير [ط].

ب	د	ح	يو	لب	سد	قكح
٢	٤	٨	١٦	٣٢	٦٤	١٢٨

ولهذا العدد خاصِّيَّةٌ أخرى أنّه إذا جُمع من واحدٍ إلى حيث ما بلغ يكون أقلَّ من ذلك العدد الذي انتهى إليه بواحد، مِثالُ ذلك إذا أُخذ واحدٌ واثنان وأربعةٌ يكون جمُلتُها أقلَّ من ثمانيةٍ بواحد، وإن زيدت الثمانيةُ عليها، تكون الجملةُ(١) أقلَّ من سِتَّةَ عشرَ بواحدٍ، وإن زيدت السّتّةَ عشرَ عليها تكون الجملةُ أقلَّ من اثنين وثلاثين بواحد، وعلى هذا القياس(٢) توجد(٣) مراتبُ هذا العدد، بالغاً ما بلغ، وهذه صورتُها(٤):

أ	ب	د	ح	يو	لب	سد	قكح	رنو
١	٢	٤	٨	١٦	٣٢	٦٤	١٢٨	٢٥٦

وكلّ واحد من هذه الأعداد نِصفٌ لما فوقه من العدد(٥).

وأمّا زوجُ الفردِ فهو كلُّ عددٍ ينقسم بنصفين مرّة واحدة، ولا ينتهي في القسمة إلى الواحد، مِثالُ ذلك: ستّةٌ، وعشرةٌ، وأربعةَ

---

(١) «الجملة»: سقطت من [ل]، ثمّ وضعت في الهامش.

(٢) «القياس» [ن، S]: المثال [ع]

(٣) «توجد»: سقطت من [ل]، ثمّ وضعت في الهامش.

(٤) < حاول ناسخ [ن] تصحيح هذه الصورة والتي تلتها من بعد عدّة مرات في هوامش المخطوط > .

(٥) «وكلّ واحد من هذه الأعداد نصف لما فوقه من العدد»: زيادة في [ع]، سقطت من [ن، ط، ف، ك، ل، Ṣ]. < وردت هذه الجملة في طبعة صادر [Ṣ] في غير موضعها في سياق النص، وقد صحّحناها تبعاً لما ذكر في مخطوط [ع] > .

عشرَ، وثمانيةَ عشرَ، واثنان وعشرون، [٦٣] وستّةٌ وعشرون؛ فإنّ كُلَّ واحدٍ من هذه وأمثالها من العدد ينقسم مرّة واحدة ولا ينتهي إلى الواحد، ونشوءُ مثلِ هذا العدد من ضربِ كلِّ عددٍ فردٍ في اثنين وهذه صورتُها:

و	ي	يد	يح	كب	كول	لو	لج	مب	مو

وأمّا زوجُ الزوجِ والفردِ فهو كلُّ عددٍ ينقسم بنصفين[١] أكثر من مرّة واحدة، ولا ينتهي في القسمة إلى الواحد، مثل اثني عشرَ، وعشرين، وأربعة وعشرين، وثمانية وعشرين، وأمثالِها من[٢] الأعداد، وهذه صورتُها:

يب	ك	كد	كح	لو	مد	نب	س	سح
١٢	٢٠	٢٤	٢٨	٣٦	٤٤	٥٢	٦٠	٦٨

ونشوءُ هذا العددِ من ضربِ زوجِ الفردِ في اثنين مرّةً أو مِراراً كثيرةً، ولها خواصُّ تركنا ذِكرها مخافة[٣] التطويل.

وأمّا العددُ الفردُ فإنّه على[٤] قسمين: فردٌ أوّلٌ، وفردٌ مركّبٌ. والفردُ المركّبُ نوعان: مُشترِكٌ ومُتبايِنٌ. تفصيلُ ذلك[٥]: أمّا الفردُ

(١) «بنصفين» في [ن، ف، ك، ل، S]: بقسمين [ع]. < وردت في مخطوط [ع] «بقسمين»، ولم يضف الناسخ في الهامش «متساويين» كما فعل سابقاً، حسب ما أشرنا أعلاه. والمعنى هنا يُرَجِح صواب «بنصفين» > .

(٢) «من» [ع، ن]: في [S].

(٣) «مخافة»: كراهية [ك]. < وضعها ناسخ [ل] في متن النصّ «مخافة» ثمّ نقّحت بين السطرين بذكر «كراهية» > .

(٤) «فإنه على»: يتنوّع [ك، ل، S]، فتتنوّع [ن].

(٥) زيادة في [ف]: «فنقول».

الأوّلُ فهو كلُّ عددٍ لا يعدُّه غيرَ الواحدِ عددٌ آخرُ مِثلُ[1]: ثلاثةٍ، خمسةٍ، سبعةٍ، أحدَ عشرَ، ثلاثةَ عشرَ، سبعةَ عشرَ، تسعةَ عشرَ، ثلاثة وعشرين، وأشباهِ ذلك من العدد.

وخاصِّيَّةُ هذا العدد أنّه ليس له جزءٌ سوى المسمّى له، وذلك أنّ الثلاثةَ ليس لها إلاّ الثُّلثُ، والخمسةَ ليس لها إلا الخُمس، وكذلك السبعةُ ليس لها إلا السُّبعُ، وهكذا الأحدَ عشرَ والثلاثةَ عشرَ والسبعةَ عشرَ. وبالجُملةِ جميع الأعداد الصُّمّ |ع ١٠ ظ| لا يعدّها إلاّ الواحدُ، فإنّ اسم جُزئها مشتقٌّ منها.

وأمّا الفردُ المركّب فهو كلُّ عددٍ يعدُّه غيرَ الواحدِ عددٌ آخرُ مِثلُ تسعةٍ، وخمسةٍ وعشرين، وتِسعةٍ وأربعين، وواحدٍ وثمانين، وأمثالُها من العدد، وهذه صورتُها [٦٤]:

ط　　　كه　　　مط　　　فاقكا　　　قسط

وأمّا الفردُ المشتركُ فهو كلُّ عددين يعدُّهما غيرَ الواحدِ عددٌ آخرُ مثلُ تسعةٍ، وخمسةَ عشرَ، وإحدى وعشرين[2]، فإنّ الثلاثة[3] تعدُّها كلَّها. وكذلك خمسةَ عشرَ، وخمسةٌ وعشرون، وخمسةٌ وثلاثون، فإنّ الخمسةَ تعدُّها كلَّها. فهذه الأعدادُ وأمثالُها تُسمّى «مشتركة في العدد الذي يعدُّها»، وهذه صورتُها:

ط　　　يه　　　كا　　　كه　　　له

وأمّا الأعدادُ المتباينَةُ فهي كلُّ عددين يعدُّهما عددان آخران

---

(١) «مثل»: مثال ذلك [ف].

(٢) زيادة في [ع]: «وخمسة وثلاثين».

(٣) زيادة في [ع]: «والخمسة والسبعة».

غير الواحد، ولكن الذي يعدُّ أحدَهما لا يعدُّ الآخرَ مِثلُ التسعةِ، والخمسةِ والعشرين، فإنّ الثلاثةَ تعدُّ التسعةَ، ولا تعدُّ الخمسةَ والعشرين. والخمسةُ تعدُّ الخمسةَ والعشرين ولا تعدُّ التسعةَ، فهذه الأعدادُ وأمثالُها يقال لها المتبايِنَة[١].

< فصل ١٣ >

واعلَمْ يا أخي[٢]، أيّدك الله وإيّانا بروح منه[٣]، بأنّ مِن خاصِّيَّةِ كلِّ عددٍ فردٍ[٤] أنّه إذا قُسِّمَ بقسمين كيف ما كان، فأحدُ القسمين يكون زوجاً، والآخرُ فرداً. ومن خاصِّيَّة كلِّ عددٍ زوجٍ أنّه إذا قُسِمَ كيف ما كان، فيكونُ كِلا قِسمَيه[٥] إمّا زوجاً، وإمّا فرداً، وهذه صورتُها:

زوجيّ			فرديّ		
٤	١٠	٤	١	١١	١٠
٧	١٠	٧	٢	١١	٩
٢	١٠	٢	٣	١١	٨
١	١٠	١	٤	١١	٧
٥	١٠	٥	٥	١١	٦

(١) زيادة في [ف، ك]: «وكلّ عددين فردين مجذورين هذا حكمهما». <سقطت من متن النصّ في [ل] ولكنها أضيفت في الهامش>.

(٢) «واعلم يا أخي»: أيّها الأخ البارّ [ف].

(٣) «أيّدك الله وإيّانا بروح منه»: سقطت من [ن، ك].

(٤) زيادة في [ك]: «أيّ عدد كان». <وأضيفت أيضاً في هامش [ل]>.

(٥) «فيكون كلا قسميه»: على قسمته [ن]، فإن أحد القسمين [ك].

فصل < ١٤ > في التامِّ والناقصِ والزائدِ[١]

واعلَمْ يا أخي[٢]، أيّدك الله وإيّانا بروح منه[٣]، بأنّ العددَ ينقسِمُ من جهة [٦٥] أخرى ثلاثةَ أنواعٍ: إمّا تامّاً، وإمّا زائداً، وإمّا ناقصاً.

فالتامّ هو كلُّ عددٍ إذا جُمعت أجزاؤه كانت الجملةُ مثلَه سواءً مِثلُ ستّةٍ، وثمانيةٍ وعشرين، وأربعمائةٍ وستّةٍ وتسعين، وثمانيةِ[٤] آلافٍ ومائةٍ وثمانيةٍ وعشرين. فإنّ كلَّ واحدٍ من هذه الأعداد إذا جُمعت أجزاؤه كانت الجملَةُ مثلَه سواءً. ولا يوجد من هذا العدد إلاّ في كلِّ مرتبة من مراتب العدد واحدٌ كالستّةِ في الآحاد، والثمانيةِ والعشرين في العشرات، والأربعمائةِ وستّةٍ وتسعين في المئات، والثمانيةِ[٥] آلافٍ ومائةٍ وثمانيةٍ وعشرين في الأُلوف، وهذه صورتُها:

٦ ٢٨ ٤٩٦ ٨١٢٨

(١) «في التام والناقص والزائد»: سقطت من [ع، ن، ط، ف، ك، ل]، وردت في [Ṣ]. <سياق النص ومحتواه يتطلّبان ذكر عنوان «الفصل ١٤» في هذا الموضع بدلاً من وضعه أعلاه في «الفصل ١٣»، أيّ في المكان الذي ذكر فيه في طبعة دار صادر [Ṣ]>.

(٢) «واعلم يا أخي»: أيّها الأخ [ف].

(٣) «أيدك الله وإيّانا بروح منه»: سقطت من [ع، ن، ط، ف، ك، ل]، وردت في [Ṣ].

(٤) «ثمانية»: سبعة [ف، ط ، ك، ل]. <والتصويب العدديّ هنا يكون منه: «ثمانية»>.

(٥) «الثمانية»: السبعة [ف، ط ، ك، ل]. <والتصويب العدديّ هنا: «الثمانية»>.

وأمّا العددُ الزائدُ فهو كلُّ عددٍ إذا جُمعت أجزاؤه كانت أكثرَ منه، مِثلُ الاثني عشرَ، والعشرين والستّين[(١)]، وأمثالِها من العدد. وذلك أنّ الاثني عشرَ نِصفُها ستّةٌ وثُلثُها أربعةٌ ورُبعُها ثلاثةٌ وسُدسُها اثنان ونِصفُ سُدسِها واحدٌ، فجُملةُ هذه الأجزاءِ ستّةَ عشرَ وهي أكثرُ من اثني عشرَ[(٢)].

وأمّا العدد |ع ١١ و| الناقص فهو كلُّ عددٍ إذا جُمعت أجزاؤه كانت أقلَّ منه مثلُ الأربعةِ والثمانيةِ والعشرةِ وأمثالِها من العدد[(٣)]. وذلك أنّ الثمانيةَ نِصفُها أربعةٌ ورُبعُها اثنان وثُمنُها واحِدٌ، وجُملتُها تكون سبعةً، وهي أقلُّ من الثمانية. وعلى هذا القياس حُكمُ سائرِ الأعدادِ الناقصة[(٤)].

فصل < ١٥ > في الأعداد المتحابّة[(٥)]

واعلَمْ يا أخي[(٦)]، أيّدك الله وإيّانا بروح منه[(٧)]، بأنّ العدد من جهةٍ أخرى ينقسم قسمين، أحدُهما يقال له أعدادٌ متحابّةٌ[(٨)]، وهي

(١) : زيادة في [ع]: «والإثنين والستّين».
(٢) «من اثني عشر»: سقطت من [ف].
(٣) «والعشرة وأمثالها من العدد»: وردت في هامش [ل].
(٤) «وذلك أن الثمانية نصفها أربعة وربعها اثنان وثمنها واحد، وجملتها تكون سبعة، وهي أقلّ من الثمانية. وعلى هذا القياس حكم سائر الأعداد الناقصة»: وذلك أن الأربعة نصفها اثنان وربعها واحد وجملتها ثلاثة، وهو أقلّ من أربعة. وهذا حكم الثمانية وغيرها من الأعداد الناقصة [ن].
(٥) «في الأعداد المتحابّة»: سقطت من [ع، ن، ك، ل]، وردت في [Ṣ].
(٦) «يا أخي»: أيّها الأخ البارّ [ف].
(٧) «أيّدك الله وإيّانا بروح منه»: سقطت من [ن، ط، ف، ل].
(٨) «متحابّة»: متجانسة [ك]. < ذكرت «متحابّة» في متن نصّ [ل] ولكن وضع اللفظ «متجانسة» في الهامش > .

كلُّ عددين أحدُهما زائدٌ والآخرُ ناقصٌ، وإذا جُمعت أجزاءُ العددِ الزائدِ كانت مساويةً لجُملةِ العددِ الناقصِ، وإذا جُمعت أجزاءُ العددِ الناقصِ كانت مساويةً لجمُلةِ العددِ الزائدِ. مِثالُ ذلك مائتان وعشرون وهو عدد زائد، ومائتان وأربعة وثمانون وهو عدد ناقص، فإذا جُمعت أجزاء مائتين وعشرين كانت مساويةً لمائتين وأربعةٍ [٦٦] وثمانين، وإذا جُمعت أجزاءُ هذا العدد[١] تكون جُملتها مائتين وعشرين . فهذه الأعدادُ وأمثالُها تُسمّى متحابّة[٢] وهي قليلة الوجود، وهذه صورتُها[٣]:

عدد زائد: ٢٢٠

نصفه: ١١٠ + رُبعه: ٥٥ + خمُسه: ٤٤ + نِصف الخُمس: ٢٢ + رُبع الخُمس: ١١ + مخرج رُبع الخُمس: ٢٠ + مخرج نِصف الخُمس: ١٠ + مخرج الخُمس: ٥ + مخرج الرُّبع: ٤ + مخرج النِّصف: ٢ + جزؤه: ١

= جُملته: ٢٨٤

عدد ناقص: ٢٨٤

نِصفه: ١٤٢ + رُبعه: ٧١ + مخرج الرُّبع: مخرج الخُمس: ٤ + مخرج النِّصف: ٢ + جزؤه: ١

= جُملته: ٢٢٠

(١) زيادة في [ك]: «الناقص».

(٢) «متحابّة»: متجانسة [ك]. < ذكرت «متحابّة» في متن نصّ [ل] ولكن وضع اللفظ «متجانسة» في الهامش > .

(٣) < أشار إليها ناسخ [ل] ولكنها لم تُرسَم في المخطوط > .

< فصل ١٦ في > تَضْعيف العدد[١]

واعلَمْ يا أخي[٢] بأنّ من خاصِّيَّةِ العدد أنّه يقبل التَضْعيفَ والزيادةَ بلا نهاية، ويكون ذلك على خمسة أنواع:

فمنها «على النظم الطبيعيّ» مِثلُ هذا بالغاً ما بلغ:

١ ٢ ٣ ٤ ٥ ٦ ٧ ٨ ٩ ١٠ ١١

ومنها «على نظم الأفراد» بالغاً ما بلغ مِثلُ هذا:

١ ٣ ٥ ٧ ٩ ١١ ١٣ ١٥

ومنها «على نظم الأزواج» بالغاً ما بلغ مِثلُ هذا[٣]:

٢ ٤ ٦ ٨ ١٠ ١٢ ١٤

ومنها «بالطرح» كيفما اتّفق، كما يوجد في سائر الحساب.

ومنها «بالضرب»، كما نُبيّن من بعدُ[٤].

فصل < ١٧ > في خواصّ الأنواع[٥]

واعلَمْ يا أخي[٦]، أيّدك الله وإيّانا بروح منه[٧]، بأنَّ لكلِّ نوعٍ

---

(١) «في تضعيف العدد»: سقطت من [ع، ن، ط، ك، ل]، وردت في [Ṣ].

(٢) «يا أخي»: سقطت من [ك].

(٣) < ورد نظم الأزواج في [ع] بعد نظم الأفراد، على عكس الترتيب المعتمد في [Ṣ] >.

(٤) «نبيّن من بعد»: سنبيّن [ن، ف، ك، ل].

(٥) «في خواص الأنواع»: سقطت من [ع، ن، ط، ف، ك، ل]، وردت في [Ṣ].

(٦) «يا أخي»: سقطت من [ف].

(٧) «أيّدك الله وإيّانا بروح منه»: سقطت من [ع، ن، ف، ك، ل]، وردت في [ط، Ṣ].

من هذه الأنواعِ عِدّةَ خواصّ، وقد ذِكر ذلك في[1] كتاب الأرِثماطيقي بشرح طويل، ولكن نذكر منها طرفاً في هذا الفصل فنقول:

[٦٧] إنّ من خاصِّيَّةِ النظم الطبيعيّ أنّه إذا جُمع من واحدٍ إلى حيث ما بلغ[2] يكون المجموعُ مساوياً لضرب ذلك العدد الأخير بزيادة واحدٍ عليه في نِصفه، مِثالُ ذلك إذا قيل: كم من واحدٍ إلى عشرةٍ مجموعاً على النظم الطبيعيّ؟

فقِياسُه أن يزاد على[3] العشرة واحد، ثم يُضرب في نصف العشرة، فيكون خمسةً وخمسين، وذلك بابُه[4]، فقِس عليه[5]. أو تُضرب الخمسة في نفسها، فيكون خمسةً وعشرين، ثم في النصف الآخر الذي هو ستّةٌ فيكون ثلاثين: الجملةُ خمسةٌ وخمسون، وذلك بابُه المطلوب وقياسُه[6].

|ع ١١ ظ| وأمّا نظمُ الأزواجِ فهو مِثلُ واحدٍ، اثنين، أربعة، ستّة، ثمانية، عشرة، اثني عشرَ، وعلى هذا المِثال[7] بالغاً ما

---

(١) زيادة في [ف]: «علم».

(٢) «أنه إذا جمع من واحد إلى حيث ما بلغ»: خطّت في [ن] ومن ثمّ شطبت بعد ذلك من النصّ.

(٣) زيادة في [ن]: «هذه».

(٤) «بابه»: بيانه [ك، ل].

(٥) «فقس عليه»: وقياسه [ط].

(٦) «أو تضرب الخمسة في نفسها، فيكون خمسة وعشرين، ثم في النصف الآخر الذي هو ستة فيكون ثلاثين: الجملة خمسة وخمسون، وذلك بابه المطلوب وقياسه»: سقطت من [ع، ن، ف، ك، ل]، وردت في [S].

(٧) «عشرة، اثني عشر، وعلى هذا المثال»: على هذا القياس [ن].

بلغ.[١] ومن خاصّية هذا النظم[٢] أن يكون المجموع أبداً فرداً، ومن خاصّيته أيضاً أنّه إذا جُمع على نظمه الطبيعيّ[٣] من واحدٍ إلى حيث ما بلغ يكون المجموعُ مساوياً لضرب نصفِ ذلك العدد في النصف الآخر بزيادة واحد، ثم يزاد على الجملة واحد.

مِثالُ ذلك إذا قيل لك: كم من واحد إلى العشرة مجموعاً على نظمِ الأزواج؟ فقياسُه أن تأخذ نصفَ العشرة، فتزيد عليه واحداً، ثم تضربه في النصف الآخر، ثم تزيد على الجملة واحداً، فذلك أحدٌ وثلاثون، وعلى هذا القياس سائر الأعداد[٤].

وأمّا نظمُ الأفرادِ[٥] فمِثلُ واحدٍ، ثلاثة، خمسة، سبّعة، تسعة، أحد عشر، بالغاً ما بلغ. فمن خاصِّيَّتهِ أنّه إذا جُمع على نظمِه الطبيعيّ يكون المجموعان: الواحد زوج والآخر فرد، يتلو بعضُها بعضاً، بالغاً ما بلغ، وتكون كلُّها مجذورات. ومن خاصِّيَّتِه أيضاً أنّه إذا جُمع على نظمِه الطبيعيّ من واحدٍ إلى حيثُ ما بلغ، فإنّ المجموعَ يكون مُساوياً لضربِ نصفِه مجذوراً[٦]

---

(١) «أو تضرب الخمسة في نفسها، فيكون خمسة وعشرين، ثم في النصف الآخر الذي هو ستة فيكون ثلاثين: الجملة خمسة وخمسون، وذلك بابه المطلوب وقياسه. وأما نظم الأزواج فهو مثل واحد، اثنين، أربعة، ستّة، ثمانية، عشرة، اثني عشر، وعلى هذا المثال بالغاً ما بلغ»: سقطت من [ط].

(٢) «النظم»: نظم الأزواج [ط].

(٣) «الطبيعيّ»: سقطت من [ط].

(٤) زيادة في [ف]: «الأزواج». <وأضيف بين السطور في [ل]: «الزوج»> .

(٥) «الأفراد»: سقطت من [ف].

(٦) «مجذوراً»: سقطت من [ط].

مجبوراً في نفسه[١]، مِثالُ ذلك إذا قيل: كم من واحدٍ[٢] إلى أحدَ عشرَ على نظمِ الأفراد؟ فقياسُه[٣] أن تأخذ نِصفَ الأحدَ عشرةَ[٤]، وهو خمسةٌ ونِصفٌ[٥]، وتجبره[٦] فيصير ستّة، فتضربه في نفسه[٧]، فيكون ستّةً وثلاثين، وذلك بابُه فقِس عليه.

< فصل ١٨ >

[٦٨] واعلَمْ يا أخي[٨]، أيّدك الله وإيّانا بروح منه[٩]، بأنّ معنى[١٠] الضربِ هو تضعيفُ أحدِ العددين بقدرِ ما في الآخر من الآحاد، مِثالُ ذلك إذا قيل: كم ثلاثةٌ في أربعة؟ فمعناه كم جملةُ ثلاثةٍ أربعَ مرّاتٍ؟

---

(١) «نفسه»: مثله [ك].

(٢) «مثال ذلك إذا قيل لك: كم من واحد إلى العشرة مجموعاً على نظم الأزواج؟ . . . وأما نظم الأفراد فمثل واحد، ثلاثة، خمسة، سبعة، تسعة، أحد عشر، بالغاً ما بلغ . . . فإن المجموع يكون مساوياً لضرب نصفه مجذوراً مجبوراً في نفسه، مثال ذلك إذا قيل: كم من واحد»: ورد هذا المقطع الطويل في هامش ظهر الورقة ١١ من المخطوط [ع]، وهو متطابق مع ما ورد في المخطوط [ط] وطبعة دار صادر [Ṣ].

(٣) «الأحد عشرة» [ع، ن، ف، ل]: العدد [Ṣ].

(٤) «فقياسه» [ن، ف]: فبابه [Ṣ].

(٥) «وهو خمسة ونصف»: زيادة في[ع]، سقطت من [ن، Ṣ].

(٦) زيادة في [ف]: «وتضربه في نفسه».

(٧) «فتضربه في نفسه»: سقطت من [ف].

(٨) «واعلم يا أخي»: سقطت من [ف].

(٩) «أيّدك الله وإيّانا بروح منه»: سقطت من [ع، ن، ط، ف، ك، ل]، ووردت في [Ṣ].

(١٠) «بأن معنى»: المعرفة [ن، ل]، الحساب [ف]، ملاك الحساب [ط، ك].

واعلَمْ يا أخي[١] بأنّ العدد نوعان، صحيحٌ وكُسورٌ، كما بيّنا قبلُ. فصار أيضاً ضرب العدد بعضُه في بعض نوعين: مُفرداً ومُركّباً. فالمفردُ ثلاثةُ أنواعٍ: الصحيحُ في الصحيح، مثلُ اثنين في ثلاثةٍ، وثلاثةٍ في أربعةٍ، وما شاكله. ومنها الكسورُ في الكسورِ، مِثلُ نِصفٍ في ثُلثٍ، وثُلثٍ في رُبعٍ وما شاكله. ومنها الصحيحُ في الكسور، مِثلُ اثنين في ثُلثٍ، أو ثُلثٍ في أربعةٍ وما شاكله[٢].

وأمّا المركّبُ فهو أيضاً ثلاثةُ أنواع: فمنها الكسورُ والصحيحُ في الصحيح، مِثلُ اثنين وثُلثٍ في خمسةٍ وما شاكلها. ومنها الصحيحُ والكسورُ في الصحيح والكُسور، مِثلُ اثنين وثُلثٍ في ثلاثةٍ ورُبعٍ وما شاكلها. ومنها الصحيحُ والكسورُ في الكسور[٣]، مثل اثنين وثُلثٍ في سُبعٍ.

فصل < ١٩ > في العدد الصحيح[٤]

واعلَمْ يا أخي بأنّ ضربَ العددِ الصحيحِ على أربعةِ أنواعٍ، وجُملتُها عشرةُ أبوابٍ وهي: آحادٌ وعشراتٌ ومئاتٌ وألُوفٌ[٥].

---

(١) «يا أخي»: سقطت من [ع، ن، ف، ك، ل]، وردت في [Ṣ].

(٢) «شاكلها»: شاكله [ف].

(٣) «ومنها الصحيح والكسور في الصحيح والكسور ... ومنها الصحيح والكسور في الكسور»: وردت على هذا الترتيب الأدقّ في [ط، ك، ل، Ṣ]، أما في [ع] فقد أشير إلى «الصحيح والكسور في الكسور» من الأعداد المركّبة قبل «الصحيح والكسور في الصحيح والكسور».

(٤) «في العدد الصحيح»: سقطت من [ع، ن، ط، ف، ك، ل]، وردت في [Ṣ].

(٥) «آحاد وعشرات ومئات وألوف»: أضافها ناسخ [ل] في الهامش.

فالآحاد في الآحاد، واحِدُها واحدٌ، وعشرتُها[١] عشرةٌ. والآحادُ في العشراتِ، واحِدُها عشرةٌ، وعَشرتها مئةٌ. والآحادُ في المئاتِ، واحِدُها مِئةٌ، وعشرتُها ألفٌ. والآحادُ في الأُلوفِ، واحِدُها ألفٌ، وعشرتُها عشرةُ آلافٍ. فهذه أربعةُ أبواب.

وأما العشراتُ في العشراتِ، فواحِدُها مئةٌ، وعشرتُها ألفٌ. والعشراتُ في المئاتِ، واحِدُها ألفٌ، وعشرتُها عشرةُ آلافٍ. والعشراتُ في الأُلوفِ، واحِدُها عشرةُ آلافٍ، وعشرتُها [٦٩] مِئةُ ألفٍ. فهذه ثلاثةُ أبواب.

وأمّا المئاتُ في المئاتِ، فواحِدُها عشرةُ آلافٍ، وعشرتُها مِئةُ ألفٍ. والمئاتُ |ع ١٢ و| في الألوفِ، فواحِدُها مِئةُ ألفٍ، وعشرتُها ألفُ ألفٍ. فهذان بابان.

وأمّا الألوفُ في الألوفِ، فواحِدُها ألفُ ألفٍ، وعشرتُها عشرةُ آلافِ ألفٍ. وهذا بابٌ واحد.

فصار جُملةُ الجميع عشرةَ أبواب، وهذه صورتُها[٢]:

آحاد في آحاد: واحدها واحد، وعشرتها عشرة.

آحاد في عشرات: واحدها عشرة، وعشرتها مئة.

آحاد في مئات: واحدها مئة، وعشرتها ألف.

آحاد في أُلوف: واحدها ألف، وعشرتها عشرة آلاف.

---

(١) «وعشرتها»: وعشراتها [ع]. < وردت على هذه الشاكلة في بقيّة هذا الفصل >.

(٢) < وضعت خانة خالية في [ن] لرسم الصورة ولكنها بقيت فارغة. وكذلك الأمر في [ف، ك] >.

عشرات في عشرات: واحدها مئة، وعشرتها ألف.

عشرات في مئات: واحدها ألف، وعشرتها عشرة آلاف.

عشرات في أُلوف: واحدها عشرة آلاف، وعشرتها مئة ألف.

مئات في مئات: واحدها عشرة آلاف، وعشرتها مئة ألف.

مئات في أُلوف: واحدها مئة ألف، وعشرتها ألف ألف.

أُلوف في أُلوف: واحدها ألف ألف، وعشرتها عشرة آلاف ألف.[١]

فصل < ٢٠ > في الضرب[٢] والجذور والمكعّبات وما يستعمله الجبريّون والمهندسون من الألفاظ ومعانيها

فنقول[٣]: كلُّ عددين، أيّ عددين كانا، إذا ضُرب أحدُهما في الآخر، فإنَ المُجتَمِعَ منهما[٤] يُسمّى عدداً مُربّعاً. فإن كان العددان متساويين يُسمّى المُجتَمِعُ من ضربِهما عدداً مُربَّعاً مجذوراً، أو العددان يُسمّيان جذريْ ذلك العدد، مِثالُ ذلك إذا ضُرب اثنان في اثنين يكون أربعةٌ، وثلاثةٌ في ثلاثةٍ تسعةٌ، وأربعةٌ في أربعةٍ ستّةَ عشرَ. فالأربعةُ والتسعةُ والستّةَ عشرَ وأمثالُها من العدد يُسمَّى كُلُّ واحدٍ منها مُربَّعاً مجذوراً، والاثنان والثلاثة والأربعة يُسمّى جذراً، لأنّ الاثنين هو جَذرُ الأربعةِ، والثلاثةَ جَذرُ التسعةِ، والأربعةَ جَذرُ الستّةَ عشرَ، وعلى هذا القياس يُعتبر

---

(١) < وضعناها في صورتها هذه من أجل التنبيه إلى جدولتها بوضوح، بينما وردت في [ع] في رسم مؤلف من مربّعات صغيرة متراصّة >.

(٢) «الضرب»: سقطت من [ط، ك].

(٣) «فنقول»: سقطت من [ن، ط، ك].

(٤) «منهما»: من ذلك [ك، S].

سائر المُربّعات المجذورات وجُذورُها. وهذه صورتُها[١]:

٢	٣	٤	٥	٦	٧	٨	٩
د	ط	يو	كه	لو	مط	دس	فا
< ٤	٩	١٦	٢٥	٣٦	٤٩	٦٤	٨١ >[٢]

[٧٠] وكلُّ عددين مختلفين، أيّ عددين كانا، إذا ضُرب أحدُهما في الآخر، فإنّ المجتمعَ من ذلك يُسمّى عدداً مُربّعاً غيرَ مجذورٍ، والعددان المختلفان[٣] يُسمّيان جُزأين[٤] له، ويُسمّيان ضِلعين لذلك المُربّع، وهي من ألفاظ المهندسين. مِثالُ ذلك اثنان في ثلاثةٍ، أو ثلاثةٌ في أربعةٍ، أو أربعةٌ في خمسةٍ، وأشباهُ ذلك، فإنّ المُجتَمِعَ من مِثلِ هذه الأعداد المضروبة[٥] بعضِها في بعضٍ تُسمَّى مُربَّعات غير مجذورات[٦].

---

(١) «وهذه صورتها»: سقطت من [ف]، ووردت في هامش [ل].

(٢) < أضفنا الصف الأخير من الأعداد هنا للدلالة على المربّعات المجذورات على الرغم من أن صورتها لم ترد في المخطوط [ع] ولا في طبعة دار صادر [Ṣ] ولا في بقيّة المخطوطات >.

(٣) «المختلفان»: سقطت من [ع]، وردت في [ن، ف، Ṣ].

(٤) «جزأين»: جذرين [ك]. < وردت في هذا الشكل في متن نصّ [ل] وبعدها أضيف في الهامش «جذرين» >.

(٥) «المضروبة»: المضروب [ك].

(٦) < وردت صورة شبه مطموسة وغير واضحة لهذه الأعداد في المخطوطات [ع، ن، ف، ك، ل]، وربّما المراد من رسمها تصوير هيئة الأعداد المربّعات غير المجذورات على هذا الشكل: ٢ ضرب ٣ = ٦ | ٣ ضرب ٤ = ١٢ | ٤ ضرب ٥ = ٢٠. وهنا تترابط الأعداد المربّعات غير المجذورات مع حروف الجمل المركّبة على هذه الصورة: ٦ دط | ١٢ طيو | ٢٠ يوكه >.

فصل < ٢١ > في العدد المربّع[١]

[٢] كُلُّ عددٍ مُربَّعٍ |ع ١٢ ظ|، أيّ عددٍ[٣] [٤] كان مجذوراً أو غيرَ مجذور، ضُرب في عدد آخر، أيّ عدد كان، فإنّ المُجتَمِعَ من ذلك يُسمّى عدداً مجسّماً. فإن كان العددُ المربّعُ مجذوراً وضُرب في جذرِه، يُسمّى العددُ[٥] المُجتَمِعُ من ذلك عدداً مجسّماً مكعّباً. مِثالُ ذلك أربعةٌ، فإنّه عددٌ مربّعٌ مجذورٌ، ضُرب في الاثنين الذي هو جذرُها، فخرج منه ثمانيةٌ. وكذلك أيضاً التسعةُ، وهو أيضاً عدد مربّعٌ مجذورٌ ضُرب في الثلاثةِ الذي هو جذرُها، فخرج[٦] منه سبعةٌ وعشرون.

وكذلك الستّةَ عشَر فإنّه عددٌ مجذورٌ، ضُرب في الأربعةِ التي هي جذرُها فخرج منه أربعةٌ وسِتّون، فالثمانيةُ، والسبعةُ والعشرون، وأربعةٌ وستون، وأمثالُها من العدد يُسمّى مجسّماً مكعّباً[٧]. والمكعّبُ جِسمٌ طولُه وعرضُه وعمقُه متساويةُ الأضلاع[٨]، وله ستّةُ سطوحٍ مربّعات، متساوية الأضلاع، قائمة الزوايا، وله اثنا عشرَ ضِلعاً متوازية، وثماني زوايا مجسّمة،

---

(١) «في العدد المربّع»: سقطت من [ن، ف، ك، ل]، وردت في [ع، Ṣ].

(٢) زيادة في [ك، ل]: «إعلم بأن».

(٣) «أيّ عدد»: زيادة في [ع]، سقطت من [Ṣ].

(٤) زيادة في [ك]: «مربّع».

(٥) «العدد»: زيادة في [ع]، سقطت من [ن، ف، ك، Ṣ].

(٦) «فخرج» في [ع، ف، ك]: كانت [Ṣ]، فكان منه [ن، ل].

(٧) «العدد يسمّى مجسّماً مكعّباً»: الأعداد تسمّى أعداداً مجسّمة مكعّبة [Ṣ].

(٨) «الأضلاع»: سقطت من [ن، ف، Ṣ].

وأربعةٌ وعشرون زاويةً مسطّحة[١].

وإن ضُرب العددُ المربّعُ المجذورُ في عددٍ أقلَّ من جذرِه يُسمّى المُجتَمِعُ من ضربِه عدداً مجسّماً لبنيّاً. والمجسّمُ اللبنيُّ هو الذي طولُه وعرضُه متساويان[٢]، وأكبرُ من سَمْكِه[٣]، وله ستّةُ سطوحٍ مربّعات، متوازية الأضلاع، قائمة [٧١] الزوايا، لكنّ له سطحين متقابلين مربّعين، متساويي الأضلاع، قائمي الزوايا[٤]، ولها أربعة سطوح مستطيلات، ولها اثنا عشرَ ضِلعاً وكلُّ اثنين منها متوازيان، وثماني زوايا مجسّمة، وأربعٌ وعشرون زاويةً مسطّحة. وإن ضُرب المربّعُ المجذورُ في أكثرَ من جذرِه يُسمّى المُجتَمِعُ منه عدداً مجسّماً بيريّاً. مِثالُ ذلك أربعةٌ، فإنّه عددٌ مجذورٌ ضُرب في الثلاثةِ التي هي أكثرُ من جذرِها، فخرج[٥] منه اثنا عشرَ. وكذلك التسعةُ إذا ضُربت في الأربعة التي هي أكثرُ من جذرِها خرج منها ستّةٌ وثلاثون. فالاثنا عشرَ، والستّةُ والثلاثون، وأمثالُها من العدد تُسمّى مجسّمات بيريّات[٦].

والمجسّمُ[٧] البيريُّ هو الذي سَمْكهُ أكثرُ من طولِه وعرضِه،

---

(١) <أشكال هذه المجسّمات الهندسيّة صوِّرت في الرسالة الثانية في الجومَطْريا>.

(٢) «متساويان»: سقطت من [ع]، وردت في [ن، ف، ك، Ṣ].

(٣) «وأكبر من سمكه»: وسمكه أقلّ منهما [ن، ف، ك، Ṣ].

(٤) «لكن له سطحين متقابلين مربّعين، متساويي الأضلاع، قائمي الزوايا»: سقطت من [ع]، وردت في [ن، ف، ك، ل، Ṣ].

(٥) «فخرج»: فكان [ن، ف، ك، Ṣ].

(٦) «تسمّى مجسّمات بيريّات»: يسمّى مجسّماً بيريّاً [ن، ف، ل، Ṣ]، تسمّى مجسّمة بيريّة [ك].

(٧) «والمجسّم»: سقط من [ن].

وله ستّةُ سطوحٍ مربّعات: اثنان منها مربّعان متقابلان، متساويا الأضلاع، قائما الزوايا، وأربعةٌ منها مستطيلة، متوازية الأضلاع، قائمة الزوايا. وله اثنا عشرَ ضِلعاً كلُّ اثنين منها متوازيان متساويان، وله ثماني زوايا مجسّمة، وأربعٌ وعشرون زاوية مسطّحة. وكلُّ عددٍ مربّعٍ غيرِ مجذورٍ ضُرب في ضِلعهِ الأصغرِ، فإنّ المُجتَمِعَ منه يُسمّى مُجسّماً لبنيّاً، وإن ضُرِب في ضِلعهِ الأطولِ فإنّ المُجتَمِعَ منه يُسمّى مجسّماً بيريّاً[(١)]، وإن ضُرب في عددٍ أقلَّ منه أو أكثرَ منه، فإنّ المُجتَمِعَ منه يُسمّى مجسّماً لوحِيّاً. مِثالُ ذلك الاثنا عشرَ، فإنّه عددٌ مربّعٌ غيرُ مجذورٍ، وأحدُ ضِلعيه ثلاثةٌ، والآخرُ |ع ١٣ و| أربعةٌ، فإن ضُرب اثنا عشرَ في ثلاثةٍ خرج منه ستّةٌ وثلاثون، وهو مجسّمٌ لبنيٌّ. وإن ضُرب في أربعةٍ خرج منه ثمانيةٌ وأربعون، وهو مجسّمٌ بيريٌّ[(٢)]. وإن ضُرب في أقلَّ من الثلاثةِ أو أكثرَ من الأربعةِ يُسمّى مجسّماً لوحِيّاً.

والمجسّمُ اللوحيُّ هو الذي طولُه أكثرُ من عرضِه، وعرضُه أكثرُ[(٣)] من سَمْكِه، وله ستّةُ سطوح، كلُّ اثنين منها متساويان متوازيان، وله اثنا عشرَ ضِلعاً، كلُّ اثنين منها متوازيان، وثماني زوايا مجسّمة، وأربعٌ وعشرون زاويةً مسطّحة.

---

(١) «وإن ضرب في ضلعه الأطول فإن المُجتَمِع منه يسمى مجسماً بيرياً»: سقطت من [ع، ف]، وردت في [ن، ط، Ṣ].

(٢) «وهو مجسم بيريّ»: وذلك يسمّى مجسّماً بيريّاً [ن].

(٣) «أكثر»: أكبر [ف، ك، ل].

[٧٢] فصل < ٢٢ > في خواصِّ الأعداد المجذورة[١][٢]

فنقول[٣]: وكلُّ عددٍ مجذورٍ، إذا زيد عليه جذراه وواحِدٌ، كان المجتمعُ من ذلك مجذوراً[٤]. وكلُّ عددٍ مجذورٍ إذا انتُقِصَ منه[٥] جذراه إلّا واحداً يكون الباقي مجذوراً. وكلُّ عددين مجذورين على الوِلاء، إذا[٦] ضُرب جذرُ أحدِهما في جذر الآخر، وزيد عليه رُبعُ الواحد، يكون الجملةُ مجذوراً. مِثالُ ذلك: جذرُ أربعةٍ وهو اثنان، في جذر تسعةٍ وهو ثلاثة، فيكون ستّةً، وزيد عليهُ ربعٌ، يكون ستةً ورُبعاً[٧][٨]، جذرها اثنان ونِصفٌ[٩]. وكلُّ عددين مجذورين على الوِلاء إذا ضُرب جذرُ

---

(١) «الأعداد المجذورة»: العدد المجذور [Ṣ].

(٢) «في الأعداد المجذورة»: سقطت من [ن، ك، ل].

(٣) «فنقول»: سقطت من [ن، ط، ف، ك، ل].

(٤) زيادة في [ك]: «مثاله تسعة وهو عدد مجذور إذا زيد عليه جذران وهو ستّة وواحد فالمجتمع منه ستّة عشر وهو عدد مجذور». <أضيفت كذلك إلى هامش [ل]>.

(٥) «إنتقص منه»: ينقص [ك].

(٦) «إذا»: سقطت من [ع]، وردت في [Ṣ]. <وهي على هذه الشاكلة في بقيّة الفصل>.

(٧) «جذر أربعة وهو اثنان، في جذر تسعة وهو ثلاثة، فيكون ستة، وزيد عليه ربع، يكون ستة وربعاً»: تسعة وستّة عشر وهما عددان مجذوران على الولاء جذرهما ثلاثة وأربعة إذا ضرب الثلاثة في الأربعة يكون إثنا عشر ... <والبقيّة مطموسة> [ف]، ستّة ويزاد عليه ربع فتصير ستّة وربع [ك]، إثنان في ثلاثة فيكون ستّة [ل].

(٨) «جذر أربعة وهو اثنان، في جذر تسعة وهو ثلاثة، فيكون ستة، وزيد عليه ربع، يكون ستة وربعاً»: سقطت من [ن].

(٩) «مثال ذلك: جذر أربعة وهو اثنان، في جذر تسعة وهو ثلاثة، فيكون ستة،

أحدِهما في جذر الآخر خرج بينهما عددٌ وسطٌ، وتكون ثلاثتُها في نِسبة واحدة. مِثالُ ذلك: أربعةٌ وتسعةُ، فإنّهما عددان مجذوران، وجذراهما اثنان وثلاثةٌ، واثنان في ثلاثةٍ: ستّةٌ، فنسبةُ الأربعة إلى الستّة كنسبة الستّة إلى التسعة، وعلى هذا القياس يُعتبر سائرها.

فصل < ٢٣ > في مسائل(١) من المقالة الثانية من كتاب أوقليدِس في الأصول(٢)

كلُّ عددين قُسِّمَ أحدُهما بأقسامٍ كم كانت، فإنّ ضربَ أحدِهما في الآخر مساوٍ لضربِ الذي لم يُقْسَم في جميع أقسام العدد(٣) المقسوم قِسماً قِسماً. مِثالُ ذلك عشرةٌ وخمسةَ عشرَ. وقُسِّمَ الخمسةَ عشرَ ثلاثةَ أقسام: سبعة وثلاثة وخمسة، فنقول:

« أ »: إنّ ضربَ العشرةِ في خمسةَ عشرَ مساوٍ لضربِ العشرةِ في سبعةٍ وفي ثلاثةٍ وفي خمسةٍ. [٧٣]

« ب »: كُلُّ عددٍ قُسِّمَ بأقسامٍ كم كانت، فإنّ ضربَ ذلك العددِ في مثلِه مساوٍ لضربهِ في جميع أقسامه. مِثالُ ذلك عشرةٌ قُسِّمَت بقسمين: سبعة وثلاثة، فأقول: إنّ ضربَ العشرةِ في نفسِها مساوٍ لضربِها في سبعةٍ وفي ثلاثةٍ.

وزيد عليه ربع، يكون ستة وربعاً، جذرها اثنان ونصف»: سقطت من [ع]، وردت في [Ṣ].

(١) «في مسائل»: سقطت من [ك].

(٢) «في الأصول»: سقطت من [ع، ط، ن، ل]، وردت في [Ṣ].

(٣) «العدد»: سقطت من [ن].

« ج »: كلُّ عددٍ قُسِّمَ بقسمين، فأقول: إنّ ضربَ ذلك العددِ في أحدِ قِسميه مساوٍ لضربِ تلك القسمةِ[١] في نفسِها وفي القسمِ الآخر. مِثالُ ذلك عشرةٌ قُسِّمَت بِقسمين: ثلاثة وسبعة، فأقول: إنّ ضربَ العشرةِ في سبعةٍ مساوٍ لضربِ السبعةِ في نفسِها وثلاثةٍ في سبعة[٢].

|ع ١٣ ظ| « د »: وكلُّ عددٍ قُسِّمَ بقسمين، فإنّ ضربَ ذلك العددِ[٣] في نفسِه[٤] مساوٍ لضربِ كلِّ قسمٍ في نفسِه مجموعين[٥]، وأحدهما في الآخر مرّتين، مِثالُ ذلك عشرةٌ قُسِمَت قِسمين: سبعة وثلاثة. فضربُ السبعة في نفسه، وثلاثة في نفسه، يكون ثمانيةً وخمسين، ثمّ ضُرب كلُّ واحدٍ منهما في الآخر مرّتين، فيكون اثنين وأربعين، وزدناه على الثمانية والخمسين، فصارت مائةً مساوِيةً لضربِ العشرة في نفسها[٦].

« هـ »: كلُّ عددٍ قُسِّمَ بنصفين ثم بقسمين مختلفين، فإنّ ضربَ أحدِ المختلفين في الآخر، وضربَ التفاوتِ في نفسِه

---

(١) «تلك القسمة»: ذلك القسم [ن، S].

(٢) «فأقول: إن ضرب العشرة في سبعة مساوٍ لضرب السبعة في نفسها وثلاثة في سبعة»: أضافها مدقق [ل] في الهامش.

(٣) «العدد»: سقطت من [ع]، وردت في [ك، S].

(٤) «نفسه»: مثله [ف].

(٥) «مجموعين»: زيادة في [ع] ، سقطت من [S].

(٦) «السبعة في نفسه، وثلاثة في نفسه، يكون ثمانية وخمسين، ثمّ ضرب كل واحد منهما في الآخر مرّتين، فيكون اثنين وأربعين، وزدناه على الثمانية وخمسين، فصارت مائة مساوية لضرب العشرة في نفسها»: فأقول إن ضرب العشرة في نفسها مساوٍ لضرب سبعة في نفسها، وثلاثة في نفسها، وسبعة في ثلاثة [S].

مساوٍ لضربِ نِصفِ ذلك العدد في نفسِه. مثالُه عشرةٌ قُسِّمَت بنصفين ثم بقسمين مختلفين: ثلاثة وسبعة، فأقول[١]: إنّ ضربَ السبعةِ في ثلاثةٍ والتفاوتِ في نفسِه وهو اثنان مجموعين مساوٍ لضربِ الخمسةِ في نفسِها.

« و »: كلُّ عددٍ قُسِّمَ بنصفين ثم يُزاد فيه زيادة ما، فأقول: إنّ ضربَ ذلك العددِ مع الزيادة في تلك الزيادةِ ونصفِ العدد في نفسِه مجموعاً يكون مساوياً لضربِ نِصفِ ذلك العدد[٢] مع الزيادة في نفسِه، مِثالُ ذلك[٣] عشرةٌ قُسِّمَت بنصفين ثم زيد عليها اثنان. فأقول[٤]: إنّ ضربَ الاثني عشرَ في الاثنين وخمسةٍ في نفسها مجموعاً يكون مساوياً لضربِ الاثنين والخمسةِ مجموعاً في نفسه.

[٧٤] « ز »: كلُّ عددٍ قُسِّمَ بقسمين، فأقول: إنّ ضربَ ذلك العددِ في نفسِه وضربَ أحدِ القسمين في نفسِه مجموعاً مساوٍ لضربِ ذلك[٥] العدد في ذلك القسم مرّتين، وضربِ القسمِ الآخر في نفسه مجموعاً. مِثالُ ذلك عشرةٌ قُسِّمَت بقسمين: سبعة وثلاثة، فأقول: إنّ ضربَ العشرةِ في نفسِها، وسبعةٍ في نفسِها مجموعاً مساوٍ لضربِ العشرةِ في سبعةٍ مرّتين، وثلاثةٍ في نفسِها مجموعاً.

« ح »: كلُّ عددٍ ينقسم بقسمين، ثم زيد عليه مثلُ أحدِ

---

(١) «فأقول» [ع، ك]: فنقول [ن، ل، ṣ].

(٢) «لضرب نصف ذلك العدد»: بالضرب [ع].

(٣) «مثال ذلك»: مثاله [ṣ]. < وردت على هذا الشكل في بقيّة الفصل > .

(٤) «فأقول»: فنقول [ن، ṣ].

(٥) «ذلك»: سقطت من [ع]، وردت في [ن، ṣ].

القسمين، فأقول[1]: إنّ الذي يكون من ضربِ جميعِ ذلك في نفسِه مساوٍ لضرب ذلك العدد قبل الزيادة في تلك الزيادة أربع مرّات، و ضربِ القسمِ الآخرِ في نفسِه مرّةً واحدة[2]. مِثالُ ذلك عشرةٌ قُسِّمَت بقسمين: سبعة وثلاثة، ثم زيد عليها ثلاثةٌ، فأقول[3]: إنّ ضربَ الثلاثةَ عشرَ في نفسها مساوٍ لضربِ العشرةِ في الثلاثةِ أربعَ مرّات، وضربِ سبعةٍ في نفسِها مرّةً واحدة.

« ط »: كل عدد قُسِّمَ بنصفين ثم بقسمين مختلفين[4]، فإنّ الذي يكون من ضربِ القسمين المختلفين كلٍّ واحدٍ منهما في نفسِه مجموعاً، مِثلُ ما يكون من ضربِ نِصفِ ذلك العدد[5] في نفسه، وضربِ التفاوتِ ما بين العددين في نفسه مجموعاً مرّتين[6]. مِثالُ ذلك عشرةٌ قُسِّمَت بنصفين ثم بقسمين مختلفين: ثلاثة وسبعة، فأقول: إنّ الذي |ع ١٤ و| يكون من ضربِ سبعةٍ في نفسِها، وثلاثةٍ في نفسها مجموعاً، مثلُ ما يكون من ضربِ خمسةٍ في نفسها، ومن ضربِ اثنين الذي هو التفاوت ما بين القسمين في نفسه مجموعاً مرّتين[7].

« ي »: كلُّ عددٍ قُسِّمَ بنصفين، ثم زيد فيه زيادةٌ ما، فإنّ الذي يكون من ضربِ ذلك العددِ مع الزيادة في نفسه، وضربِ

---

(١) «فأقول»: فنقول [ن، ṣ].

(٢) «وضرب القسم الآخر في نفسه مرّة واحدة»: والقسم الآخر في نفسه [ن، ṣ].

(٣) «فأقول»: فنقول [ن، ṣ].

(٤) «قُسِّمَ بنصفين ثم بقسمين مختلفين»: بقسمين معتدلين [ع].

(٥) «العدد»: زيادة في [ع]، سقطت من [ن، ṣ].

(٦) «مرّتين»: زيادة في هامش [ع]، سقطت من [ن، ṣ].

(٧) «مرّتين»: زيادة في [ع]، سقطت من [ن، ṣ].

الزيادةِ في نفسها مجموعاً، يكون مِثلَي ما يكون من ضربِ نِصفِ العددِ الأوّلِ[١] مع الزيادةِ في نفسه، وضربِ[٢] نِصفِ العددِ الأوّلِ[٣] في نفسِه بلا زيادة[٤]. مِثالُ ذلك عشرة قُسِّمَت بنصفين، ثم زيد عليها اثنان. [٧٥] فأقول: إنّ ضربَ الاثني عشرَ في نفسه، والاثنين في نفسه مجموعاً، يكون مِثلَي ما يكون من ضربِ سبعةٍ في نفسها، وخمسةٍ في نفسها مجموعاً.

فصل < ٢٤ > عِلم العدد والنفس[٥]

واعلَمْ أيّها الأخ البارّ الرحيم[٦]، أيّدك الله وإيّانا بروح منه[٧]، أنّه إنّما قدَّم الحكماءُ النظرَ في عِلم العدد قبل النظر في سائر العلوم الرياضيّة، لأنّ هذا العلمَ مركوزٌ في كلِّ نفسٍ بالقوّة، وإنّما يحتاج الإنسان إلى التأمُّل بالقوّة الفكريّة حَسْبُ، من غير أن يأخذ لها مِثالاً من عِلمٍ آخرَ، بل منه يؤخذ المِثالُ على كلِّ علمٍ[٨] ومعلوم. فأمّا ما أشرنا إليه من المثالات التي بالخطوط الهنديّة[٩]

---

(١) «الأوّل»: زيادة في [ع]، سقطت من [ن، Ṣ].

(٢) «ضرب»: سقطت من [ع]، وردت في [ن، Ṣ].

(٣) «الأوّل»: زيادة في [ع]، سقطت من [Ṣ].

(٤) «بلا زيادة»: زيادة في [ع]، سقطت من [Ṣ].

(٥) «علم العدد والنفس»: سقطت من [ن، ط، ف، ك، ل]، وردت في [Ṣ].

(٦) «أيها الأخ البارّ الرحيم»: يا أخي [ك، ل].

(٧) «أيها الأخ البارّ الرحيم، أيّدك الله وإيّانا بروح منه»: سقطت من [ع، ف]، وردت في [ن، ط، Ṣ].

(٨) «علم»: زيادة في [ع، ن، ك]، سقطت من [Ṣ].

(٩) «الهنديّة»: زيادة في [ط، ف، ك]. <أضافها كذلك منقّح [ل] في الهامش>. <وقد عني بها هنا «الأرقام الهنديّة»>.

في هذه الرسالة[١] فإنّما تلك للمتعلّمين المبتدئين الذين قوّةُ أفكارِهم ضعيفةٌ، فأمّا من كان منهم فهيماً ذكياً[٢] فغيرُ محتاج إليها إذا تأمّل[٣].

واعلَمْ[٤] أيّها الأخ البارّ الرحيم، أيّدك الله وإيّانا بروح منه[٥]، [٦] أنّ أحدَ أغراضِنا من هذه الرسالة ما قد بَيّنا في أوّلها، وأمّا الغرضُ الثاني[٧] فهو التنبيهُ على عِلم النفس والحثُّ على معرفةِ جوهرِها، وذلك أنّ العاقلَ الذهينَ إذا نظر في عِلم العدد وتفكّر في كميّة أجناسه وتقاسيم أنواعه وخواصّ تلك الأنواع، عَلِمَ أنّها كلَّها أعراضٌ[٨] وجودُها وقِوامُها بالنفس[٩]، فالنفسُ إذاً جوهرٌ، لأنّ العَرضَ لا يكون[١٠] له قِوامٌ إلّا بالجوهر ولا يوجد إلّا فيه[١١] [١٢].

---

(١) «في هذه الرسالة»: سقطت من [ع]، وردت في [ن، Ṣ].

(٢) زيادة في [ف]: «حيّ القلب».

(٣) «إذا تأمّل»: زيادة في [ع]، سقطت من [ن، ك، ل، Ṣ].

(٤) «واعلم»: سقطت من [ن].

(٥) «أيها الأخ البارّ الرحيم، أيّدك الله وإيّانا بروح منه»: سقطت من [ع، ك]، وردت في [ن، Ṣ].

(٦) «أيها الأخ البارّ الرحيم، أيّدك الله وإيّانا بروح منه»: يا أخي [ك].

(٧) «الثاني»: الآخر [ن، ط، ف، ك، ل، Ṣ].

(٨) «أعراض»: رسوم ومثالات [ط، ف].

(٩) زيادة في [ف]: «لا في الجسد». <أضافها منقّح [ل] في الهامش>.

(١٠) «يكون»: يقوم [ف].

(١١) «فيه»: منه [ف].

(١٢) <وهذا القول ورد تبعاً لما درج عليه الحكماء من تحصيل ما ذُكر عند أرِسطوطاليس حول الجوهر والأعراض>.

< فصل ٢٥ في > الغرض من العلوم[١]

واعلَمْ يا أخي، أيّدك الله وإيّانا بروح منه[٢]، بأنّ غرضَ الفلاسفةِ الحكماءِ[٣] الموحّدين[٤] من النظر في العلوم الرياضيّة، وتخريجهم تلامذتهم[٥] بها، إنّما هو السلوكُ [٧٦] والتطرّقُ منها إلى علوم الطبيعيّات. وأمّا غرضُهم من النظر[٦] في الطبيعيّات فهو الصعودُ منها والتّرقي إلى العلوم الإلٰهية[٧]،[٨] الذي[٩] هو أقصى غرضِ الحكماءِ، والنهايةُ التي إليها يُرتقى بالمعارف الحقيقيّة.[١٠]

---

(١) «الغرض من العلوم»: سقطت من [ن، ف، ك، ل].

(٢) «يا أخي، أيّدك الله وإيّانا بروح منه»: سقطت من [ع، ف]، وردت في [ن، ك، Ṣ].

(٣) «الحكماء»: العلماء الفضلاء [ط].

(٤) «الموحّدين»: زيادة في [ع]، سقطت من [Ṣ]. < أضاف منقّح [ع] في الهامش: «العلماء الفضلاء» >.

(٥) «تلامذتهم»: أولادهم وتلامذتهم [ط].

(٦) «من النظر»: سقطت من [ل].

(٧) «الإلهية»: الإلهيات [ل].

(٨) «وأما غرضهم من النظر في الطبيعيّات فهو الصعود منها والتّرقي إلى العلوم الإلهية»: والغرض والمراد من العلوم الطبيعيّات هو التطرّق والسلوك إلى العلوم الإلاهيّات والصعود إليها [ك].

(٩) < كما ورد في حاشية طبعة دار صادر، إن لفظ «الذي» هو كناية عن صفة للتّرقي >.

(١٠) «إنما هو السلوك والتطرّق منها إلى علوم الطبيعيّات. وأما غرضهم من النظر في الطبيعيّات فهو الصعود منها والتّرقي إلى العلوم الإلهية الذي هو أقصى غرض الحكماء، والنهاية التي إليها يُرتقى بالمعارف الحقيقيّة»: إن النظر في الطبيعيّات والغرض والمراد من العلوم الطبيعيّات هو التطرّق والسلوك إلى العلوم الإلاهيّات والصعود إليها، وهو أقصى غرض الحكماء من التّرقي بالمعارف [ط].

ولمّا كان أوّل درجة من النظر في العلوم الإلٰهيّة هي معرفةُ جوهر النفس، والبحثُ عن مبدئها من أين كانت قبل تعلّقها بالأجساد[١]، والفحصُ عن معادها إلى أين يكون بعد فراقها الجسدَ، الذي يُسمّى[٢] الموتَ، وعن كيفيّة[٣] ثواب المحسِنين كيف يكون في عالم الأرواح[٤]،[٥] وعن جزاء المسيئين[٦] كيف يكون في دار الآخرة، وخصلةٌ أُخرى أيضاً، |ع ١٤ ظ| لمّا كان الإنسان مدعوّاً[٧] إلى معرفة ربّه تعالى[٨]،[٩] ولم يكن له طريق إلى معرفته إلّا بعد معرفة نفسه، كما قال الله تعالى: ﴿وَمَن يَرْغَبُ عَن مِّلَّةِ إِبْرَٰهِـۧمَ إِلَّا مَن سَفِهَ نَفْسَهُۥ﴾[١٠]، أي جَهِلَ النفس[١١]، وكما قيل[١٢]: «من عَرَفَ نَفسَه، عَرَفَ رَبَّه»[١٣]، وقد

---

(١) «بالأجساد»: بالجسد [ل، Ṣ].

(٢) «يسمّى»: هو [ط].

(٣) «وعن كيفيّة»: والبحث أيضاً [ط].

(٤) «عالم الأرواح»: دار الآخرة [ل]. < ثمّ أضافها منقحّ [ل] في الهامش >.

(٥) زيادة في [ط، ف]: «يسمّى الدار الآخرة».

(٦) «كيف يكون في عالم الأرواح، وعن جزاء المسيئين»: زيادة في [ع، ن]، سقطت من [Ṣ] // «وعن جزاء المسيئين»: سقطت من [ك].

(٧) «مدعوّا»: مندوباً [ن، ط، Ṣ].

(٨) «تعالى»: زيادة في [ع]، سقطت من [ن، ك، Ṣ].

(٩) زيادة في [ط]: «الذي خلقه وسوّاه ورزقه» // زيادة في [ف]: «الذي خلقه».

(١٠) < القرآن: البَقَرة ١٣٠ >.

(١١) «كما قال الله تعالى: ﴿وَمَن يَرْغَبُ عَن مِّلَّةِ إِبْرَٰهِـۧمَ إِلَّا مَن سَفِهَ نَفْسَهُۥ﴾، أي «جَهِلَ النفس»: سقطت كلّها من [ع، ن، ط، ف]، وردت في [Ṣ].

(١٢) زيادة في [ك]: «في الخبر عن النبيّ صلّى الله عليه وسلّم أنه قال» // زيادة في [ط]: «كما ورد من الخبر عن النبيّ صلّى الله عليه وسلّم أنه قال».

(١٣) «من عرف نفسه، عرف ربه» [ع، ف، ك]: من عرف نفسه فقد عرف ربه [ن، Ṣ].

قيل أيضاً[١]: «أعرَفُكُم بِنَفسِه أعرَفُكُم بِرَبِّه»، وجبَ على كلِّ عاقلٍ طلبُ[٢] عِلمِ النفسِ ومعرفةُ جوهرِها وتهذيبِها[٣]، وقد قال الله عزّ وجلّ[٤]: ﴿وَنَفۡسٖ وَمَا سَوَّىٰهَا (٧) فَأَلۡهَمَهَا فُجُورَهَا وَتَقۡوَىٰهَا (٨) قَدۡ أَفۡلَحَ مَن زَكَّىٰهَا (٩) وَقَدۡ خَابَ مَن دَسَّىٰهَا (١٠)﴾[٥].[٦] وقال الله تعالى حكايةً عن امرأةِ العزيز في قِصّة يوسُفَ الصِدِّيقِ[٧]، عليه السلام[٨]: ﴿ٱلنَّفۡسَ لَأَمَّارَةُۢ بِٱلسُّوٓءِ إِلَّا مَا رَحِمَ رَبِّيٓۚ﴾[٩].

وقال الله[١٠] تعالى: ﴿وَأَمَّا مَنۡ خَافَ مَقَامَ رَبِّهِۦ وَنَهَى ٱلنَّفۡسَ عَنِ

---

(١) «وقد قيل أيضاً»: وقال أيضاً [ط] <النبيّ عليه السلام>. سقطت من [ع]، وردت في [Ṣ].

(٢) زيادة في [ف]: «معرفة».

(٣) زيادة في [ف]: «ومداها ومعادها».

(٤) «عزّ وجلّ»: تعالى [Ṣ].

(٥) <القرآن: الشَّمْس ٧-١٠>.

(٦) <كما ذكر في طبعة دار صادر [Ṣ]، القول «دسّاها» عُنيَ به أنه «أخفاها بالمعصية»>.

(٧) «الصِدّيق»: زيادة في [ن، ك].

(٨) «وقال الله تعالى حكاية عن إمرأة العزيز في قصة يوسُف، عليه السلام»: وقال يوسُف عليه السلام [ع]، وقال يوسُف الصدّيق [ن]، وقال حكاية عن يوسُف الصدّيق [ط]. <أبقينا هنا على ما ورد في طبعة دار صادر [Ṣ] بدلاً مما خُطَّ في [ع، ن، ط]، لعلّ ذلك فيه تصويب أفضل لكيفيّة ذكر الآية القرآنيّة الواردة أعلاه>.

(٩) <القرآن: يوسُف ٥٣>.

(١٠) «وقال الله»: وقوله [ع]. <أبقينا هنا على ما ورد في طبعة دار صادر [Ṣ] بدلاً عما خُطَّ في [ع] وذلك للإبقاء على نفس الألفاظ المعتمدة في وتيرة السرد>.

ٱلْهَوَىٰ (٤٠) فَإِنَّ ٱلْجَنَّةَ هِىَ ٱلْمَأْوَىٰ (٤١)﴾[١]. وقال تعالى: ﴿يَوْمَ تَأْتِى كُلُّ نَفْسٍ تُجَٰدِلُ عَن نَّفْسِهَا﴾[٢]. وقال تعالى: ﴿يَٰٓأَيَّتُهَا ٱلنَّفْسُ ٱلْمُطْمَئِنَّةُ (٢٧) ٱرْجِعِىٓ إِلَىٰ رَبِّكِ رَاضِيَةً مَّرْضِيَّةً (٢٨) فَٱدْخُلِى فِى عِبَٰدِى (٢٩) وَٱدْخُلِى جَنَّتِى (٣٠)﴾[٣][٤]. وقال سُبحانَه وتعالى[٥]: ﴿ٱللَّهُ يَتَوَفَّى ٱلْأَنفُسَ حِينَ مَوْتِهَا وَٱلَّتِى لَمْ تَمُتْ فِى مَنَامِهَا﴾[٦]؛ وآياتٌ كثيرةٌ في[٧] القُرآن، دالّاتٌ[٨] على وجود النّفس وعلى تعرُّفِ حالاتها[٩]، وهي حُجّة على الخُرَّمِيّين[١٠] المنكِرين أمرَ النّفسِ ووِجْدانها[١١].

---

(١) <القرآن: النازِعات ٤٠-٤١>.

(٢) <القرآن: النَّحْل ١١١>.

(٣) <القرآن: الفَجْر ٢٧-٣٠>.

(٤) «﴿فَادْخُلِي فِي عِبَادِي وَادْخُلِي جَنَّتِي﴾»: زيادة في [ع، ن، ط]، سقطت من [ف، Ṣ].

(٥) «سُبحانَه وتعالى» في [ن]: سُبحانَه [ع]، تعالى [Ṣ].

(٦) <القرآن: الزُّمَر ٤٢>.

(٧) «في»: من [ع].

(٨) «دالّاتٍ» في [ع، ن]: ودلالات [Ṣ].

(٩) «تعَرُّفِ حالاتها»: تَصرُّفِ حالاتها [ن، Ṣ]، تَصَرُّفِ حالاتها [ط].

(١٠) «على الخُرّميين»: للعارفين بأمر النّفس [ط]، على الجُرميين [Ṣ]. <ورد ذكر «الخُرّميين» في مخطوط [ع]، أي للدلالة على مذهب الخُرميّة. أما ذكر «الجُرميين» في طبعة دار صادر [Ṣ]، فيه نسبة إلى الأجرام السماوية، أي النجوم، والمراد من ذلك الدلالة على مذهب الماديّة>. <سقطت كلّها من [ف]>.

(١١) «أمر النّفس ووجدانها»: وجودها وحقيقة جوهرها [ط].

[٧٧] وأمّا أولئك الحكماءُ[١] الذين تكلّموا[٢] في عِلم النفس قبل نزول القرآن والإنجيل والتوراة[٣]،[٤] فإنّهم لمّا بحثوا[٥] عن عِلم النفس وألّفوا فيه الكتُبَ[٦] بقرائح قلوبهم الصّافيةِ[٧]، واستخرجوا معرفةَ جوهرِها[٨] بنتائج عقولهم[٩]، دعاهم ذلك إلى تصنيف الكتب الفلسفيّة التي تقدّم ذِكرُها في أوّل[١٠] هذه الرسالة، ولكنّهم لما طوّلوا الخُطَبَ فيها، ونَقلها من لغة إلى لغة مَن لم يكن فهِمَ معانِيَها ولا عرفَ[١١] أغراضَ مؤلّفيها، انغلقَ[١٢] على الناظرين[١٣] في تلك الكتب فهمُ معانيها، وثَقُلَ[١٤] على الباحثين أغراضُ مُصنّفيها. ونحن قد أخذنا لُبَّ معانيها وأقصى

---

(١) زيادة في [ط]: «والفلاسفة الذين كانوا قبل الإسلام».

(٢) «تكلّموا»: كانوا يتكلمون [Ṣ].

(٣) «التوراة» في [ف، ل، Ṣ]: التوريّة [ع، ن، ك]. < وردت الإشارة إلى التوراة باللفظ «التوريّة» في مجمل مخطوط [ع] و[ن، ك]، ولكننا فضّلنا التسمية الأكثر تداولاً في اللغة العربيّة، أي «التوراة»> .

(٤) «القرآن والإنجيل والتوراة»: القرآن والتوريّة والإنجيل [ن، ك]. سقطت من [ط].

(٥) «فإنهم لما بحثوا»: كأنهم بحثوا [ع].

(٦) «وألّفوا فيه الكتب»: زيادة في [ع]، سقطت من [ط، Ṣ].

(٧) «قلوبهم الصّافية» في [ن]: صافية [ع]، قلوبهم [Ṣ].

(٨) «معرفة جوهرها»: سقطت من [ط].

(٩) زيادة في [ط]: «وبيّنوا حقيقة جوهرها».

(١٠) «أوّل»: سقطت من [ط].

(١١) «عرف»: فهم [ن، ل].

(١٢) «انغلق»: إنغلقت [ن].

(١٣) «الناظرين»: الناظر [ل].

(١٤) «ثَقُلَ»: ثقلت [Ṣ].

أغراض واضعيها، وأوردناها بأوّجز ما يمكن من[١] الاختصار في اثنتين وخمسين[٢] رسالةً، أوّلُها هذه[٣]، ثمّ يتلوها أخواتُها على الوِلاءِ[٤] كترتيب العدد تجدُها إن شاء الله تعالى[٥].[٦]

والحَمدُ لواهِبِ العَقل ولملَقّن الصواب[٧].[٨]

تمّت الرسالة الأولى[٩] من القسم الأوّل في الرياضيّات وهو ثلاثَ عشرةَ رسالةً[١٠].[١١]

---

(١) زيادة في [ط، ف]: «الألفاظ».

(٢) «اثنتين وخمسين»: إحدى وخمسين [ع، ن، ط، ف، ك، ل]. <أبقينا هنا على ما ورد في طبعة دار صادر [Ṣ] بدلاً عما خُطَّ في [ع]، وذلك للدلالة على الترتيب المعتمد في التحقيق النقديّ لمجمل رسائل إخوان الصفاء في سلسلتنا هذه، والذي يتوافق أيضاً مع [Ṣ]>.

(٣) زيادة في [ط، ف]: «الرسالة».

(٤) زيادة في [ف]: «إلى آخرها».

(٥) «كترتيب العدد تجدها إن شاء الله تعالى»: سقطت من [ف]. سقطت فقط اللّفظة «تعالى» من [ط].

(٦) زيادة في [ف]: «وهي الأرِثماطيقي».

(٧) «والحمد لواهب العقل ولملقّن الصواب»: زيادة في [ع]، سقطت من [ن، ط، ف، Ṣ].

(٨) «لواهب العقل ولملقّن الصواب»: ربّ العالمين [ك].

(٩) «تمّت الرسالة الأولى»: تمّت رسالة الأرِثماطيقي وهي الأوّلى من إحدى وخمسين رسالة [ك]. <أضاف منقّح [ل] هذه الجملة في الهامش>.

(١٠) «من القسم الأوّل في الرياضيّات وهو ثلاث عشرة رسالة»: زيادة في [ع]، سقطت من [ن، ف، ك، Ṣ].

(١١) «تمّت الرسالة الأولى من القسم الأوّل في الرياضيّات وهو ثلاث عشرة رسالة»: تمّت الرسالة الأولى المسمّاة الأرِثماطيقي من رسايل إخوان السّعادة والحمد لله ربّ العالمين والصلاة على رسوله محمّد وآله أجمعين وحَسبُنا الله ونِعمَ الوكيل [ط]؛ تمّت الرسالة والحمد لله ربّ العالمين وصلّى الله على سيّدنا محمّد وآله الطاهرين [ل].

والحمد لله ربّ العالمين، وصلّى الله على رسوله محمّد النّبيّ وآله الطاهرين، وسلّم تسليماً[١] .[٢]

(١) «ربّ العالمين، وصلّى الله على رسوله محمّد النبي وآله الطاهرين، وسلّم تسليما»: وحده [ن].

(٢) «والحمد لله ربّ العالمين، وصلّى الله على رسوله محمّد النّبيّ وآله الطاهرين، وسلّم تسليماً»: سقطت من [ع، ف، ك]، وردت في [Ṣ]. < خُتم مخطوط [ل] بشكل خاطىء من خلال تقديم ناسخه للرسالة الثانية تحت عنوان «النِسَبّ العدديّة والهندسيّة» بدلاً من الإشارة إلى الرسالة التالية على أنها «رسالة الهندسة» > .

|ع ١٥ و| [٧٨](*) الرِّسالَةُ الثانيةُ مِنَ القِسْمِ الأَوَّلِ في العُلومِ الرِّياضيَّةِ التَعْليميَّة

(*) تمّ الاستناد إلى مخطوط عاطف أفندي، المشار إليه هنا بالرمز [ع]، في التحقيق النقديّ لهذه الرسالة وهو الأقدم من مجمل مخطوطات رسائل إخوان الصفاء. وقد اعتُمِد على ما ورد في هذا المخطوط كما لو أنه المصدر الأساسي في وضع متن النص في هذا المجلّد من دون الأخذ به بشكل مطلق على أنه المخطوط الأمّ. وقد قوبل هذا المخطوط على طبعة دار صادر، التي أُشير إليها هنا بالرمز اللاتينيّ [Ṣ]، كما أن تحقيقه استند إلى مقارنته مع بقيّة المخطوطات الأخرى. وقد وردت الإشارة إلى رموزها المعتمدة في التحقيق في توطئة هذا الكتاب المكتوبة باللّغة الإنكليزيّة (Foreword, pp. xx-xxi). وللدلالة على أوراق مخطوط عاطف أفندي، تمّ الاعتماد على هذا الشكل الرمزي في الرجوع إلى محتواها: |ع ١٥ و|. الحرف «ع» يدلّ هنا على مخطوط عاطف أفندي، ثمّ الرقم «١٥» يشير إلى الورقة (folio) الخامسة عشرة مثلاً من المخطوط، وإذا كانت الكتابة على وجه (recto) مثل هذه الورقة، أُشير إلى ذلك عندها بالحرف «و»، أما إذا وردت الكتابة على ظهر (verso) الورقة، فيشار إليها بالحرف «ظ». أما بقيّة المخطوطات فإن الاختلاف في محتوياتها عمّا تضمّنه مخطوط عاطف أفندي قد تمّت الإشارة إليه في الحواشي، مع الدلالة عليه من خلال الرموز المعتمدة في التحقيق، حسب ما ذكرنا أعلاه، وتبعاً لما ورد في توطئة هذا الكتاب. أما الإشارة إلى صفحات طبعة دار صادر في متن النصّ فترد بين [...] على هذه الصورة: [٧٨]؛ أي أن الرقم الوارد هنا يشير إلى الصفحة ٧٨ على سبيل المثال. ولقد أشرنا أيضاً إلى بعض التنّبيهات والتّعليقات المقتضبة حول النصّ العربيّ في الحاشية، ووضعناها هنا بين < ... > للدلالة عليها بوضوح، ولإبراز تباين محتواها عن الرموز المعتمدة في التحقيق النقدي للمخطوطات في الحواشي. أمّا الشروح المفصّلة فقد أوردناها في مقدّمة (Introduction) الكتاب باللّغة الإنكليزيّة، حيث ذكرنا أيضاً حيثيات المنهج المتّبع في التحقيق النقديّ والترجمة في المقدمات التقنيّة (Technical Introduction). نتقدم أيضاً بالشكر للأستاذ صالح الأشمر على تدقيقه النص العربي وضبط شكله.

في الهَندَسَةِ

< الموسومة بجومَطْريا في الهَندَسَةِ[١] وبيانِ ماهيّتها > [٢]

|ع ١٥ ظ| بسمِ اللهِ الرَّحمٰنِ الرَّحيمِ[٣]

< فصل ١ > [٤]

إعلَمْ أيّها الأخ البارّ الرحيم، أيّدنا الله وإيّاك بروح منه، أنّا قد فرغنا من رسالة العدد في الأرِثماطيقي وبيّنا من خواصّ العدد قدرَ الكِفاية والجهد، وانتقلنا من تلك الرسالة إلى هذه الرسالة

(١) ورد في [ط، ك]: «في المدخل إلى علم الهندسة»، ثمّ كمّل العنوان على الشكل التالي الذي تطابق ههنا مع [ن ،ل]: «وبيان ماهيّتها وكميّة أنواعها والغرض المقصود منها هو التهدّي للنّفوس من المحسوسات إلى المعقولات»، وأضاف ناسخ [ك]: «وتُسمّى برسالة الجومطريا» وأكمل بعدها كما في [ن ،ل] بالقول: «وكيفيّة رؤية النفس الصّور المجرّدة عن الهيولى».

(٢) وردت الزيادة التالية في مخطوط [ع]: «من جملة إحدى وخمسين رسالة لإخوان الصفاء من القسم الأوّل في الرياضيّات في تهذيب النفس وإصلاح الأخلاق من كلام الصوفيّة». < وقد ذكرت على هذا الشكل أيضاً في [ل] مع إسقاط الإشارة إلى «كلام الصوفيّة». وقدّمها ناسخ [ل] عن طريق الخطأ تحت عنوان شامل كرسالة «النّسب العدديّة والهندسيّة» ثمّ عاد بعد ذلك وصحّحها في الهامش على أنها «رسالة الهندسة» > .

(٣) وردت الزيادة التالية في مخطوط [ع]: «ربِّ وفّق. رسالة في المدخل إلى علم الهندسة من جملة إحدى وخمسين رسالة لإخوان الصفاء أكثرهم في تهذيب النفس وإصلاح الأخلاق».

(٤) وردت الزيادة التالية في طبعة دار صادر [S]: «الحمد لله وسلام على عباده الذين اصطفى، ألله خيّر أم ما يشركون؟».

التي هي الثانية من رسائل الرياضيّات في المدخل إلى علم الهندسة، فنقول:[١]

إعلَمْ أيّها الأخ البارّ الرحيم، أيّدنا الله وإيّاك بروح منه[٢]، بأنّ العلومَ التي كان القُدماءُ[٣] يخرّجون[٤] أولادَهم بها[٥]، ويروضون بها تلامذتهم، أربعةُ أجناسٍ: أوّلُها علومُ[٦] الرياضيّات، والثاني علومُ المنطقيّات، والثالثُ علومُ[٧] الطبيعيّات، والرابعُ علومُ الإلهيّات. فالرياضيّات أربعة أنواع: أوّلُها الأرِثماطيقي، وهو معرفة العدد، وكميّةِ أجناسه وخواصّه[٨] وأنواعِه، وخواصّ تلك الأنواع. ومبدأ هذا العلم من الواحِد الذي قبل الاثنين. والثاني الجومَطريا، يعني[٩] الهندسةَ. وهي معرفةُ المقاديرِ والأبعاد[١٠] وكميّةِ أنواعِها وخواصّ تلك الأنواع.

---

(١) «إعلم أيها الأخ البارّ الرحيم . . . في المدخل إلى علم الهندسة، فنقول»: سقطت من [ع، ط، ن، ف، ك، ل]، وردت في [Ṣ].

(٢) «أيها الأخ البارّ الرحيم، أيّدنا الله وإيّاك بروح منه»: زيادة في [ع]، سقطت من [Ṣ]. وقد وردت في [ن، ط، ف، ك، ل]: «أيّدك الله وإيّانا».

(٣) «كان القدماء»: كانوا [ن]، كانت [ك]. < وسقطت من [ل]> .

(٤) «يخرّجون»: يستخرج [ك].

(٥) «أولادهم بها»: أولاد الفلاسفة [ن، ك، ل].

(٦) <أسقطنا هنا «ألْ» التعريف الواردة في مخطوط [ع] عند ذكر هذه «العلوم». و قد اتّبعنا إسقاط «ألْ» التعريف أو الإبقاء عليها تبعاً لسياق النصّ وحسب ما يتطلّبه تيسير السلاسة في القراءة وبالتوافق مع ما يظهر عبر مقابلة المخطوطات بعضها ببعض> .

(٧) «علوم»: وردت بين السطرين في [ل].

(٨) «أجناسه وخواصه»: سقطت من [ن، ف، ك، ل]. سقطت «خواصه» من [Ṣ].

(٩) «يعني»: وهو علم [Ṣ].

(١٠) «الأبعاد»: ذي الأبعاد [ن، ل]، ذوات الأبعاد [ك].

ومبدأ هذا العلم من النُّقطة التي هي طرفُ[١] الخطّ أي نهايته[٢]. [٣] والثالثُ الأسطُرُنوميا. يعني علمَ النجوم، وهو معرفةُ [٧٩] تركيبِ الأفلاكِ وتخطيطِ البروج وعددِ الكواكب وطبائعها[٤] ودلائلها على الأشياء[٥] الكائنات في هذا العالم،[٦] من الشمس وحركتها[٧]. [٨] والرابعُ[٩] الموسيقى[١٠]، وهو معرفة التأليف[١١] والنِسَب بين الأشياء المختلفة والجواهر المتضادّة القوى. ومبدأ هذا العلم من نسبة المساوات والكيفيّات[١٢]،

---

(١) «طرف»: رأس [ط، ك، ل].

(٢) «أي نهايته»: سقطت من [ن، ط، ك، ل].

(٣) «طرف الخطّ أي نهايته»: نهاية الخطّ [ع].

(٤) «طبائعها» في [ك، Ṣ]: طبايع [ع، ن، ف، ل]. <الكلمات المشابهة لهذا اللّفظ وردت بمجملها في المخطوطات على هذا الشكل، وذلك من خلال استخدام «ي» (الياء في: «. . .يع») بدلاً عن «ئ» (الهمزة المتوسّطة في: «. . .ئع»). ولقد إستخدمنا هنا في سياق تحقيق هذه الرسالة الهمزة المتوسّطة بدلاً من الياء> .

(٥) «على الأشياء»: في [ع].

(٦) زيادة في [ن] «ويبدأ هذا العلم».

(٧) «حركتها»: حركاتها [ط، ف، ك، ل].

(٨) «الشمس وحركتها»: حركة الشمس [Ṣ].

(٩) زيادة في [ل]: «علم».

(١٠) <كلمة «الموسيقى» وردت في المخطوط [ع، ن] على شاكلة: «الموسيقي»، وهي بذلك أقرب إلى أصل الكلمة اليونانية من المصطلح العربي المتداول في شكله الحديث، ولكن لأجل المحافظة على قراءة سلسة للنص أورَدّتُ هذه الكلمة بشكلها المعاصر تماشياً من خلال ذلك مع مفردات اللغة العربيّة. وتجدر الإشارة هنا أيضاً إلى أن اللّفظ «موسيقى» قد ظهر كذلك في مخطوط [ط]> .

(١١) «التأليف»: التأليفات [Ṣ].

(١٢) «من نسبة المساوات والكيفيّات»: من النسب المتساوية الكمّيّات [ك]، من النسب المتساوية إما الكمّيّات [ل].

وهي نسبةُ[(١)] الثلاثةِ إلى الستّة، كنسبةِ الاثنين إلى الأربعة.

حدُّ المنطقِ أنّه عِلمٌ يُتوَصَّل به إلى اكتساب المجهولات من التصوّرات والتّصديقات[(٢)] بمعلوماتٍ هي مبادئ لها[(٣)].

وأمّا المنطقيّاتُ فهي معرفةُ معاني الأشياء الموجودة التي هي مصوّرةٌ في أفكار النفوس. ومبدأها من العقل والنفس[(٤)] [(٥)].

وأمّا الطبيعيّاتُ فهي معرفةُ جواهرِ الأجسام وما يعرض لها من الأعراض. ومبدأ هذا العلم من الحركة والسكون.

وأمّا عِلمُ[(٦)] الإلٰهيّاتِ فهي معرفةُ الصّور المجرّدة المفارقة للهَيُولى في هذا العالم[(٧)]. ومبدأ هذا العِلم[(٨)] من معرفةِ الله عزّ وجلّ[(٩)]، وجوهرِ العقلِ[(١٠)] والنفسِ[(١١)] والملائكةِ[(١٢)] والشياطينِ والجنّ[(١٣)]، وهُم أرواحٌ[(١٤)] بلا أجسام. وأنّ الأجسام عندهم

---

(١) «المساوات والكيفيّات، وهي نسبة» زيادة في [ع]، سقطت من [ن، Ṣ].

(٢) «والتّصديقات»: سقطت من [ف].

(٣) «حدّ المنطق أنه علم يَتوصّل إلى اكتساب المجهولات . . . بمعلومات هي مبادئ لها»: زيادة في [ع، ف]، سقطت من [ن، ط، ك، ل، Ṣ].

(٤) «العقل والنفس»: الجوهر [ن، Ṣ]، المقولات [ط، ك، ل].

(٥) «ومبدأها من العقل والنفس»: ومبدأ هذا العلم من الجوهر [ف].

(٦) «علم»: زيادة في [ع]، سقطت من [Ṣ].

(٧) «في هذا العالم»: زيادة في [ع]، سقطت من [Ṣ].

(٨) «ومبدأ هذا العلم»: سقطت من [ع]، وردت في [Ṣ].

(٩) «الله عزّ وجلّ»: زيادة في [ع]، سقطت من [Ṣ].

(١٠) «العقل»: زيادة في [ع]، سقطت من [Ṣ].

(١١) «الله عزّ وجلّ وجوهر العقل والنفس»: جواهر الأنفس [ن، ك، ل].

(١٢) زيادة في [ن، ك، ل]: «النفوس».

(١٣) «والشياطين والجنّ»: والجنّ والشياطين [ل]. سقطت كلّها من [ع]، ووردت في [Ṣ].

(١٤) «وهُم أرواح»: أرواح [ن، ف]، والأرواح [Ṣ]، والتي عندهم [ك].

ذوات أبعادٍ ثلاثة.[١] ومبدأ معرفة[٢] هذا العلم من معرفة[٣] جوهر النفس[٤]. وقد عَمِلنا في كلِّ نوعٍ من هذه العلوم رسالةً شِبْهَ المدخلِ والمقدِّمات. فأوّلُها رسالةٌ في العدد قبل هذه، وقد بيّنا فيها طرفاً من خواصّ العدد[٥] وكمّية[٦] أنواعها وكيّفية نشوئها من الواحد الذي قبل الاثنين. ونريد أن نذكر ونبيّن[٧] في هذه الرسالة أصلَ الهندسة التي هي أصلُ[٨] المقادير الثلاثة، وكميّةَ أنواعِها وخواصَّ تلك الأنواع، وكيّفيةَ نشوئها من النُّقطةِ التي هي نهايةُ[٩] الخطّ، وأنّها في صناعة الهندسة مثلُ الواحدِ في صناعة العدد.

|ع ١٦ و| واعلَمْ أيّها الأخ البارّ الرحيم، أيّدك الله وإيّانا بروح منه[١٠]،[١١] أنّ الهندسة، يقال على نوعين: عقّليّة وحسّية. فالحسّيةُ[١٢] هي معرفةُ المقاديرِ وما يعرِض فيها من المعاني، إذا

(١) «وهُم أرواح بلا أجسام. وأن الأجسام عندهم ذوات أبعاد ثلاثة»: سقطت من [ف].

(٢) «معرفة»: هذا العلم معرفة [ل]. زيادة في [ع]، سقطت من [ن، Ṣ].

(٣) «معرفة»: زيادة في [ع]، سقطت من [Ṣ].

(٤) زيادة في [ل]: «كالملائكة».

(٥) «العدد»: الأعداد [Ṣ].

(٦) زيادة في [ن]: «جميع».

(٧) «نذكّر ونبيّن»: نبيّن ونذكر [Ṣ]. سقطت «نذكّر» من [ف، ك، ل].

(٨) «أصل»: سقطت من [ع، ن، ف، ك، ل]، وردت في [Ṣ].

(٩) «نهاية»: رأس [ن، ف، ك، ل، Ṣ].

(١٠) «البارّ الرحيم، أيّدك الله وإيّانا بروح منه»: سقطت من [ن، ط، ك].

(١١) «أيها الأخ البارّ الرحيم، أيّدك الله وإيّانا بروح منه»: سقطت من [ع، ف، ل]، وردت في [Ṣ].

(١٢) «فالحسّية»: زيادة في [ع]، سقطت من [ن، Ṣ].

أضيف بعضُها إلى بعض، وهي ما يُرى بالبصر، ويُدرَك باللمس. والعقليُّ بضِدِّ ذلك، وهو ما يُعرف ويُفهم. فالذي يُرى بالبصر [٨٠] هو الخطُّ والسطحُ[١] والجسمُ ذو الأبعاد وما يعرِض فيها[٢]، كما أنّ الثِّقلَ في الثقيل لا يُعرف إلّا بالعقل، والثِّقلُ عينُ[٣] الثقيل. والمقاديرُ ثلاثةُ أنواعٍ وهي: خطوط، وسطوح، وأجسام.

وهذه الهندسة تدخل في الصنائع[٤] كلّها، وذلك أنّ كلَّ صانعٍ إذا قَدَّرَ في صناعته قبل العمل، فهو ضربٌ من الهندسة العقّليّة[٥]، [٦] فهي معرفة الأبعاد، وما يعرض فيها من المعاني، إذا أُضيف بعضُها إلى بعض، وهي متصوَّرة[٧] في النفس بالفكر، وهي تنقسم[٨] ثلاثةَ أنواعٍ: الطول والعرض والعمق.

وهذه الأبعادُ العقّليّةُ صفاتٌ لتلك المقاديرِ الحسِّية. وذلك أنّ الخطَّ هو أحدُ المقادير، وله صِفةٌ واحدة، وهي الطولُ حَسْبُ. وأمّا السطح فهو مِقدارٌ ثانٍ، وله صفتان وهما الطول والعرض. وأمّا الجسم فهو مقدارٌ ثالثٌ، وله ثلاثُ صِفاتٍ وهي الطول والعرض والعمق.

---

(١) «والسطح»: سقطت من [ع]، وردت في [Ṣ].

(٢) «ذو الأبعاد وما يعرض فيها»: سقطت من [ع]، وردت في [Ṣ].

(٣) «عين»: غير [ط].

(٤) «الصنائع»: الصناعات [ن].

(٥) «العقّليّة»: سقطت من [ن، ط، ف، ك، ل].

(٦) زيادة في [ن، ف، ك، ل]: «فأما الهندسة العقّليّة».

(٧) «متصوّرة»: تُتَصَوّر [ط]، ما يتصوّر [Ṣ].

(٨) «تنقسم»: زيادة في [ع]، سقطت من [ن، ط، ف، ك، ل، Ṣ].

واعلَمْ أنّ[١] النظر في هذه الأبعاد[٢] مجرّدةً عن الأجسام من صناعة المتفلسِفين[٣]، فنبدأ أولاً بوصف الهندسة الحسّية لأنّها أقربُ إلى فَهمِ المتعلّمين، فنقول:

إنّ الخطَّ الحسيَّ، الذي هو أحدُ المقادير، أصلُه النقطةُ كما بيّنا قَبْلُ في الرسالة التي في خواصّ العدد بأنّ الواحدَ أصلُ العدد، وذلك أنّ النقطةَ الحسّيةَ إذا انتظمت ظهر الخطُّ بحاسّة البصر[٤] مِثلَ هذا:

. . . .

ولسنا[٥] نقول إنّ هذه النقطة هي التي لا جُزءَ لها[٦]، لكنّ النّقطةَ العقليَّة هي التي لا جُزءَ لها. ونقول أيضاً إنّ الخطَّ الحِسِّيَّ[٧] أصلُ السطح[٨] كما أنّ النّقطةَ أصلُ الخطّ، وكما أنّ الواحدَ أصلُ الاثنين، والاثنان أصلٌ لعدد الزوج كما بيّنا قبل ذلك في رسالة العدد[٩]، وذلك أنّ الخطوطَ إذا تجاورت ظهر السطحُ لحاسّةِ البصر مِثلَ هذا[١٠]:

(١) «واعلم أن»: سقطت من [ع]، وردت في [Ṣ].
(٢) «الأبعاد»: الأشياء [ط].
(٣) «المتفلسفين»: المحقّقين [Ṣ].
(٤) «البصر»: النظر [Ṣ].
(٥) «ولسنا»: فإنّا لا [Ṣ].
(٦) «هي التي لا جزء لها»: شيء لا جزء له [Ṣ].
(٧) «الحسيّ»: زيادة في [ع]، سقطت من [Ṣ].
(٨) «السطح»: الهندسة للسطح [ل].
(٩) «في رسالة العدد»: زيادة في [ع، ن، ل]، سقطت من [ك، Ṣ].
(١٠) «ولسنا نقول إن هذه النقطة هي التي لا جزء لها لكن النقطة العقليّة هي التي لا

[٨١] ونقول أيضاً[١] إنّ السطحَ أصلٌ للجسم، كما أنّ الخطَّ أصلٌ للسطح، والنقطةُ أصلٌ للخطّ، كما أنّ الواحدَ أصلُ الاثنين، والاثنان والواحدُ أصلان لعدد الأفراد[٢] كما بيّنا قبل ذلك[٣]. وذلك أنّ السطوحَ إذا تراكمت بعضُها فوق بعضٍ ظهر عُمقُ[٤] الجسم لحاسّةِ البصر[٥] مثلَ هذا:

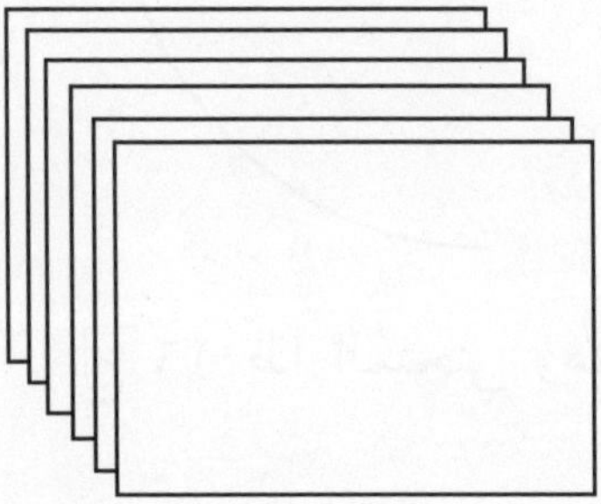

جزء لها . . . وذلك أن الخطوط إذا تجاورت ظهر السطح لحاسة البصر مثل هذا»: زيادة في [ع]، سقطت من [ن].

(١) «أيضاً»: زيادة في [ع]، سقطت من [S̤].

(٢) «لعدد الأفراد»: الفرد [ن، ف]، للعدد [ك، ل]، لأوّل الفرد [S̤].

(٣) «قبل ذلك»: في رسالة الأرِثماطيقي قبل ذلك [ك، وفي هامش ل].

(٤) «عمق»: زيادة في [ع]، سقطت من [S̤].

(٥) «البصر»: النظر [S̤].

فصل < ٢ > في أنواع الخطّ[١]

فنقول[٢]: الخطوطُ[٣] ثلاثةُ أنواعٍ: أوّلُها الخطُّ المستقيمُ، وهو مِثلُ الذي يُخطُّ بالمِسطَرة[٤]، على ما يُرى في هذه الصورة[٥]، مِثلَ هذا[٦]:

والثاني الخطُّ المقوَّسُ وهو[٧] مِثلُ الذي يخطُّ بالبِركار مثلَ هذا:

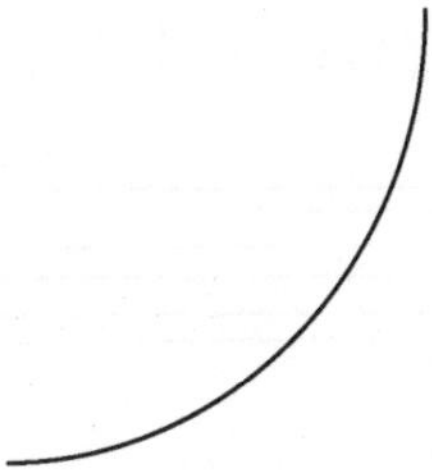

والثالثُ الخطُّ[٨] |ع ١٦ ظ| المُنحني وهو المركّبُ[٩] منهما مِثلُ هذا[١٠]:

(١) «في أنواع الخطّ»: في أنواع الخطوط [ط، ك، ل]. سقطت من [ن].
(٢) «فنقول»: سقطت من [ع، ن]، وردت في [Ṣ].
(٣) «الخطوط»: الخطّ [ل].
(٤) «بالمِسطَرة» في [ط، ف، ك، ل]: بالمِسطَر [ع]، بالمِستَرة [ن].
(٥) «على ما يرى في هذه الصورة»: سقطت من [ع، ن، ط، ل]، وردت في [Ṣ].
(٦) «مثل هذا»: سقطت من [ل].
(٧) «وهو»: سقطت من [ع]، وردت في [Ṣ].
(٨) «الخطّ»: الذي هو [ل]. سقطت من [ع]، وردت في [ن، Ṣ].
(٩) «هو المركّب»: يسمّى المكوّكب [ع].
(١٠) «منهما مثل هذا»: سقطت من [ع]، وردت في [ط، Ṣ].

فهذه أنواع الخطوط الثلاثة[١] .[٢]

[٨٢] فصل < ٣ > في ألقاب الخطوط المستقيمة[٣]

فنقول[٤]: إنّ الخطوطَ المستقيمةَ[٥] إذا أُضيف بعضُها إلى بعضٍ، إمّا أن تكون متساويةً أو متوازيةً أو متلاقيةً أو متماسّةً أو متقاطعةً.

فالمتساويةُ هي التي طولُها واحِدٌ مِثالُ هذا:

والمتوازيةُ هي التي إذا كانت في سطحٍ واحدٍ وأُخرجت في كلتا الجهتين إخراجاً دائماً، لا يلتقيان أبداً مِثلَ هذا[٦] :[٧]

(١) «فهذه أنواع الخطوط الثلاثة»: سقطت من [ع، ن، ط، ف، ك، ل]، وردت في [S].

(٢) < أضاف ناسخ [ك] بعض العبارات لوصف الخطوط المتلاقية والمتماسّة والمتقاطعة التي وردت لاحقاً في السياق العام للنصّ تبعاً لما ذكر في المخطوطات الأخرى > .

(٣) «المستقيمة»: سقطت من [ف].

(٤) «فنقول»: سقطت من [ع، ط]، وردت في [S].

(٥) «فنقول إن الخطوط المستقيمة»: سقطت من [ن، ل].

(٦) «مثل هذا»: مثله [ع].

(٧) «فنقول: إن الخطوط المستقيمة إذا أضيف بعضها إلى بعض، إما أن تكون متساوية أو متوازية أو متلاقية أو متماسّة أو متقاطعة . . . والمتوازية هي التي إذا كانت في سطح واحد وأخرجت في كلتا الجهتين إخراجاً دائماً، لا يلتقيان أبداً مثل هذا»: سقطت هذه الجمل الثلاث كلّها من [ك].

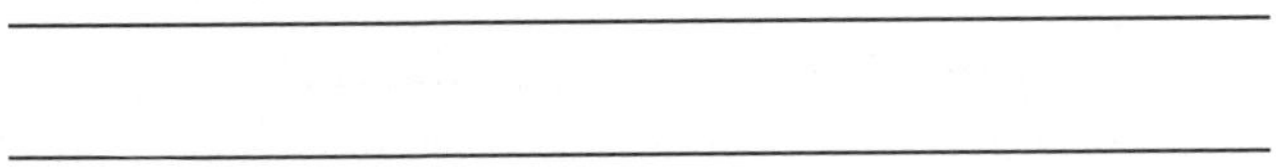

والمتلاقيةُ هي التي تلتقي في إحدى[١] الجهتين، وتحيط بزاوية واحدة[٢] مِثلَ هذا:

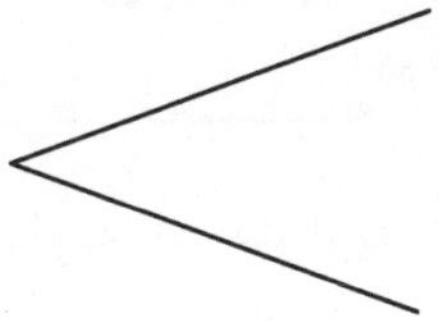

والمتماسّةُ هي التي تماسُّ إحداهما الأخرى، وتُحدث زاويتين أو زاويةً[٣] مِثلَ هذا[٤]:

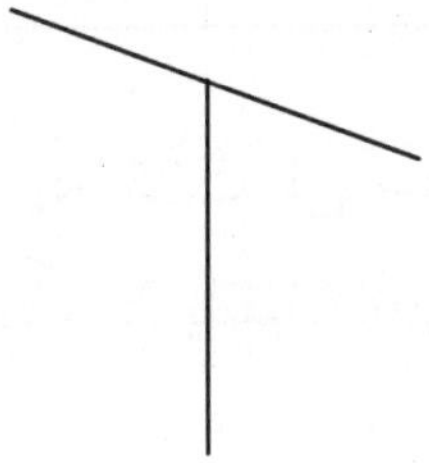

[٨٣] والمتقاطعةُ التي تقطع إحداهما الأخرى[٥] وتُحدث من تقاطعهما أربعَ زوايا مِثلَ هذا[٦]:

(١) < وردت في هامش [ل] >.

(٢) «واحدة»: سقطت من [ع]، ووردت في [Ṣ].

(٣) «أو زاوية»: سقطت من [ع، ن، ط، ف، ك، ل]، ووردت في [Ṣ].

(٤) «المثال»: زيادة في [Ṣ].

(٥) «تقطع إحداهما الأخرى»: سقطت من [ع، ن، ك، ل]، ووردت في [ف، Ṣ].

(٦) «والمتقاطعة التي تقطع إحداهما الأخرى وتحدث من تقاطعهما أربع زوايا مثل هذا»: سقطت من متن النصّ في [ن]، ثمّ وردت بعد ذلك في الهامش.

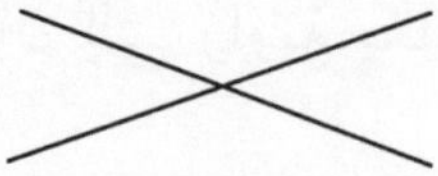

فهذه ألقابُ الخطوطِ المستقيمة[١] .

فصل < ٤ > في أسماء الخطوط المستقيمة[٢]

إذا قام خطٌّ مستقيمٌ على خطٍّ آخرَ قِياماً مستوياً من غير مَيلٍ إلى طرفٍ[٣] ، يقال عند ذلك للخطّ القائم العْمودُ، وللقائم عليه القاعدةُ[٤] ، مِثلُ هذا :

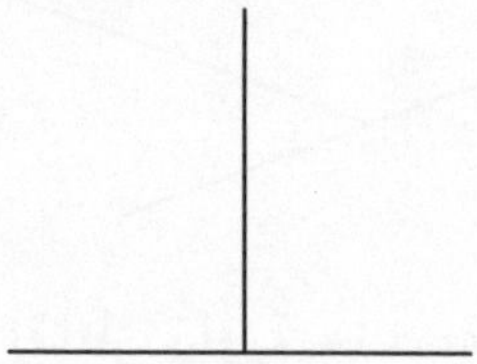

كلُّ خطٍّ يقابل زاويةً ما يقال له وتَرُ تلك الزاوية التي تقابلها مِثل هذا[٥] :

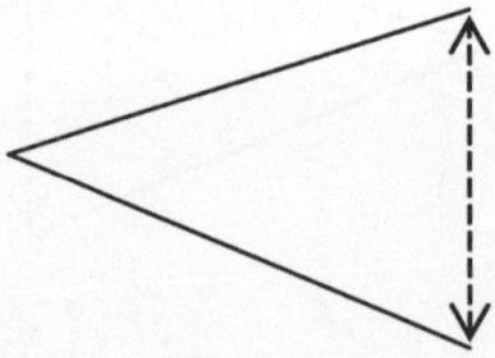

(١) «فهذه ألقاب الخطوط المستقيمة»: سقطت من [ع، ن، ط، ك، ل]، وردت في [ṢB].

(٢) «في أسماء الخطوط المستقيمة»: في أسماء الخطّ المستقيم [ن، ط، ف، ṢB]. سقطت من [ل].

(٣) «من غير ميل إلى طرف»: سقطت من [ع، ن، ط، ك، ل]، وردت في [ṢB].

(٤) «وللقائم عليه القاعدة»: وللقايم القاعدة [ك، وفي هامش ل].

(٥) «كلّ خطّ يقابل زاوية ما يقال له وتَر تلك الزاوية التي تقابلها مثل هذا»: زيادة في [ع، ن، ط، ك، ل]، سقطت من [ف، ṢB].

وإذا أُضيف الخطّانِ إلى زاويةٍ يقال لهما الساقان لتلك الزاوية، مثل هذا[1]:

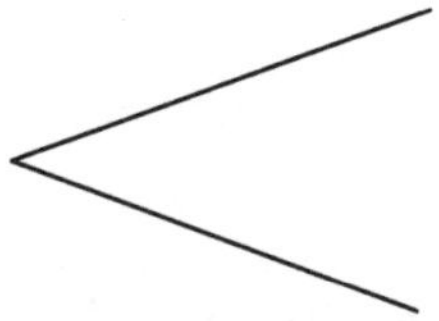

وإذا قام خطٌّ مستقيمٌ على خطٍّ، وللخطِّ والقائمِ ميلٌ إلى أحد الطرفين، يحصل زاويتان إحداهما أكبر يقال لها المنفرجة، والأخرى أصغر يقال لها الحادّة:

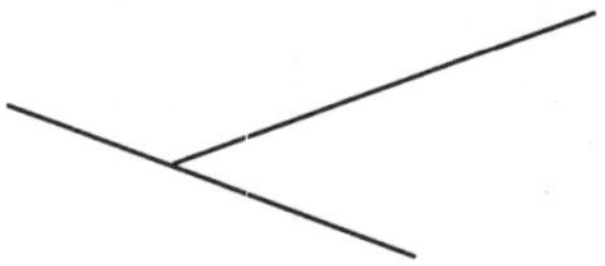

وكلُّ خطٍّ مستقيم يقابل زاويةً ما، يقال له وترُ تلك الزاوية التي يقابلها، مثل هذا[2]:

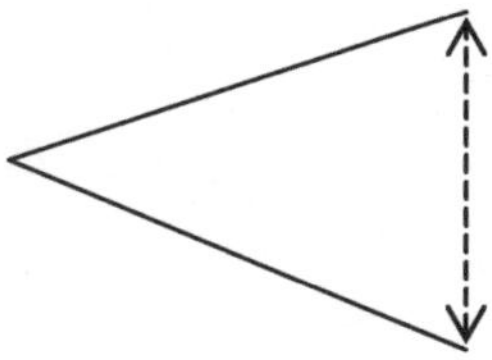

[٨٤] والخطوطُ إذا أُضيفت بعضُها البعض[3] إلى سطح ما،

---

(١) «وإذا أضيف الخطّان إلى زاوية . . . مثل هذا»: سقطت من [ع، ل]، وردت في [ن، ف، ك، ṣ]. <وقد ذكرت في [ن] قبل ذكر وتر الزاوية الذي ورد في النصّ أعلاه> .

(٢) «وإذا قام خطّ مستقيم على خطّ، وللخطّ والقائم ميل إلى أحد الطرفين . . . يقال له وتر تلك الزاوية التي يقابلها، مثل هذا»: سقطت من [ع، ن، ف، ك، ل]، وردت في [ṣ].

(٣) «بعضها البعض»: زيادة في [ع]، سقطت من [ل، ṣ].

يقال لها أضلاعُ ذلك السطح، مثل هذا:

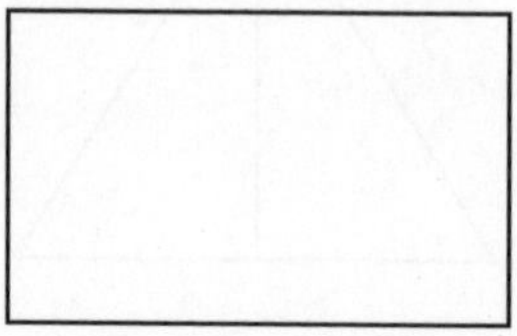

وكلُّ خطٍّ يخرج من زاويةٍ وينتهي إلى أخرى[1] يقال له قَطرُ المربّعِ[2] مِثل هذا[3]:

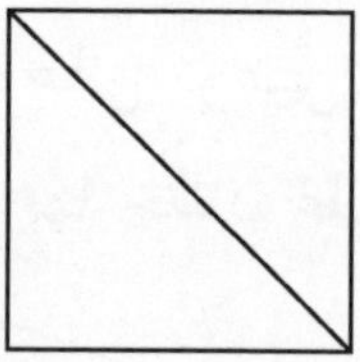

وكلُّ خطٍّ يخرج من زاويةِ المثلّثِ[4] وينتهي إلى الضِّلعِ المقابلِ لها، ويقوم على الخطّ المقابل لها على زاوية قائمة[5]، يقال لذلك الخطّ مَسقَطُ الحَجَر، ويسُمّى[6] العَمودُ أيضاً، ويُسمّى الذي[7] يقع[8] عليه مَسقَطُ الحَجَر القاعدةُ، مثل هذا:

(١) «أخرى»: زاوية أخرى [ط، ف، ك، ل].

(٢) «المربّع»: المربّعات [ك].

(٣) «وكل خطّ يخرج من زاوية وينتهي إلى أخرى يقال له قطر المربع مثل هذا»: سقطت من [ع]، سقطت من [ن] ثمّ أضيفت في الهامش، وردت في [S̱].

(٤) «المثلّث»: سقطت من [ك]، وردت في هامش [ن].

(٥) «ويقوم على الخطّ المقابل لها على زاوية قائمة»: سقطت من [ل].

(٦) «يسمّى»: يقال له [ف، ل، ك، S̱].

(٧) «ويسمّى الذي»: ويقال لذلك الخطّ الذي [ل]، ويقال للخطّ الذي وقع [S̱].

(٨) «يسمّى»: ويقال للخطّ الذي وقع [ك، S̱].

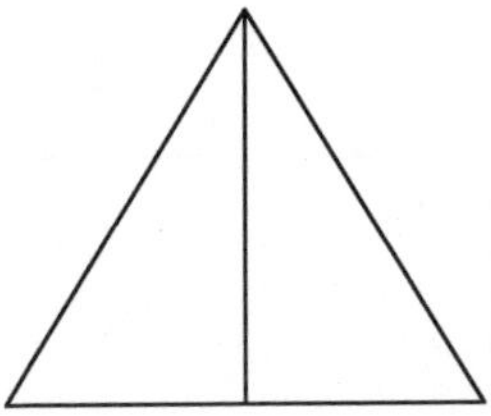

فهذه أسماء الخطوط المستقيمة[١].

[٨٥] فصل < ٥ > في أنواع الزوايا[٢]

نقول[٣]: إنّ الزوايا على نوعين: مسطّحة ومجسّمة[٤]. والمسطّحةُ هي التي يحيط بها خطّان على غير استقامة مثل هذا:

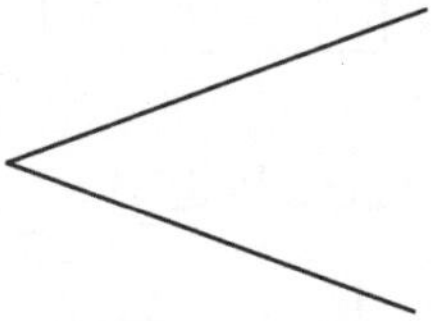

والمجسّمةُ هي التي تحيط بها ثلاثةُ خطوطٍ في زاوية[٥]، كلُّ اثنين منها[٦] على غير استقامة[٧].

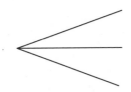

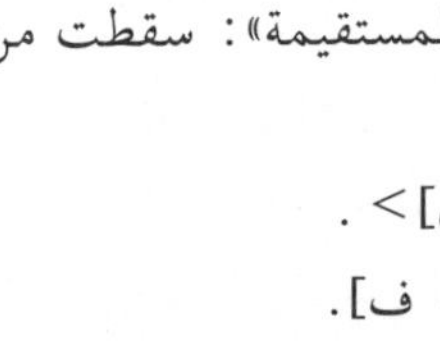

(١) «فهذه أسماء الخطوط المستقيمة»: سقطت من [ع، ن، ك، ل]، وردت في [S̤].

(٢) < أضيفت إلى هامش [ل] >.

(٣) «نقول»: سقطت من [ط، ف].

(٤) «مسطّحة ومجسّمة»: مسطّح ومجسّم [S̤].

(٥) «في زاوية»: سقطت من [ع، ف]، وردت في [ن، ك، ل، S̤].

(٦) «منها»: زاوية [ن، S̤].

(٧) زيادة في [ن، ل]: «مثل هذا».

فصل < ٦ > في أنواع الزوايا المسطّحة من لونٍ آخر[(١)]

تتنوّعُ من جهة الخطوط ثلاثةُ أنواع، إمّا من خطّين مستقيمين مثل هذا:

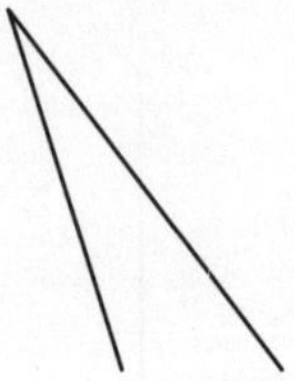

أو خطّين مقوّسين مثل هذا:

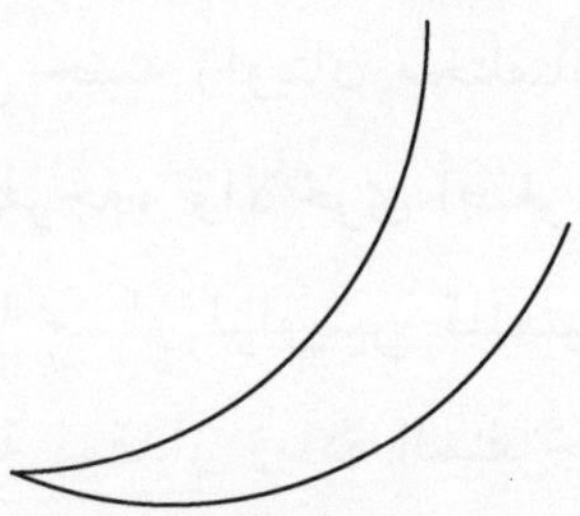

أو أحدهما مستقيم والآخر مقوّس[(٢)]:

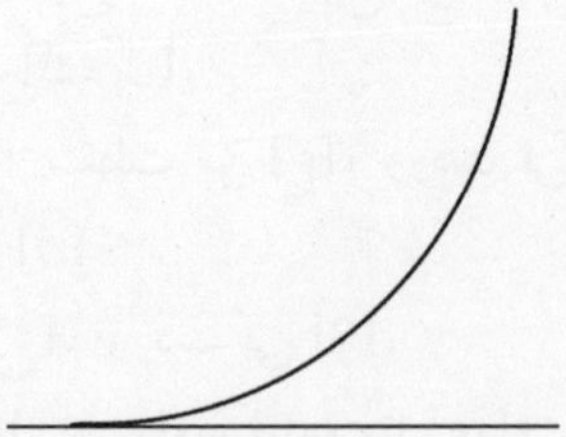

(١) «في أنواع الزوايا المسطّحة من لونٍ آخر»: الزاوية المسطّحة [ف]. زيادة «من لونٍ آخر» في [ع]، «من لونٍ آخر» سقطت من [ط، Ṣ].

(٢) «مستقيم والآخر مقوّس»: مقوّس والآخر مستقيم [Ṣ].

والزوايا التي تحيط بها خطوطٌ مستقيمةٌ تتنوّعُ من جهة الكيفيّة ثلاثةَ أنواعٍ[١]: قائمة ومنفرجة وحادّة. فالقائمة هي التي إذا قام خطٌّ مستقيمٌ[٢] على خطٍّ آخرَ مستقيمٍ قِياماً مستوياً[٣] حدث عن جنبيه[٤] زاويتان متساويتان، وكلٌّ واحدة منهما يقال لها زاوية[٥] قائمة مثل هذا[٦]:

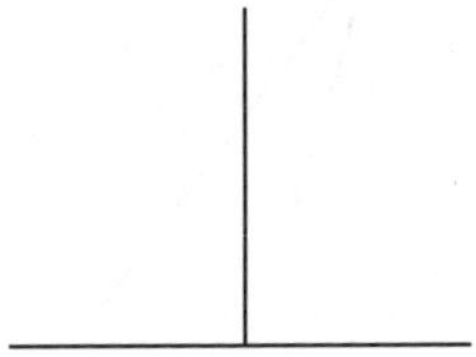

[٨٦] وإذا قام ذلك الخطُّ |ع ١٧ و| قياماً غير مستوٍ على خطٍّ مستقيمٍ[٧]، حدث عن جنبيه زاويتان مختلفتان، إحداهما أكبر من القائمة، يقال لها المنفرجة، والأخرى أصغر من القائمة، يقال لها الحادّة، ومجموعهما مساوٍ لزاويتين قائمتين[٨]، لأنّ نقصان[٩] الحادّة[١٠] عن القائمة بمقدار زيادة المنفرجة على القائمة، مثل هذا[١١]:

(١) «من جهة الكيفيّة ثلاثة أنواع»: ثلاثة أنواع من جهة الكيفيّة [ك].
(٢) «مستقيم»: سقطت من [ك، ل].
(٣) «مستقيم قياماً مستوياً»: سقطت من [ع]، وردت في [ن، ل، Ṣ].
(٤) <أضيفت إلى هامش [ل]>.
(٥) «زاوية»: سقطت من [ع]، وردت في [Ṣ].
(٦) «وكل واحدة منهما يقال لها زاوية قائمة مثل هذا»: أضيفت إلى هامش [ل].
(٧) «على خطّ مستقيم»: سقطت من [ع، ن]، وردت في [ل، Ṣ].
(٨) «لزاويتين قائمتين»: لقائمتين [Ṣ].
(٩) زيادة في [ن، ك، ل]: «الزاوية».
(١٠) «نقصان الحادّة»: الزاوية الحادّة تنقص [ط، Ṣ].
(١١) «مثل هذا»: على هذا المثال [Ṣ].

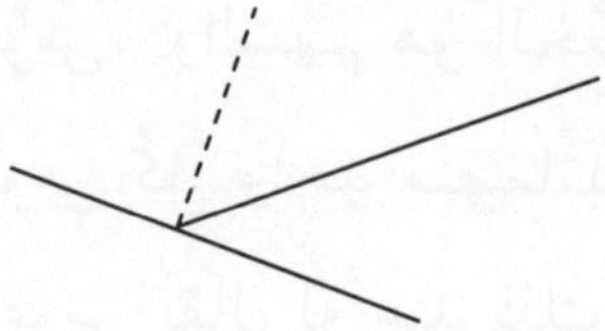

فهذا عددُ أنواعِ الزوايا[١].

فصل < ٧ > في أنواع الخطوط المقوِّسة[٢]

فنقول[٣]: إنّ الخطوطَ المقوِّسةَ أربعةُ أنواعٍ: منها محيطُ الدائرةِ، ومنها نِصفُ الدائرةِ، ومنها أكثرُ[٤] من نصف الدائرة، ومنها أقلُّ من نصف الدائرة.

ومركز الدائرة[٥] هي النقطةُ التي داخلها[٦] في وسط الدائرة، وكلُّ الخطوط الخارجة منها إلى المحيط متساوية[٧]. وقُطر الدائرةِ[٨] هو الخطُّ المستقيمُ الذي يقطع الدائرةَ بنصفين ويمرُّ على المركز[٩]. والوترُ هو الخطُّ المستقيمُ[١٠] الذي يصل بين

---

(١) «فهذا عدد أنواع الزوايا»: سقطت من [ع، ن، ك، ل]، وردت في [Ṣ].

(٢) «المقوّسة»: القوسيّة [Ṣ]. <وهذا ينطبق على ذكر هذه اللفظة في بقيّة هذا الفصل>.

(٣) «فنقول»: سقطت من [ع، ن، ط، ف، ك، ل]، وردت في [Ṣ].

(٤) «أكثر»: أكبر [ط، ك، ل].

(٥) «الدائرة»: الدواير [ل]. <وردت على هذا الشكل حتى نهاية الفصل>.

(٦) «داخلها»: زيادة في [ع]، سقطت من [ن، Ṣ].

(٧) «وكلّ الخطوط الخارجة منها إلى المحيط متساوية»: زيادة في [ع]، سقطت من [ن، Ṣ].

(٨) «قطر الدائرة»: قطرها [ع].

(٩) «ويمرّ على المركز»: زيادة في [ع، ن، ف، ك، ل]، سقطت من [Ṣ].

(١٠) «المستقيم»: سقط من [ن].

طرفَيّ الخطّ[1] المقوِّس. والسهم هو الخطُّ المستقيمُ[2] الذي يفصل[3] الوتَر والقوسَ كُلَّ واحدٍ منهما بنصفين. وهو[4] إذا أُضيف إلى نصف القوس يقال له عند ذلك الجَيبُ المعكوس. وإذا أُضيف نِصفُ الوتر إلى نصف القوس[5]، يقال له عند ذلك الجَيبُ المستوي.

والخطوط المقوِّسة المتوازية هي التي مركزُها واحِدٌ مِثلُ هذا:

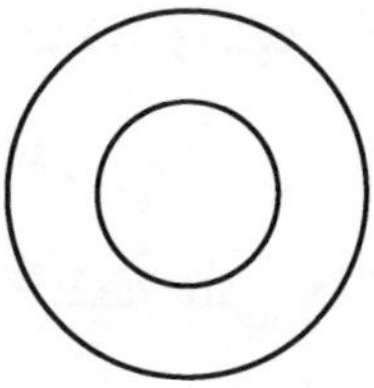

[٨٧] والخطوط المقوّسة المتقاطعة هي التي مراكزُها مختلفةٌ مِثلُ هذا:

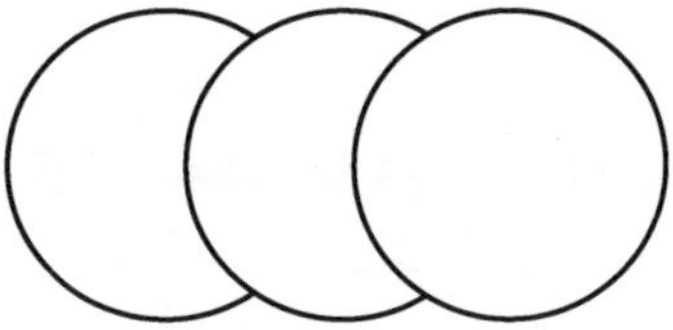

والخطوط المقوِّسة المتماسَّة هي التي تماسّ[6] بعضها

---

(١) «بين طرفيّ الخطّ»: سقط من [ن].
(٢) «المستقيم»: سقطت من [ع]، وردت في [ن، ف، ك، ل، Ṣ].
(٣) «يفصل»: يصل [Ṣ].
(٤) «وهو»: السهم [ن].
(٥) «القوس»: المقوّس [ل].
(٦) «تماسّ»: يماسّ [ن، ك].

بعضاً[١] إمّا من داخلٍ أو خارجٍ ولا تقاطُعَ بينها[٢]، مِثلُ هذا:

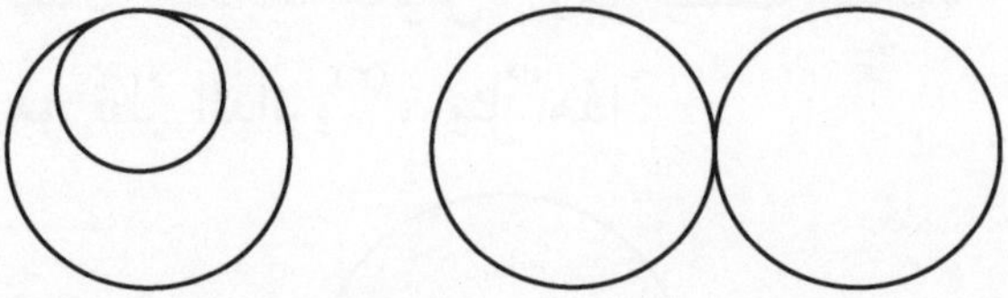

وأمّا الخطوط المنحنية فتُرِكَ[٣] ذِكرُها لأنّها غيرُ مستعملة. إنقضى ذكرُ الخطوطِ[٤]. فاعلَمْ جميعَ ذلك[٥].

فصل < ٨ > في ذكر السطوح

فنقول[٦]: السطحُ هو شكلٌ[٧] يحيط به خطٌّ أو خطوطٌ. والدائرةُ هي شكلٌ يحيط به خطٌ واحِدٌ مِثلُ هذا:

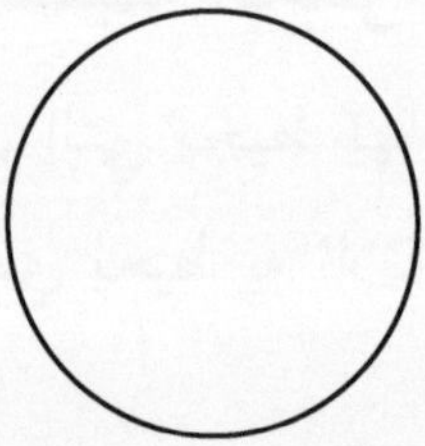

وفي داخله نُقطةٌ كلِّ الخطوطِ المستقيمةِ التي تخرج منها،

(١) <أضيفت إلى هامش [ن]>.

(٢) «تقاطع بينها»: يتقاطع [ك، ل، Ṣ].

(٣) «فترك»: فقد تركنا [Ṣ].

(٤) «إنقضى ذكر الخطوط»: زيادة في [ع، ن، ل]، سقطت من [Ṣ].

(٥) «فاعلم جميع ذلك»: سقطت من [ع، ن، ط، ك]، وردت في [Ṣ].

(٦) «فنقول»: سقطت من [ع، ن، ط، ف، ك، ل]، وردت في [Ṣ].

(٧) «السطح هو الشكل»: الشكل هو السطح [ع ن، ط، ك، ل].

وينتهي إلى جهتين مساوٍ بعضُهما لبعضٍ[١]. ونِصفُ الدائرةِ شكلٌ يحيط به خطّان أحدُهما مقوِّس، وهو نِصفُ الدائرة[٢]، والآخرُ مستقيمٌ، وهو قُطرُ الدائرةِ[٣]، مِثلُ هذا:

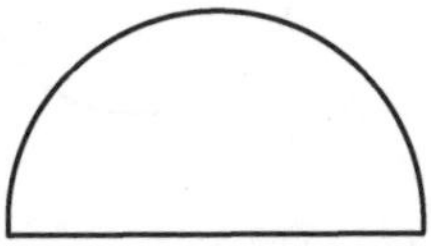

[٨٨] وقطعةُ الدائرةِ هو شكلٌ يحيطُ به خطٌّ مستقيمٌ وقوسٌ من محيط الدائرة، إمّا أكبرُ من نِصفِه، وإمّا أصغرُ حسبَ ما بيّنا وأوردنا مِثالَها قبل هذا[٤].

فصل < ٩ > في أنواع الأشكال المستقيمة الخطوط[٥]

فنقول[٦]: الأشكالُ التي يحيط بها خطوطٌ مستقيمةٌ أوّلُها الشكلُ المثلّثُ، وهو الذي يحيط به ثلاثةُ خطوطٍ متساوية[٧]، وله ثلاثُ زوايا مِثلُ هذا:

---

(١) «وفي داخله نقطة كلّ الخطوط المستقيمة التي تخرج منها، وينتهي إلى جهتين مساوٍ بعضهما لبعض»: سقطت من [ف]، وداخله نقطة كلّ الخطوط الخارجة منها إليه متساوية [ل، وفي هامش ن].

(٢) «وهو نصف الدائرة»: زيادة في [ع]، سقطت من [ن، ط، ل، Ṣ].

(٣) «وهو قطر الدائرة»: زيادة في [ع]، سقطت من [ن، ط، ل، Ṣ].

(٤) «وقطعة الدائرة هو شكل يحيط به خط مستقيم وقوس من محيط الدائرة . . . وأوردنا مثالها قبل هذا»: سقطت من [ع، ن، ف، ك، ل]، وردت في [Ṣ].

(٥) «أنواع الأشكال المستقيمة الخطوط»: الأشكال المستقيمة الخطوط وأنواعها [Ṣ].

(٦) «فنقول»: سقطت من [ع، ن، ط، ف، ك]، وردت في [Ṣ].

(٧) «متساوية»: زيادة في [ع، ن]، سقطت من [ط، ف، ك، ل، Ṣ].

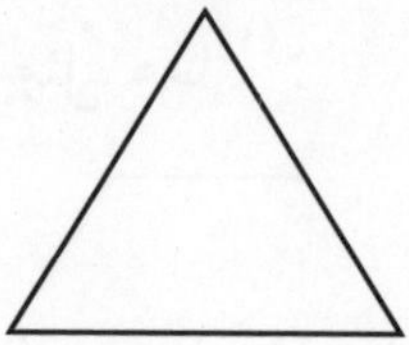

وبعده[1] المربّعُ، وهو شكلٌ[2] يحيط به أربعةُ خطوطٍ مستقيمةٍ[3]، وأربعُ زوايا قائماتٍ[4] مِثلُ هذا:

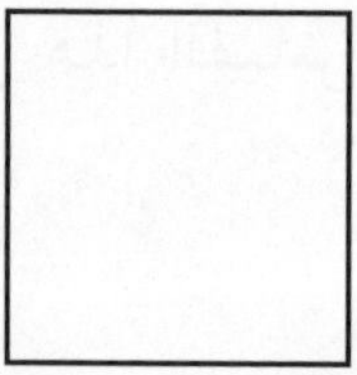

وبعده المُخمَّسُ وهو شكلٌ يحيط به خمسةُ خطوطٍ، وله خمسُ زوايا مِثلُ هذا:

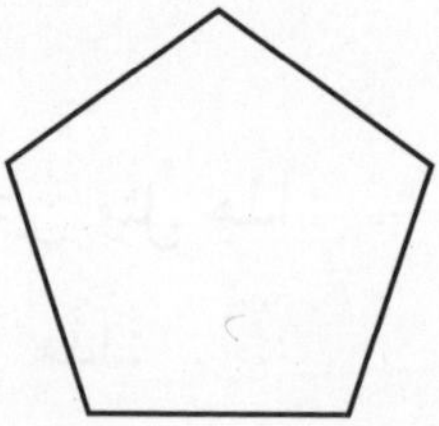

وبعده المُسدَّسُ وهو الذي يحيط به سِتّةُ خطوطٍ، وله ستُّ زوايا مِثلُ هذا:

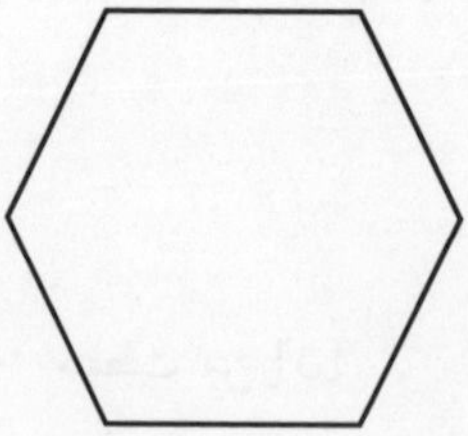

(١) «وبعده»: ثم [Ṣ]. < وردت على هذا الشكل حتى نهاية الفصل >.

(٢) «شكل»: الشكل الذي [ن]، الذي [Ṣ].

(٣) «مستقيمة»: سقطت من [ع، ن، ل]، وردت في [Ṣ].

(٤) «وأربع زوايا قائمات»: وأربع زوايا قائمة [ط]. سقطت من [ع، ن]، وردت في [Ṣ].

[٨٩] وبعده المُسبَّعُ مِثلُ هذا[١]:

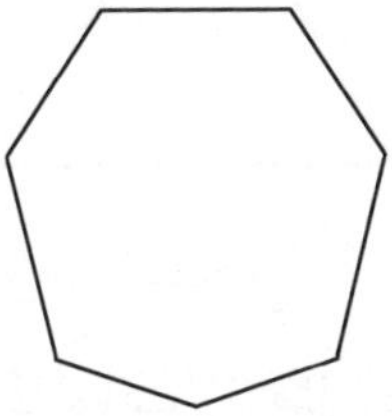

|ع ١٧ ظ |[٢] وعلى هذا القياس تتزايد الأشكالُ كتزايُدِ العدد[٣].

فصل < ١٠ > من النُّقطِ لحاسّةِ البصر[٤]

وقد بيّنا أنّ الخطوطَ[٥] يظهر طولُها لحاسّةِ البصرِ من النُّقطِ[٦] إذا انتظمت.

فأقصرُ خطٍّ من نُقطتين مِثلُ هذا: . .

ثمّ من ثلاثٍ[٧] مِثلُ هذا: . . .

ثمّ من أربعٍ مِثلُ هذا: . . . .

ثمّ من خمسٍ مِثلُ هذا: . . . . .

---

(١) «وبعده المسبّع مثل هذا»: سقطت من [ن].

(٢) زيادة في [ف]: «والمثمّن».

(٣) زيادة في [ط]: «على النظم الطبيعيّ».

(٤) «من النّقط لحاسّة البصر»: سقطت من [ن، ط، ف، ك].

(٥) «الخطوط»: الخطّ [ن، ط، ك، ل].

(٦) «النّقط»: النّقطة [ك، S].

(٧) زيادة في [ن]: «أجزاْ».

ويتزايد واحداً بعد واحد[١] كتزايُدِ العدد على النّظمِ الطبيعي.[٢]

وأصغرُ شكل المثلّث من ثلاثةِ أجزاءٍ مِثلُ هذا:

وبعده[٣] من ستّةِ[٤] أجزاءٍ مِثلُ هذا:

وبعده من عشرةِ أجزاءٍ مِثلُ هذا[٥]:

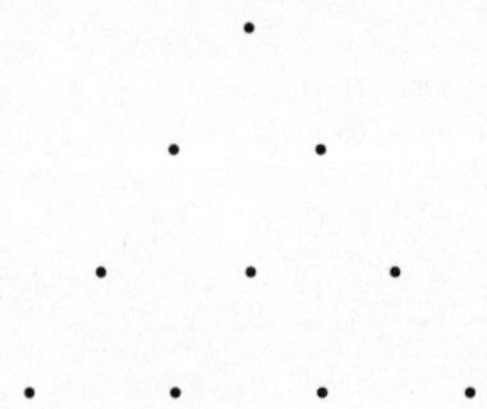

[٩٠] وعلى هذا القياس يتزايد دائماً[٦] كتزايُدِ[٧] جمعِ[٨]

---

(١) «واحداً بعد واحد»: سقطت من [ع]، وردت في [ṢṢ].

(٢) <أضاف ناسخ مخطوط [ع] فصل جديد هنا من دون عنوان، ولكن ذلك يقطع حُسن التسلّسل في محتويات النصّ>.

(٣) «وبعده»: ثم [Ṣ]. <وتتبع كذلك على هذا المنوال في هذا الفصل>

(٤) «ستّة»: أربعة [Ṣ].

(٥) زيادة في [ن، ط، ف، ك، ل]: «وبعده من خمسة عشر جزءًا».

(٦) «دائماً»: زيادة في [ع، ن، ل]، سقطت من [Ṣ].

(٧) «كتزايد»: كما يتزايد [Ṣ].

(٨) «جمع»: سقطت من [ع، ن]، وردت في [Ṣ].

العددِ على النّظم الطبيعيّ.

وأمّا الأشكالُ المربّعاتُ فأوّلُها يظهر في أربعةِ أجزاءٍ مِثلُ هذا:

وبعده من تسعةِ أجزاءٍ مِثلُ هذا:

وبعده من سِتَّةَ عشرَ مِثلُ هذا:

وبعده من خمسةٍ وعشرين جُزءاً مِثلُ هذا:

وعلى هذا القياس تتزايد المربّعاتُ دائماً[1] كتزايُدِ جمعِ[2] العددِ[3] على نظم طبيعة الأفراد[4] وتكون كلُّها مجذوراتٍ.

[٩١] فصل > ١١ < في بيان أنّ المثلّثَ أصلٌ لجميع الأشكال

فنقول: إنّ الشكلَ المثلّثَ أصلٌ لجميع الأشكالِ المستقيمةِ الخطوطِ، كما أنّ الواحدَ أصلٌ لجميع العدد، والنقطةُ أصلٌ للخطوط، والخطُّ أصلٌ للسطوح، والسطوحُ[5] أصلٌ للأجسام، كما بيّنا قبلُ[6]. وذلك أنّه إذا أُضيف شكلٌ مثلّثٌ إلى شكلٍ آخرَ مِثلِه[7]، حدث[8] من جُملتهما شكلٌ مربّعٌ مِثلُ هذا:

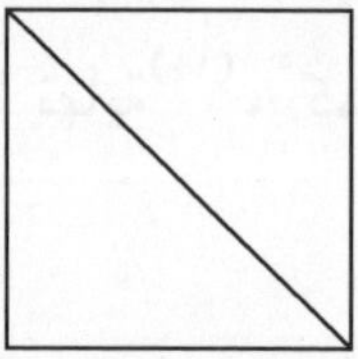

(١) «دائماً»: سقطت من [ع]، وردت في [S].

(٢) «جمع»: جميع [ن، ف، ل]. سقطت من [ع]، وردت في [S].

(٣) «العدد»: الأعداد [ل].

(٤) «نظم طبيعة الأفراد»: نظم الطبيعة للأفراد [ن، ل]، النظم الطبيعيّ للأفراد [ف].

(٥) «السطوح»: السطح [ك، ل، S].

(٦) «كما بيّنا قبل» في [S]: كما بيّنا قبل ذلك [ن]، كما بيّنا قبل هنا [ك]، كما بيّنا قبل هذا [ل]. سقطت من [ع].

(٧) «مثله»: سقطت من [ع]، وردت في [ن، S، وفي هامش ل].

(٨) «حدث»: وجدت [ن].

وإن[١] أُضيف إليهما شكلٌ آخرُ مثلّثٌ[٢]، حدث منه الشكلُ المخمَّسُ[٣]. وإن أُضيف إلى ذلك[٤] شكلٌ آخرُ مثلّثٌ[٥]، حدث من ذلك[٦] شكلٌ مُسدَّسٌ، وإن أُضيف إليها شكلٌ آخرُ، حدث من ذلك شكلٌ مُسبَّعٌ مِثلُ هذا[٧]:[٨]

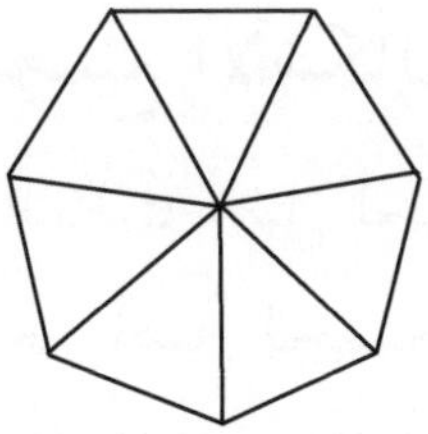

وعلى هذا القياس تحدث الأشكالُ المستقيمةُ الخطوطِ الكثيرةُ الزوايا من الشكل المثلّث إذا ضُمَّ بعضُها إلى بعضٍ، وتتزايد[٩] دائماً بلا نهاية كتزايُدِ العددِ[١٠] من الآحاد، إذا ضُمَّ بعضُها إلى بعضٍ دائماً بلا نهاية[١١]، كما بيّنا قبلُ.

---

(١) «إن»: إذا [Ṣ].
(٢) «مثلّث»: سقطت من [ع]، وردت في [Ṣ].
(٣) «منه الشكل المخمّس»: من ذلك شكل مخمّس [Ṣ].
(٤) «إلى ذلك»: إليها [Ṣ].
(٥) «مثلّث»: سقطت من [ع]، وردت في [Ṣ].
(٦) «من ذلك»: زيادة في [ع]، سقطت من [Ṣ].
(٧) زيادة في [ن]: «وعلى هذا إذا ما زيد زاد في الهندسة».
(٨) زيادة في [ط]: «وعلى هذا إذا زيد زاد في الهيئة».
(٩) «تتزايد»: سقطت من [ع]، وردت في [Ṣ].
(١٠) «العدد»: الأعداد [ل].
(١١) «كتزايد العدد من الآحاد، إذا ضمّ بعضها إلى بعض دائماً بلا نهاية»: سقطت من [ن].

وقد تبيّن أنّ من الشكل المثلّث تتركّب[١] الأشكالُ المستقيمةُ الخطوطِ، وأنّ من السطح تتركّبُ الأجسامُ، وأنُ من الخطوط تتركّب السطوحُ، وأنّ من النُّقطةِ تتركّب الخطوطُ، كما أنّ من الواحد يتركّب العددُ.

[٩٢] فإنّ النُّقطةَ في صِناعة الهندسة كالواحد في صناعة العدد، وكما أنّ الواحِدَ لا جُزءَ له فكذلك النُّقطةُ العقّليّةُ لا جُزءَ لها.

فصل < ١٢ > في أنواع السطوح[٢]

السطوحُ من جهة الكيفيّة تتنوعُ ثلاثةَ أنواعٍ: مُسطَّحاً ومُقعَّراً ومُقبَّباً. فالمسطَّحُ كوجوهِ الألواحِ[٣]،|ع ١٨ و| والمقعَّرُ كقعْرِ الأواني، والمقبَّبُ كظهرِ القِباب.

ومن الأشكال ما يُسمّى البَيضيَّ مِثلَ هذا:

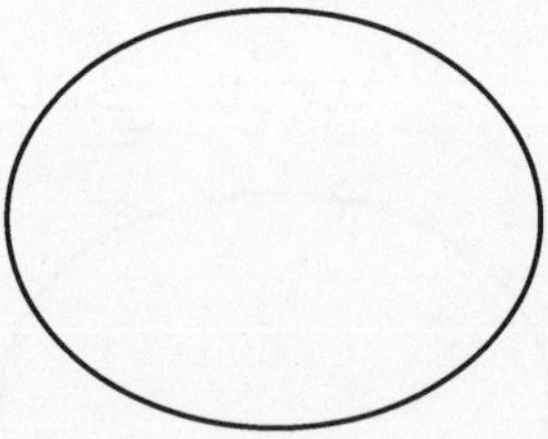

(١) «تتركّب»: يتركّب [ن]، إذا ضمّ يتركّب [ك]، مركّب [ل]. < وعلى هذا المنوال إلى نهاية الفصل >.

(٢) «في أنواع السطوح»: سقطت من [ع، ن، ف، ك، ل]، وردت في [S].

(٣) «الألواح»: اللّوح [ل].

ومنها الهِلاليُّ مِثلُ هذا:

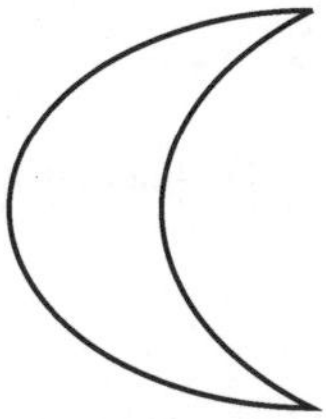

ومنها المخروطُ الصنوبريُّ[١] مِثلُ هذا:

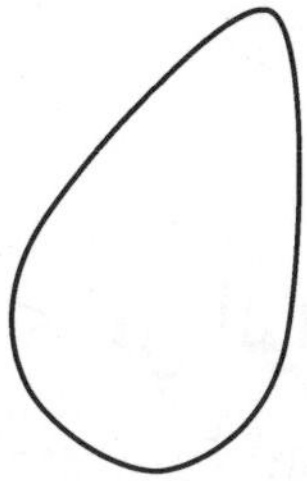

ومنها الإهليليجيُّ[٢] مِثلُ هذا:

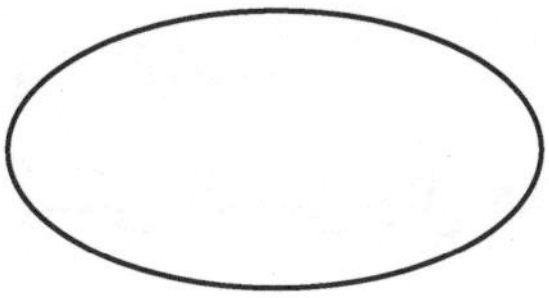

[٩٣] ومنها نيم خانجيّ مِثلُ هذا[٣]:

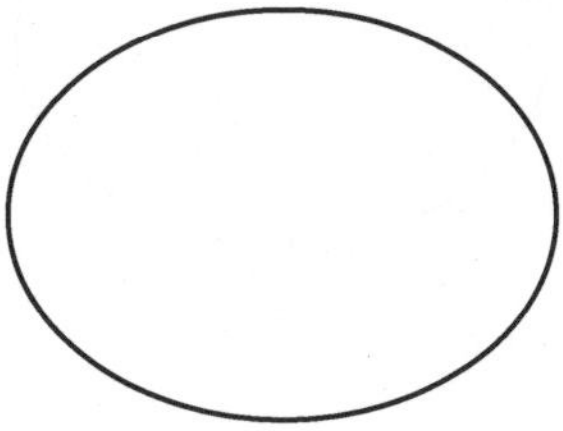

(١) < ذكر «الإهليليجيّ» قبل «المخروط الصنوبريّ» في [ف] > .

(٢) < وكما ورد في طبعة دار صادر [Ṣ]، الإهليليجي ينسب هنا إلى ثمرة الإهليلج > .

(٣) «ومنها نيم خانجيّ مثل هذا»: سقطت من [ن]. < وقد ذكر «الزيتونيّ» ثمّ «الطبليّ» ومن بعده «الإهليليجيّ» في ترتيب هذا المخطوط > .

ومنها الطَّبْليُّ مِثلُ هذا:

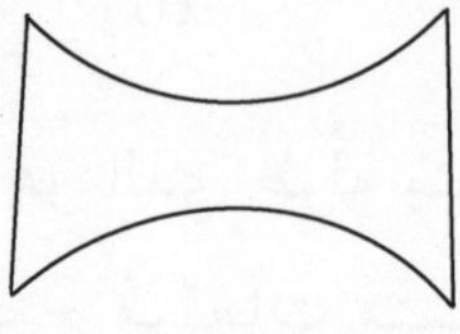

ومنها الزيتونيُّ مِثلُ هذا:

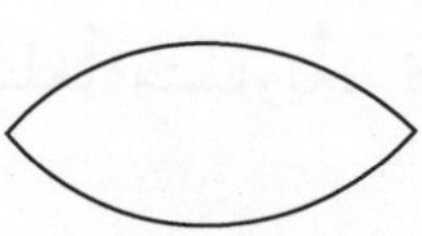

فصل < ١٣ > في ذكر الأجسام

فنقول[١]: السطوحُ هي نهاياتُ الأجسامِ، ونهاياتُ السطوحِ هي الخطوطُ[٢]، ونهاياتُ الخطوطِ هي النُّقَطُ. وذلك أنّ كلَّ خطٍّ لا بدّ له أن يبتدئ من نُقطةٍ وينتهي إلى أخرى. وكلُّ سطحٍ ينتهي إلى خطٍّ أو خطوطٍ، وكلُّ جسمٍ فلا بدّ من أن ينتهي إلى سطحٍ أو سطوح. فمن الأجسام ما يحيطُ به سطحٌ واحد[٣]، وهي الكُرةُ. ومنها ما يحيط به سطحان[٤]، وهو نِصفُ الكُرةِ، وذلك أنّ سطحاً منه مُقبَّبٌ وسطحاً مُدوَّرٌ. ومن الأجسام ما يحيط به ثلاثة سطوح، وهو رُبعُ الكُرةِ. ومنها ما يحيط به أربعة سطوح مثلّثات ويُسمّى الشّكلَ الناريَّ. ومنها ما يحيط به خمسة سطوح. ومنها

(١) «فنقول»: سقطت من [ط، ف، ك].

(٢) «الخطوط»: نهايات الخطوط [ع].

(٣) «أو سطوح. فمن الأجسام ما يحيط به سطح واحد»: وردت في هامش [ن].

(٤) «سطحان»: وردت في هامش [ن].

ما يحيط به ستّة سطوح مربّعات. فمنها المكعّبُ ومنها البئريُّ ومنها اللَّبنيُّ[١] ومنها اللوحيُّ[٢].

فالجسمُ المكعّبُ هو الذي طولُه مِثلُ عرضِه، وعرضُه مِثلُ سَمْكِه، وله ستّةُ سطوحٍ مُربّعاتٍ متساويةِ الأضلاع، قائمةِ الزوايا. وله [٩٤] ثماني زوايا مجسّمةٍ، وأربعٌ وعشرون زاويةً مسطّحةً، واثنا عشرَ ضِلعاً متساويةً، كلُّ أربعةٍ منها متوازيةٌ، وهذه صورتُها:

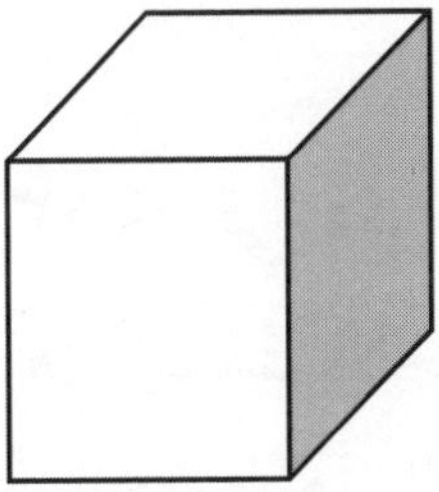

وأمّا الجسمُ البئريُّ[٣] فهو الذي طولُه مِثلُ عرضِه، وسَمْكُه أكبرُ منهما، وله ستّةُ سطوحٍ مربّعاتٍ: اثنان منها واسعان[٤] متقابلان متساويا الأضلاع، قائما الزوايا، وأربعةٌ منها مستطيلاتٌ ضيّقاتٌ[٥]، متساويةُ الأضلاعِ، قائمةُ الزوايا. وله اثنا عشرَ ضِلعاً: أربعةٌ منها طِوالٌ متساويةٌ متوازيةٌ، وثمانيةٌ قِصارٌ متساويةٌ

(١) «ومنها البئريّ ومنها اللبنيّ»: ومنها اللبنيّ ومنها البئريّ [ط، ف، ك، S].

(٢) «ومنها البئريّ ومنها اللبنيّ ومنها اللوحيّ»: سقطت من متن نصّ [ل] ثمّ أضيفت في الهامش.

(٣) < ورد ذكر مقادير «اللبنيّ» وشكله الهندسيّ قبل وصف «البئريّ» في [ن، ط، ف، ك، ل]> .

(٤) «واسعان»: زيادة في[ع]، سقطت من [S].

(٥) «مستطيلات ضيّقات»: ضيّقات مستطيلات [S].

متوازيةٌ. وله ثماني زوايا مُجسّمةٍ وأربعٌ وعشرون زاويةً مسطّحةً مِثلُ هذا[١]:

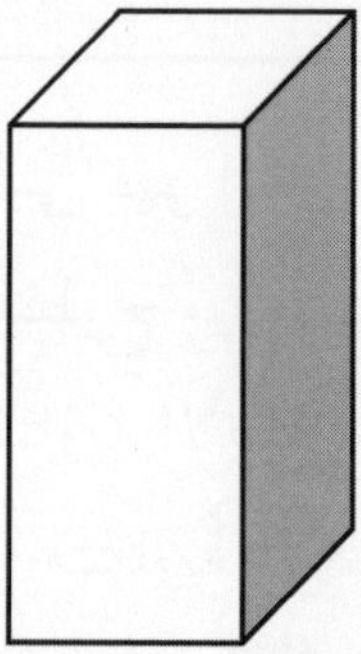

وأمّا الجسمُ اللوحيُّ[٢] فهو الذي طولُه أكبرُ[٣] من عرضِه، وعرضُه أكبرُ من سَمْكِه، وله ستّةُ سطوحٍ مُربَّعاتٍ: اثنان منها طويلان متقابلان متّسعان[٤]، ومتساويان كلُّ ضِلعينِ متقابلين[٥]، قائما الزوايا، وسطحان آخران قصيران[٦] ضيّقان[٧]، متساويا الأضلاع، قائما الزوايا. وله اثنا عشرَ ضِلعاً، أربعةٌ منها طِوالٌ، وأربعةٌ منها قِصارٌ، وأربعةٌ أقصرُ من ذلك. وله ثماني زوايا مجسّمةٍ وأربعٌ وعشرون زاويةً مسطّحةً[٨] مِثلُ هذا:

---

(١) «مثل هذا»: زيادة في[ع]، سقطت من [S].

(٢) < ورد ذكر مقادير «اللبنيّ» و«البئريّ» وشكلهما الهندسيّ قبل وصف «اللوحيّ» في [ن، ك، ل]> .

(٣) «أكبر»: أكثر [ن].

(٤) «متّسعان»: سقطت من [ن].

(٥) «ضيّقان»: طويلان [ع].

(٦) «قصيران»: طويلان [ن]. < ثمّ صحّحت في هامش [ن] إلى «قصيران»> .

(٧) «متساويان كلّ ضلعين متقابلين»: متساويا الأضلاع [S].

(٨) «وله اثنا عشر ضلعاً، أربعة منها طوال، وأربعة منها قصار، وأربعة أقصر من ذلك. وله ثماني زوايا مجسّمة وأربع وعشرون زاوية مسطّحة»: سقطت من [ع]، وردت في [ن، S].

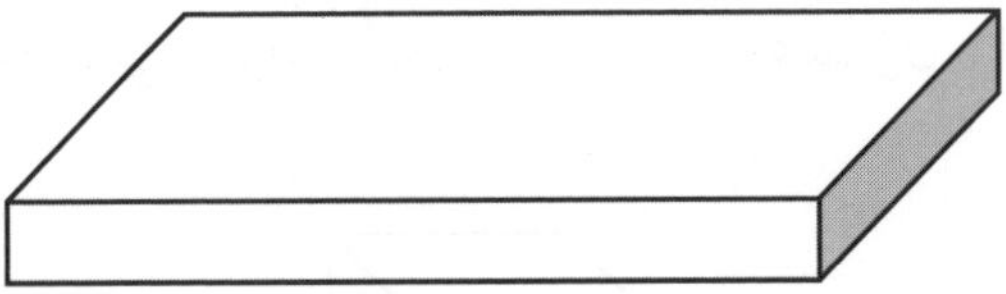

[٩٥] وأمّا الجسمُ اللبنيُّ فهو الذي طولُه مِثلُ عرضِه، وسَمْكُه أقلُّ[١] منهما. وله سِتّةُ سطوحٍ مربّعاتٍ: اثنان منها |ع ١٨ ظ| واسعان[٢] متقابلان، متساويا الأضلاع، قائما الزوايا، وأربعةٌ منها ضيّقاتٌ مستطيلاتٌ، متساويةُ الأضلاعِ، قائمةُ الزوايا. وله اثنا عشرَ ضِلعاً: أربعةٌ منها قِصارٌ[٣] متساويةٌ متوازيةٌ، وثمانيةٌ منها طِوالٌ متساويةٌ، كلُّ أربعةٍ منها متوازيةٌ[٤]، ولها ثماني زوايا مُجسّماتٍ، وأربعٌ وعشرون زاويةً مسطّحةً مِثلُ هذا:

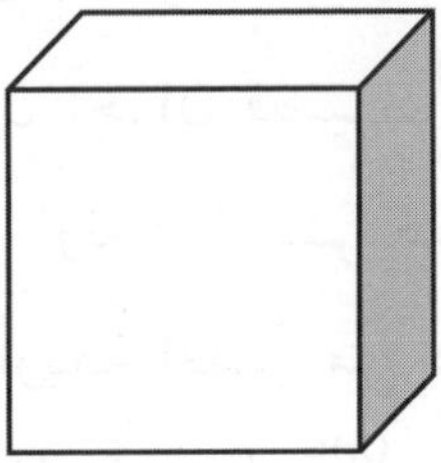

وأمّا الجسمُ الكُرِيُّ فهو الذي يحيط به سطحٌ واحِدٌ، وفي داخله نُقطة. وكلُّ الخطوط المستقيمة[٥] الخارجة من تلك النقطة إلى سطح الكرةِ متساويةٌ. والنّقطةُ[٦] يقال لها[٧] مركزُ

(١) «أقلّ»: أكثر [ن، ل].

(٢) «واسعان»: سقطت من [ع]، وردت في [Ṣ].

(٣) «قِصار»: طوال [ع]. < «قصار» في هامش [ن]> .

(٤) «كل أربعة منها متوازية»: سقطت من [ع]، وردت في [Ṣ].

(٥) «المستقيمة»: سقطت من [ع، ن]، وردت في [Ṣ].

(٦) «والنّقطة»: سقطت من [ل].

(٧) «والنّقطة يقال لها»: و يقال لتلك النّقطة [Ṣ].

الكرةِ[(١)]. وإذا دارت الكرةُ ظهر[(٢)] في سطحها نُقطتان متقابلتان[(٣)] ساكنتان يقال لهما قُطبا[(٤)] الكرةِ مِثلُ هذا:

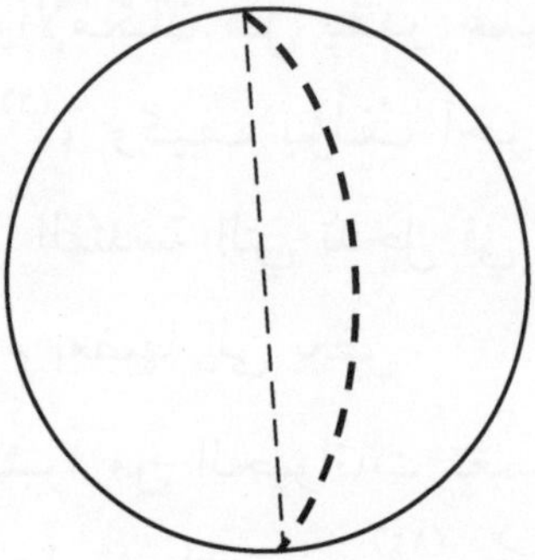

وإذا وُصِلَ بينهما بخطٍّ مستقيم وجاز ذلك الخطُّ على مركز الكرةِ يقال له مِحورُ الكرةِ، وإذا اتَّصل الخطُّ من نُقطةٍ إلى نقطةٍ، فهو المحور[(٥)].

< فصل ١٤ >

وإذ قد ذكرنا طرفاً من أصل الهندسة الحِسِّيّةِ، شِبْهَ المدخلِ والمقدِّمات، وقلنا إنّ هذا العِلمَ يحتاج إليه سائرُ[(٦)] الصُّنّاعِ، فلنبيِّنْ ذلك[(٧)]. وهو التقديرُ قبل العملِ، لأنّ كلَّ صانعٍ يؤلِّفُ

---

(١) «الكرة»: الدائرة [Ṣ].

(٢) «ظهر»: فيكون [ل، Ṣ].

(٣) «متقابلتان»: زيادة في[ع، ن، ل]، سقطت من [Ṣ].

(٤) «لها قطبا»: لهما قطب [ل، Ṣ].

(٥) «وإذا إتّصل الخطّ من نقطةٍ إلى نقطةٍ، فهو المحور»: سقطت من [ع، ن، ل]، وردت في [ط، ك، Ṣ].

(٦) «سائر»: أكثر [ن، ك، Ṣ]، كثير [ل].

(٧) «فلنبيّن ذلك»: سقطت من [ع]، وردت في [ن، ك، ل، Ṣ].

الأجسامَ بعضَها إلى بعضٍ ويُركّبها فلا بدّ له أن يُقدِّرَ أولاً المكانَ في^(١) أيِّ موضع يعملُها^(٢)، والزمانَ في^(٣) أيِّ وقتٍ يعملُها [٩٦] ويبتدئ فيها^(٤)، والإمكانَ هـل يقدر عليه أم لا، وبأيِّ آلةٍ وأدواتٍ^(٥) يعملُها^(٦)، وكيف يؤلِّفُ أجزاءها، حتى تلتئِمَ وتأتلفَ^(٧). فهذه هي الهندسةُ التي تدخل في أكثر الصنائع^(٨) التي هي تأليفُ^(٩) الأجسامِ بعضِها إلى بعضٍ.

واعلَمْ^(١٠) أنّ كثيراً من الحيوانات تعمل صنعةً طبيعيّةً قد جُبِلَت طبائعُها^(١١) عليها بلا تعليم^(١٢)، كالنّحلِ في اتّخاذِها البيوتَ، وذلك أنّها^(١٣) تَبني بيوتَها طبقاتٍ^(١٤) مستديراتِ الشكل متساويةِ الأضلاع والزوايا^(١٥) كالأتراسِ^(١٦)، بعضِها فوق بعضٍ.

---

(١) زيادة في [ل]: «في».

(٢) «يعملها»: العمل [ن].

(٣) «في»: سقطت من [ع]، وردت في [Ṣ].

(٤) «يعملها ويبتدئ فيها»: يبتدئ بعملها [ط، ك، ل].

(٥) «آلة وأدوات»: آلة وأداة [ك]، أداة وآلة [ك]، آلة [ن، ط].

(٦) «يعملها»: يعمل بها [ط].

(٧) «وتأتلف»: سقطت من [ع]، وردت في [Ṣ].

(٨) «الصنائع»: الصناعات [ن].

(٩) «هي تأليف»: يعملها مؤلفوا [ع].

(١٠) زيادة في [ف، ل]: «أيّها الأخ أيّدك الله وإيّانا بروح منه».

(١١) <زيادة في [ن] ولكن ذكرت باللّفظ «طبايعها»>.

(١٢) «صنعة طبيعيّة قد جُبِلَت طبائعها عليها بلا تعليم»: صنائعها طبعاً قد جُبِلَت عليه بلا تعليم [ط]، صنائعها طبعاً قد جُبِلَ عليها بلا تعلّم [ك].

(١٣) <أنها: أيّ النّحل>.

(١٤) «طبقات»: مطبقات [Ṣ].

(١٥) «متساوية الأضلاع والزوايا»: سقطت من [ن، ك، ل].

(١٦) «كالأتراس»: كالتِراس [ع، ن، ط].

وتجعل ثُقَبَ البيوتِ كلَّها مُسدّساتِ الأضلاعِ والزوايا، لما في ذلك من إتقانِ الحِكمةِ، لأنّ من خاصِّيَّةِ هذا الشكلِ أنّه أوسعُ من المربّعِ والمخمَّسِ، وأنّها تكشفُ[١] تلك الثُقَبَ[٢] حتى لا يكون بينها خللٌ، فيدخل[٣] الهواء، فتُفسد العسل، فيَعفن العسلُ[٤]، وهذا مِثالُه[٥]:[٦]

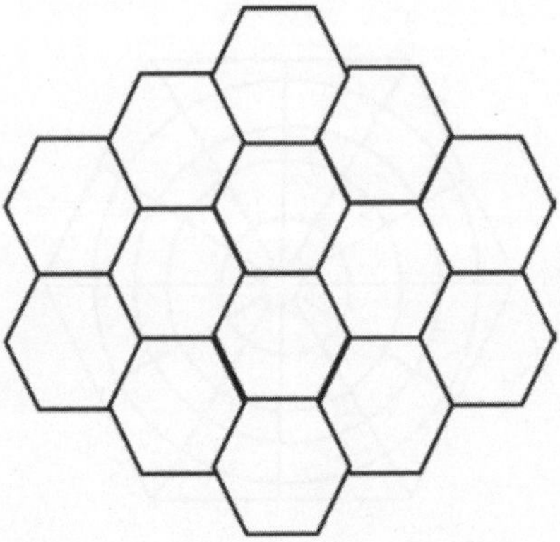

(١) «وإنها تكتشف»: وأنه تكتنف [ع].

(٢) «الثُقَب»: الثقوب [S].

(٣) «فيدخل»: وتداخل [ك].

(٤) «فتُفسد العسل، فيَعفن العسل»: العسل فيَعفن [ع، ط، ل]، فتُفسد العسل فيعفُن [ف، ك].

(٥) «مثاله»: مثال ذلك [ن، ك، S]. <ولم ترسم صورته في[ط]> .

(٦) < وذُكر هنا مقطع مطوّل في الأوراق ٢٧-٢٨ من المخطوط [ط]، ونُسِبَ إلى أقاويل «الناسخ»، حيث ورد شرحٌ مفصّل حول خواص الشكل المسدّس المتساوي الأضلاع وكيفيّة رسمه من الدائرة، وسبل تقسيم الدوائر إلى ستّة أقسام متساوية، وذلك تبَعاً لما ورد في المقالة الأولى من كتاب أوقليدِس في الأصول. ومن بعد ذلك أفرِدَ القول بأن جميع أعمال الطبيعة والحيوانات إما دوائر أو مثلّثات أو مربّعات أو مسدّسات. وبذلك انقضى شرح هذا «الناسخ» المنسوبة إليه تلك الأقاويل. وقد آثرنا عدم وضع هذا المقطع الطويل هنا، أو في متن النصّ، مخافة التطويل، ولكونه كذلك قد خرج عن سياق الرسالة

وهكذا العنكبوتُ ينسج شبكتَه[(١)] في زوايا البيوت[(٢)] والحيطان[(٣)] |ع ١٩ و| شفقة عليها من تخريق الرياح[(٤)] لها، وتزعزُعِها[(٥)]. وأمّا كيفيّةُ نسجِها فهو أن تمدَّ سَداها[(٦)] على الاستقامةِ، وخيوطَ لُحمَتِها على الاستدارةِ، لما فيه من سهولةِ العمل، وهذا مِثالُ ذلك[(٧)]:

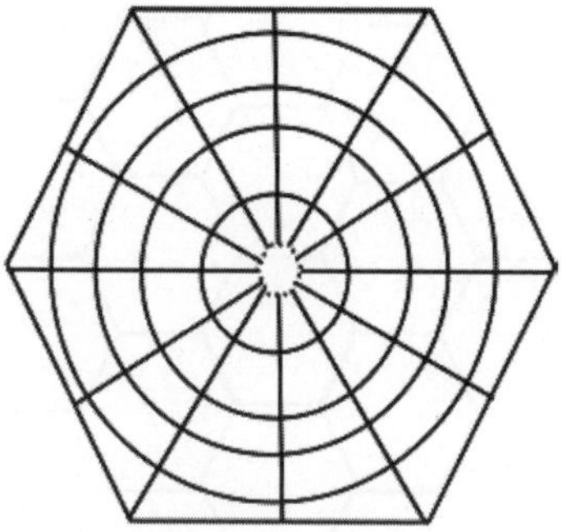

[٩٧] ومن الناس من يستخرج صناعةً بقريحته وذكاء نفسه،

برمّتها، ولم يظهر أيّ مثيل له في بقيّة المخطوطات المعتمدة في التحقيق النقدي ههنا. ولذلك ذكرناه في «ملحق» باللغة العربيّة، وضعناه بعد إنتهاء هذه الرسالة في الهندسة، وقد ترجمناه أيضاً إلى اللغة الإنكليزيّة، وكذلك أوردنا ترجمته هذه في «ملحقٍ» (Appendix) وضع بعد نهاية الترجمة الإنكليزيّة لرسالة الهندسة > .

(١) «ينسج شبكته»: تنسج شبكتها [Ṣ].

(٢) «البيوت»: البيت [ع، Ṣ].

(٣) «الحيطان»: الحائط [Ṣ].

(٤) «الرياح»: الهواء [ن، ل].

(٥) «وتزعزها»: وتمزيق حملها [ك، Ṣ]، وتزعزع الرياح لها [ن، ل].

(٦) < أي ما مدّ من خيوطه طولًا بخلاف لحمته > .

(٧) < وضعت خانات واسعة في [ن، ك، ل] لرسم الشكل المشار إليه، لكنّها بقيت فارغة > .

لم يُسبَق إليها، وأمّا أكثرُ الصُّنّاعِ فإنّهم يأخذونها توفيقاً[١] وتعليماً من الأستاذين.

فصل < ١٥ > في المِساحَة[٢]

واعلَمْ يا أخي[٣]، أيّدك الله وإيّانا بروح منه[٤]، أنّ عِلمَ الهندسةِ يدخلُ في الصنائعِ كلِّها، وخاصّةً في[٥] المِساحَة، وهي صناعة يحتاج إليها العمّال والكتّاب والدهّاقين[٦] وأصحابُ[٧] الضياعِ والعقاراتِ في معاملاتهم من جباية الخَراج وحفر الأنهار وعمل البَريدات وما شاكلها[٨].

ثمّ اعلَمْ بأنّ المقادير التي تُمسح بها الأراضي[٩] بالعراق خمسةُ مقاديرِ وهي: الأَشْلُ والبابُ والذراعُ والقبضةُ والإصبعُ. واعلَمْ بأنّ الإصبعَ الواحدةَ غِلَظُها ستُّ شَعيراتٍ[١٠] مصفوفةٍ

(١) < «توفيقاً»: أي أيضاً «تعليماً» للتوكيد >.

(٢) «في المساحة»: سقطت من [ع، ن، ط، ل]، وردت في [ف، ك، Ṣ].

(٣) «يا أخي»: أيّها الأخ [ف]. سقطت من [ع، ن، ط، ك، ل]، وردت في [Ṣ].

(٤) «أيّدك الله وإيانا بروح منه»: سقطت من [ع، ن، ط، ف، ك، ل]، وردت في [Ṣ].

(٥) «في»: سقطت من [ع، ن، ل]، وردت في [ك، Ṣ].

(٦) < «الدهاقين»، أي زعماء الفلاحين >.

(٧) زيادة في [ل]: «الصنايع».

(٨) «وما شاكلها»: وما شاكل ذلك [ط].

(٩) «الأراضي»: الأرضون [ط]،الأرض [ل].

(١٠) «شَعيراتٍ»: شعيّرات [ط]. < وردت شبه مطموسة في [ط] وربّما أراد الناسخ القول بها «شُعَيْراتٍ»، ولكن ذلك المخطوط سقط من نصّه التشكيل والتحريك >.

مضمومةٍ[١] بعضِها[٢] إلى بطون بعض[٣]. والقبضةُ الواحِدةُ أربعُ أصابع[٤]. والذراعُ الواحِدُ ثماني قبضاتٍ، وهو اثنان وثلاثون إصبعاً. والبابُ طولُه ستّةُ أذرعٍ وهي ثمانٍ وأربعون قبضةً، وهو مائةٌ واثنان وتسعون إصبعاً. والأَشْلُ حَبلٌ طولُه عشرةُ أبوابٍ، وهو سِتّون ذِراعاً، وأربعُ مئةٍ، وثمانون قبضةً، وألفٌ وتسعُ مئةٍ وعشرون إصبعاً.

واعلَمْ يا أخي[٥] بأنّك إذا ضَرَبت هذه المقاديرَ بعضَها في بعض، فالذي يخرج منها يُسمّى تكسيراً. فإذا جُمِعَت، فيكون منها جَريباتٌ وقَفيزاتٌ وعَشيراتٌ[٦]. وأمّا حِسابُها فهو أنّ القبضةَ الواحدةَ في مِثلِها تكون ستةَ عشرَ إصبعاً، والذراعَ الواحدةَ في مِثلِها تكون[٧] أربعاً وستين قبضةً مكسّرةً، وألفاً وأربعةً وعشرين إصبعاً مكسّرةً، وهو تُسعُ رُبعِ عُشرِ عُشر[٨] الجَريب. والبابُ الواحِدُ في مثلِه يكون ستةً وثلاثين ذراعاً مكسّرةً.

[٩٨] وهذه صورتُها: ٣٦ وهو ٢٣٠٤ قبضاتٍ مكسّرةً، وهو

---

(١) «مصفوفة مضمومة»: مصفوفات [ن، ل].

(٢) «بعضها»: ظهور بعضها [ك].

(٣) «مضمومة بعضها إلى بطون بعض»: سقطت «بعضها إلى بطون بعض» من [ط]. سقطت كلّها من [ع]، ووردت في [Ṣ].

(٤) «والقبضة الواحدة أربع أصابع»: سقطت من [ع]، وردت في [Ṣ].

(٥) «يا أخي»: زيادة في [ع].

(٦) «جَريباتٌ وقَفيزاتٌ وعَشيراتٌ» في [Ṣ]: جُربان وقُفزان وعُشيران [ع، ن، ف، ك، ل]، جُرباناً وقُفزاناً وعُشراناً [ط].

(٧) «ستة عشر إصبعاً، والذراع الواحدة في مثلها تكون»: سقطت من [ل].

(٨) «وهو تُسعُ رُبعِ عُشرِ عُشر»: وهي رُبع تُسع عُشر [ع، ن، ط، ك].

٣٦٨٦٤ إصبعاً مكسّرةً، وهو عُشر عُشر الجَريب[١].

وأمّا الأَشْلُ في مثلِهِ فيكون جريباً، وهو عشرة أقفِزَةٍ، وهو مائة عَشير؛ وهذه صورتُها[٢]: ٣٦٠٠ ذراعاً مكسّرة، وهو ٢٣٠٤٠٠ قبضة مكسّرة |ع ١٩ ظ|، وهو ٣٦٨٦٤٠٠ إصبعاً مكسّرة.

وأمّا القَفيز فهو عشرةُ أعشارٍ، وهو عشرةُ أبوابٍ مكسّرة، وهو من ضَربِ تسعةَ عشرَ ذراعاً إلا شيئاً يسيراً في مِثلِهِ، وهو ثلاث مائة وستون ذراعاً[٣].

وأمّا العَشيرُ فهو من ضَربِ بابٍ واحدٍ في مِثلِه نفسِه[٤]، وهو ٣٦ ذراعاً مكسّرةً، وهو ٢٣٠٤ قبضاتٍ مكسّرةً، وهو ٣٦٨٦٤ إصبعاً مكسّرةً.

والأُشولُ في الأُشولِ، واحدُها جَريبٌ وعَشرتُها عشرةُ أجربةٍ. والأُشولُ في الأبواب واحِدُها قَفيزٌ وعشرتُها جَريبٌ[٥]. والأُشولُ في الأذرعِ، واحِدُها عَشيرٌ وثُلثا عَشيرٍ، وستُّ منها[٦] قَفيز. والأُشول[٧] في القبضات واحِدُها سُدسُ عَشير ورُبْع

(١) < تركت خانة واسعة في الجانب الأيسر من الورقة ١١ في مخطوط [ف] لوضع الجدول الحسابيّ لهذه المقادير، ولكنّه بقي فارغاً. وقد صححنا حساب المقادير في متن النصّ >.

(٢) «وهو مائة عَشير؛ وهذه صورتها»: سقطت من [ع]، وردت في [S].

(٣) «وهو ثلاث مائة وستون ذراعاً»: سقطت من [ع]، وردت في [S].

(٤) «نفسه»: زيادة في [ع]، سقطت من [ن، S].

(٥) «والأشول في الأشول، واحدها جَريب وعَشرتها عشرة أجرِبة. والأشول في الأبواب واحدها قَفيز وعشرتها جَريب»: سقطت من [ع]، وردت في [S].

(٦) «واحدها عَشير وثلثا عَشير، وستّ منها»: واحدها عَشير وستّتها [ع].

(٧) «الأشول»: الأُشْلُ [S].

سُدسِ[١] عَشيرٍ، وكلُّ ثلاثةِ أخماسٍ منها عَشيرٌ، وكلُّ ستّةٍ وثلاثين منها قَفيزٌ. والأُشول[٢] في الأصابع كلُّ واحدٍ منها رُبعُ سُدسِ عَشيرٍ ورُبعُ رُبعِ سُدسِ عَشيرٍ. وكلُّ عَشرةٍ منها رُبعا عَشيرٍ، وسُدسُ ثُمنِ عَشيرٍ[٣].

والأبوابُ في الأبوابِ واحدُها عَشيرٌ، وعشرتُها قفيزٌ. والأبوابُ في الأذرُعِ واحِدُها سُدسُ عَشيرٍ، وسِتَّةٌ منها[٤] عَشيرٌ. والأبوابُ في القبضاتِ كلُّ واحِدٍ منها ثلاثةُ أرباعِ رُبعِ تُسعِ عَشيرٍ. والأبوابُ في الأصابعِ كلُّ خمسةٍ وثمانين منها ثُلثُ عَشيرٍ ورُبعُ سُدسِ عَشيرٍ وتُسعُ عَشيرٍ تقريباً[٥]. وكلُّ أربعةٍ منها ثلاثةُ أرباعٍ وتُسعُ عَشيرٍ، وكلُّ مائةٍ ثمانٍ وعشرين منها ثُلثا ثُلث

(١) «رُبع سُدس»: تُسع [ع].
(٢) «الأشول»: الأُشْلُ [S].
(٣) < وضع ناسخ مخطوط [ع] في ظهر الورقة ١٩ المقطع المطوّل الوارد أدناه. لعلّه أراد به التذكير بما ضمّنه في النصّ أعلاه. ولكن ما ذكره هنا فيه إشكاليّات حسابيّة، ومسّه التراداد الذي طغى عليه أيضاً بعض من اللغو والتطويل. ولم يظهر هذا المقطع في المخطوطات الأخرى. ولذلك آثرنا وضعه في الحاشية بدلاً من متن النصّ المحقّق. وقد خَطَّ ناسخ [ع] ما يلي: «وكلّ عشرين منها ربع تسع عشير وهو ذراعان ونصف مكسّرة. وكلّ أربعة عشر وخمسين منها يكون عشيراً. وكلّ عشرين منها ربع تسع عشير. وكلّ ثمانين تسع عشير. الأشول في الأشول واحدها جريب وعشرتها عشرة أجربة. والأشول في الأبواب واحدها قفيز وعشرتها جريب. الأشول في الأذرع واحدها عشر ثلثيّ عشير وستّتها قفيز. الأشول في القبضات كلّ خمسة منها ربع تسع عشير. وكلّ واحد وعشرين منها تسع عشير تقريباً. الأشول في الأصابع كلّ عشرين منها ربع تسع عشير» > .
(٤) «وستَّة منها»: وسِتَّتها [ع].
(٥) «عَشير ورُبع سُدس عَشير وتُسع عَشير تقريباً»: وتسع عشير على التقريب [ع].

عَشيرٍ[١]. والأذرُعُ في الأذرُعِ واحِدُها رُبعُ تُسعِ عَشيرٍ، وكلُّ أربعٍ منها تُسعُ عَشيرٍ، وكلُّ مائةٍ منها عَشيران وثُلثي عَشير وتُسع عَشيرٍ[٢].

فهذا شرحُ[٣] مِساحَةِ العرضِ والطولِ. فأمّا مِساحَةُ العُمقِ فهو أن تَضرِبَ الطولَ في العرضِ[٤] فما اجتمَعَ[٥] من ذلك فاضرِبْه في العُمق، وما يَجتَمِعُ فهو[٦] تكسيرُ المجسّم. والحاجة إلى هذا العمل عند حفر الأنهار والآبار والحفائر[٧] [٩٩] وعملِ[٨] البَريداتِ والمَسنيّات[٩] والأساسات[١٠] للديار[١١] والبُنيان[١٢] وما شاكل ذلك.

---

(١) «وكلّ أربعة منها ثلاثة أرباع وتُسع عَشير، وكلّ مائة ثمان وعشرين منها ثلثا ثلث عَشير»: سقطت من [ع]، وردت في [Ṣ].

(٢) «وتُسع عَشير»: سقطت من [ع]، وردت في [Ṣ].

(٣) «شرح»: سقطت من [ع، ن]، وردت في [Ṣ].

(٤) «الطول في العرض»: العرض في الطول [ن، ك].

(٥) «فما اجتمَعَ»: والمجتمع [ط، ف].

(٦) «من ذلك فاضربه في العمق، وما يَجتَمِعُ فهو»: سقطت من [ع]، وردت في [Ṣ].

(٧) «الآبار والحفائر»: الآبار والبِرَك [ن، ف]، الآبار والأنهار والبِرَك والحفاير والقِنا [ك]، والبِرَك والقنى [ل].

(٨) «عمل»: زيادة في [ع]، سقطت من [Ṣ].

(٩) <«المَسنيّات»: أي السدود>.

(١٠) «والأساسات »: أساسات [ك]، أساس [ف، ل]. سقطت من [ع]، وردت في [Ṣ].

(١١) «للديار»: البنيان [ك]. سقطت من [ع، ن، ط، ك، ل]، وردت في [Ṣ].

(١٢) «البنيان»: البنيانات [ل].

< فصل ١٦ >

إعلَمْ يا أخي[١]، أيّدك الله وإيّانا بروح منه[٢]، أنّه قد تدخُل الشُّبَهُ[٣] في كلّ صناعة علميّة[٤] على مَنْ يتعاطاها وليس مِنْ أهلها، أو كان ناقصاً فيها أو ساهياً عنها، مِثالُ ذلك ما ذكروا أنّ رجلاً باع من رجلٍ آخرَ[٥] قطعةَ أرضٍ بألفِ درهمٍ على أنّ طولَها مائةُ[٦] ذِراعٍ، وعرضَها مائةُ[٧] ذِراع، ثم قال له: خذْ مني عِوضاً عنها قِطعتين من[٨] أرضٍ كلٌّ واحدةٍ منهما طولُها خمسون[٩] ذِراعاً، |ع ٢٠ و| وعرضُها خمسون[١٠] ذِراعاً، وتوهّم أنّ ذلك حقُّه فتحاكما إلى قاضٍ غيرِ مُهندسٍ، فقضى له بمثل ذلك خطأً. ثم تحاكما إلى حاكمٍ[١١] من أهل صناعة الهندسة[١٢] فحكم بأنّ ذلك نِصفُ حقّه.

وهكذا أيضاً[١٣] ذكِر أنّ رجلاً استأجر رجلاً على أن يحفر له

---

(١) «يا أخي»: سقطت من [ن، ل]، وردت في [ع، ف، ك، Ṣ].

(٢) «أيّدك الله وإيانا بروح منه»: سقطت من [ع، ن، ك، ل]، وردت في [ف، Ṣ].

(٣) «تدخل الشُّبَه»: يدخل الشُبَهة [ن، ك، ل].

(٤) «علميّة»: سقطت من [ع، ن، ك، ل]، وردت في [ف، Ṣ].

(٥) «آخر»: سقطت من [ع، ن]، وردت في [ل، Ṣ].

(٦) «مائة»: ألف [ف، ل].

(٧) «مائة»: ألف [ف، ل].

(٨) «قِطعتين من»: قطعتيّ [ع، ن].

(٩) «خمسون»: خمسمائة [ف]، خمس ماية [ل].

(١٠) «خمسون»: خمسمائة [ف]، خمس ماية [ل].

(١١) زيادة في [ك]: «آخر».

(١٢) «صناعة الهندسة»: الصناعة والهندسة [ط]، الصناعة [Ṣ].

(١٣) «أيضاً»: سقطت من [ع، ل]، وردت في [ن، Ṣ].

بِركةً طولُها أربعةُ أذرُعٍ في عرضِ أربعةِ أذرُعٍ في عُمقِ أربعةِ أذرُعٍ، بثمانية دراهم. فحفر له ذراعين في ذراعين طولاً في ذراعين عرضاً في ذراعين عمقاً[١]، فطالبه بأربعة دراهم نصفِ الأُجرة، فتنازعا وتحاكما إلى مُفتٍ غيرِ مهندسٍ فحكم بأنّ ذلك حقُّه، ثم تحاكما إلى أهل صناعة الهندسة[٢] فحكموا له بدرهمٍ واحد.

وقيل لرجل يتعاطى الحسابَ ولم يكن من أهله: كم نسبة ألفِ ألفٍ إلى ألفِ ألفِ ألفٍ؟ فقال: ثُلثان. فقال أهل الصناعة إنّه عُشرُ عُشرِ العُشرِ. فعلى هذا المِثال تدخل الشُبهَةُ على كلّ من يتعاطى صناعةً وليس من أهلها. ومن قِبَلِ[٣] هذا قيلَ: استعينوا على كلِّ صِناعَةٍ[٤] بأهلها[٥].

فصل < ١٧ > في حاجة الإنسان إلى التّعاون[٦]

إعلَمْ يا أخي[٧]، أيّدك الله وإيّانا بروح منه[٨]، بأنّ الإنسانَ الواحِدَ لا يقدِر أن يعيشَ وحدَه إلا عيشاً نكْداً، لأنّه يحتاج في[٩]

---

(١) «في ذراعين عرضاً في ذراعين عمقاً»: وعرضاً وعمقاً [S].

(٢) «صناعة الهندسة»: الصناعة [ن، ف، ك، ل، S].

(٣) «قِبَلِ»: أجلِ [ن، ط، S].

(٤) «صِناعَةٍ»: صِنعَةٍ [ن، ل، S].

(٥) «بأهلها»: بصالح أهلها [ل].

(٦) «في حاجة الإنسان إلى التعاون»: سقطت من [ع، ن، ط، ف، ل]، ووردت في [ك، S].

(٧) «يا أخي»: أيّها الأخ [ف، ل]. سقطت من [ع، ط].

(٨) «أيّدك الله وإيانا بروح منه»: سقطت من [ع، ن، ط، ك].

(٩) «يحتاج في»: محتاج في [ط]، محتاج من [S].

طيبِ العيش إلى إِحكامِ[١] [١٠٠] صنائعَ شتَّى، ولا يمكن للإنسانِ الواحدِ أن يَبلُغها كلَّها، لأنّ العمرَ قصيرٌ، والصنائعَ كثيرةٌ، فمن أجل هذا اجتمع في كلِّ مدينةٍ أو قريةٍ أناسٌ كثيرون[٢] لمعاونة بعضِهم بعضاً. وقد أوجبت الحكمةُ الإلٰهيّةُ والعنايةُ الربانيّةُ بأن تشتغل جماعةٌ منهم بإِحكام الصناعات[٣]، وجماعةٌ منهم بإِحكام التجارات، وجماعةٌ بإِحكام البُنيان[٤]، وجماعةٌ بإِحكام النباتات[٥]،[٦] وجماعةٌ بتدبير السياسات، وجماعةٌ بإِحكام العلوم وتعليمها، وجماعةٌ بالخدمة للجميع والسعي في حوائجهم[٧]؛ لأنّ مَثَلهُم في ذلك كمَثَلِ أخوَةٍ من أبٍ واحدٍ في منزل واحد[٨]، متعاونين في أمر معيشتهم[٩]، كلٌّ منهم في وجهٍ منها[١٠]. فأمّا ما اصطلحوا عليه من الكيل[١١] والوزن والثمن والأُجرة، فإنّ ذلك حكمةٌ وسياسةٌ ليكون حثّاً لهم على الاجتهاد في أعمالهم

---

(١) «إحكامِ»: سقطت من [ل].
(٢) «كثيرون»: كثيرة [ن، ف].
(٣) «الصناعات»: الصنائع [ط، Ṣ].
(٤) «وجماعة بإِحكام البُنيان»: وجماعة بإِحكام البُنيانات [ك، ل]. سقطت من [ع]، وردت في [Ṣ].
(٥) «وجماعة بإِحكام النباتات»: زيادة في [ع، ن]، سقطت من [ط، ف، ك، Ṣ].
(٦) < ذكر «إحكام النباتات» قبل «إحكام البنيانات» في [ل]> .
(٧) زيادة في [ل]: «ومصالحهم».
(٨) « منزل واحد»: منزلة واحدة [ع].
(٩) «معيشتهم»: معايشهم [ع].
(١٠) «كل منهم في وجه منها»: سقطت من [ع، ط، ك]، وردت في [ن، ل، Ṣ].
(١١) «الكيل»: المكّيال [ل].

وصنائعهم ومعاوناتهم، حتّى يستحقَّ كلُّ إنسانٍ من الأُجرة[١] بحسب اجتهادِهِ في العمل[٢] ونشاطِه في الصنائع[٣].

واعلَمْ يا أخي، أيّدك الله وإيّانا بروح منه[٤]، أنّه ينبغي لك أن تتيقّن بأنّك لا تقدِر أن تَنجو وحدَك مما وقعت فيه من مِحنةِ[٥] هذه الدنيا وآفاتِها التي جناها أبونا[٦] آدمُ،[٧] عليه السلام، لأنّك محتاج في صلاحك[٨] ونجاتك وتخلُّصك[٩] من هذه الدنيا، التي هي عالمُ الكونِ والفساد، ومن عذاب نار[١٠] جهنّمَ وجِوارِ الشياطين وجنودِ إبليس أجمعين، والصُّعود إلى عالم الأفلاك وسعةِ السموات ومسكنِ العِلِّيّين وجِوار |ع ٢٠ ظ| ملائكةِ الرَّحْمٰن[١١] المقرّبين، إلى معاونةِ إخوانٍ[١٢] لك[١٣] نُصحاءَ وأصدقاءٍ لك فُضلاءَ مستبْصِرين بأمر الدّين، عُلماءَ بحقائق

---

(١) «من الأجرة»: من الآخر [ن]. سقطت من [ع]، وردت في [Ṣ].

(٢) «العمل»: العلم [ن].

(٣) «الصنائع»: الصناعة [ك].

(٤) «يا أخي، أيّدك الله وإيّانا بروح منه»: سقطت من [ع، ن، ط، ك، ل]، وردت في [Ṣ].

(٥) «محنة»: محنٌ [ف].

(٦) «التي جناها أبونا»: بالجناية التي كانت من أبينا [ن، ط، Ṣ].

(٧) «محنة هذه الدنيا وآفاتها التي جناها أبونا آدم»: محنة هذه الدنيا وآفاتها التي كانت من أبينا آدم [ف، ك]، أباك آدم [ل].

(٨) «صلاحك»: زيادة في [ع، ط، ل]، سقطت من [Ṣ].

(٩) «وتخلّصك»: ونجاحك [ل]. سقطت من [ع، ن، ف]، وردت في [Ṣ].

(١٠) «نار»: زيادة في [ع]، سقطت من [Ṣ].

(١١) «الرَحمٰن» في [ع]: الله [ن، ف، ك، ل] // وقد سقطت كلّها من [ط].

(١٢) زيادة في [ل]: «الصفاء».

(١٣) زيادة في [ل]: «وخُلّان».

الأمور، ليعرِّفوك طريقَ[١] الآخرةِ[٢] وكيفيّةَ الوصول إليها، والنجاةَ من الورطة التي وقعنا فيها كلُّنا بجنايةِ أبينا آدَمَ، عليه السلام[٣]. فاعتَبِر يا أخي[٤] بحديث الحمامةِ المطوَّقةِ المذكورةِ في كتاب «كَليلَة ودِمنَة» وكيف نجت من الشبكة لتعلمَ[٥] حقيقةَ ما قلنا[٦][٧].

---

(١) «طريق»: طرق [ل]، طرائق [S].

(٢) < ورد هنا مقطع في [ف، ل] لم يذكر في بقيّة المخطوطات أو في طبعة صادر: «منهم في الدرجة الأولى ومنهم في الدرجة الثانية ومنهم في الدرجة الثالثة ومنهم في الدرجة الرابعة. وإن القوة . . . والمعرفة بالمعاد. فإذا وفّق النفس . . . أخ ناصح فاضل فقد دلّ إلى ما فوقه حتى تصل إلى الدليل القاصد بك إلى ربّك والعرش المحيط من حوله الملائكة المقرّبين والأنبياء والمرسلين كملاً في مقامه الكريم ومحلّه الأمين، فتشاهدون كما رأى أبونا إبراهيم من ملكوت السماء وتكون من المؤمنين» >.

(٣) «عليه السلام»: سقطت من [ط].

(٤) «يا أخي»: سقطت من [ك، S].

(٥) «لتعلم»: لتعلموا [ط].

(٦) «وكيفيّة الوصول إليها، والنجاة من الورطة التي وقعنا فيها كلّنا بجناية أبينا آدم، عليه السلام . . . الحمامة المطوّقة المذكورة في كتاب «كَليلَة ودِمنَة» . . . لتعلم حقيقة ما قلنا»: سقطت من [ف، ل].

(٧) < في هذا الموضع، وردت زيادات مطوّلة في محتويات الورقة ١٢ من مخطوط [ف] والأوراق ٢٥ و ٢٦ من مخطوط [ل]، تناولت كيفيّات الصعود بالنفس إلى السماء، وأردِفَت أيضاً بالتبحّر بطبائع النفوس وتموضعها في الأجساد في هذا الكون. ثمّ وَصَفَت حسنات العبرة التي يمكن الاستدلال بها من رواية الحمامة المطوّقة المذكورة في كتاب «كَليلَة ودِمنَة». ومثيلات هذه القضايا ليس مكانها في رسالة تقنيّة مركّزة تتناول شؤون علم الهندسة. ولعل تلك الزيادات قد حصلت نتيجة لغلطٍ ما في النسخ أو في خلط تصنيف المحتويات في هذه الأوراق من المخطوطات [ف، ل]. وربّما سلك ناسخ [ل] طريق ناسخ [ف] في هذا الشأن؛ ولكن ليس لدينا الدليل القاطع الذي يبرهن مثل ذلك الأمر. ولم تظهر هذه المقاطع في المخطوطات الأخرى أو في طبعة دار صادر. ولذلك آثرنا عدم وضعها في الحاشية أو في متن النصّ

واعلَمْ أنّ الحُكماءَ إذا ضربوا مثلاً[١] لأمور الدنيا، فإنّما غرضُهم منه أمورُ الآخرةِ والإشارةُ إليها بضُروبِ[٢] الأمثالِ بحسب ما تحتمل عقولُ[٣] الناسِ في كلِّ مكانٍ وزمان.

[١٠١] فصل <١٨> في الهندسة العقليّة[٤]

وإذ قد ذكرنا طرفاً من الهندسة الحِسِّيَّةِ شِبهَ المدخلِ والمقدِّمات، فنريد أن نذكُرَ طرفاً من الهندسةِ العقليّة، إذ كانت هي أحدَ أغراضِ الحُكماءِ الراسخين في العلوم الإلهيّة، المرتاضين بالرياضاتِ الفلسفيّة، وذلك أنّ غرضَهم في تقديمِ الهندسةِ بعد عِلم العدد[٥] هو تخريجُ المتعلّمين من المحسوساتِ إلى المعقولاتِ،[٦] وترقيتهُم لأولادِهم وتلاميذِهم[٧] من الأمور الجسمانية إلى الأمور الروحانيّة.

فاعلَمْ يا أخي، أيّدك الله وإيّانا بروح منه[٨]، أنّ النظرَ في

المحقّق، واكتفينا هنا بالإشارة إليها بشكل موجز مخافة الترداد والتطويل أيضا. وعسى أن يكون في مسلكنا النقديّ هذا شيء من الصواب في التحقيق>.

(١) «مثلاً»: الأمثال [ك].

(٢) «بضُروب»: بضرب [ع].

(٣) «عقول»: عقل [ل].

(٤) «في الهندسة العقليّة»: سقطت من [ع، ط]، وردت في [Ṣ].

(٥) «العدد»: الأعداد [ك، ل].

(٦) زيادة في [ل]: «وتوفيقهم».

(٧) «لأولادهم وتلاميذهم»: لتلاميذهم وأولادهم [ن، ط، ك، Ṣ].

(٨) «يا أخي، أيّدك الله وإيّانا بروح منه»: سقطت من [ع، ن، ط، ف، ك]، وردت في [Ṣ، وفي هامش ل].

الهندسة الحِسِّيَّةِ يؤدّي إلى الحِذق في الصنائع العمليّة كلِّها، والنظرَ في الهندسة العقّليّة يؤدّي إلى الحِذق في الصنائع العلميّة[١]. لأنّ هذا العلمَ هو أحدُ الأبوابِ التي تؤدّي إلى معرفة جوهر النفس التي هي جذرُ العلومِ وعُنصرُ الحِكمةِ، وأصلُ الصنائعِ العلميّةِ والعمليّة جميعاً. أعني[٢] معرفةَ جوهرِ النفسِ. فاعلَمْ جميعَ ما قلنا[٣].

فصل < ١٩ > في توهُّم الأبعاد[٤]

الخطُّ العقليُّ لا يُرى مجرّداً إلّا بين السطحَين. فهو مثلُ الفصلِ المشتركِ الذي هو بين الشمسِ والظِّلِّ. وإذا لم يكن شمسٌ ولا ظِلٌّ[٥] لم ترَ خطّاً[٦] بنُقطتين وهميّتين. فإذا توهّمتَ أن قد تحرّكت إحدى النُّقطتين وسكنتِ الأُخرى، حتى رجعتْ إلى حيثُ ابتدأت بالحركة، حدث في فِكرك السطحُ[٧].

---

(١) «والنظر في الهندسة العقليّة يؤدي إلى الحذق في الصنائع العلميّة»: أضافها ناسخ مخطوط [ع] في الهامش ولم ترد في متن النصّ.

(٢) «جذر العلوم وعنصر الحكمة، وأصل الصنائع العلميّة والعمليّة جميعاً. أعني»: سقطت من [ن].

(٣) «فاعلم جميع ما قلنا»: سقطت من [ع، ن]، وردت في [Ṣ].

(٤) «في توهّم الأبعاد»: سقطت من [ع، ن، ك، ل]، وردت في [Ṣ].

(٥) «ظّلّ»: فيء [Ṣ].

(٦) «لم تر خطاً»: لم ير خطّ [ع].

(٧) «فإذا توهّمت أن قد تحركت إحدى النقطتين وسكنت الأخرى، حتى رجعت إلى حيث ابتدأت بالحركة، حدث في فكرك السطح»: سقطت من [ع، ف، ك]، وردت في [ن، Ṣ].

والسطحُ العقليُّ أيضاً لا يُرى بمجرّده[١] إلا بين الجِسمَين، وهو[٢] الفصلُ المشتركُ بين الدُّهنِ والماء[٣]. والنُّقطة العقليّةُ لا تُرى أيضاً بمجرّدها إلّا حيثُ ينقسم الخطُّ بنصفين[٤] بالوهمِ، أيّ موضعٍ وقعت للإشارة إليها فهي تنتهي[٥] هناك.

[١٠٢] واعلَمْ يا أخي[٦] أنّك إذا توهّمتَ حركةَ هذه النقطةِ على سَمتٍ واحدٍ، حدث في فِكرك خطٌّ وهميٌّ مستقيمٌ[٧]. وإذا توهّمتَ حركةَ هذا الخطّ في غير الجهة التي تحرّكت إليها النُّقطةُ، حدث في فِكرك وفي وهمِك[٨] سطحٌ وهميٌّ[٩] مربّعٌ[١٠]. وإذا توهّمتَ حركةَ هذا السطحِ في غير الجهة التي تحرّك إليها الخطُّ والنُّقطةُ[١١]، حدث في فِكرك[١٢] جِسمٌ وهميٌّ له ستّةُ سطوحٍ مربّعاتٍ قائمةِ الزوايا، وهو المُكعَّب.

---

(١) «بمجرّده»: مجرّداً [ع].

(٢) زيادة في [ن]: «مثل».

(٣) «الدهن والماء»: الماء والدهن [Ṣ].

(٤) «بنصفين»: سقطت من [ع]، وردت في [Ṣ]. <وقد أضاف ناسخ مخطوط [ع] في الهامش «بقسمين»>.

(٥) «تنتهي»: سقطت من [ع، ن]، وردت في [Ṣ].

(٦) «يا أخي»: سقطت من [ن]. <زيادة هنا في [ل]: «أيّها الأخ البرّ الرحيم أيّدك الله وإيّانا بروح منه»>.

(٧) «مستقيم»: سقطت من [ع]، وردت في [ن، Ṣ].

(٨) «وفي وهمك»: وفي فكرك [ن]. زيادة في [ع]، سقطت من [Ṣ].

(٩) «وهميّ»: سقطت من [ع]، وردت في [Ṣ].

(١٠) «مربّع»: زيادة في [ع، ن]، سقطت من [Ṣ].

(١١) «والنقطة»: سقطت من [ع]، وردت في [Ṣ].

(١٢) «فكرك»: وهمك [Ṣ].

وإن كانت مسافةُ حركةِ السطحِ أقلَّ من مسافةِ حركةِ الخطّ، حدث من ذلك جِسمٌ لَبِنِيٌّ، وإن كان أكثرَ من ذلك، حدث من ذلك جِسمٌ بئريٌّ[(١)]، وإن كانت متساويةً حدث مُكعَّبٌ[(٢)].

وخطُّ العنكبوتِ هو خطٌّ حِسّيٌّ، ولكنّه عند حُكماءِ الفلاسفة جِسمٌ[(٣)]. وكيف |ع ٢١ و| ما قُطّع أصغر ما يكون، فكلُّ جُزءٍ منه جِسمٌ. وعلى رأي المتكلّمين، إذا صَغُرت أجزاؤه وقُسّمت[(٤)]، فلا يكون كلُّ جُزءٍ منه جِسماً، بل يسمّونه جوهراً[(٥)]. وما دام يقبل التقسيمَ عندهم فهو جِسمٌ، وجوهرٌ إن لم يقبل ذلك. والفلاسفة بضِدِّهم[(٦)] [(٧)].

---

(١) «وإن كان أكثر من ذلك، حدث من ذلك جسم بئريٌّ»: سقطت من [ع]، وردت في [Ṣ].

(٢) < ورد هذا المقطع في [ك] بعد ذكر «رأي المتكلّمين» كما أوردناه في متن النصّ أعلاه > .

(٣) «وخطّ العنكبوت هو خطّ حسيّ، ولكنّه عند حكماء الفلاسفة جسم»: والخطّ المكتوب هو خطّ حسيّ عند حكماء الفلاسفة والجسم عندهم كيف ما قسّم في أجزاءه فإن صغر فهو جسم قابل للتجزية [ط]، والخطّ الحسيّ هو الخطّ المكتوب ولكنّه عند الحكماء والفلاسفة جسم [ك].

(٤) «إذا صَغرت أجزاؤه وقُسّمت»: القسمة تنتهي إلى أجزاء صغار ولا يقبل القسمة من غير الجسم. فكيف ما قطّع أصغر يكون كلّ جزء منه جسم [ط].

(٥) «جوهراً»: جوهراً فرداً [ك].

(٦) «بضدّهم»: يقولون بضدّ ذلك [ط].

(٧) «وخطّ العنكبوت هو خطّ حسيّ، ولكنّه عند حكماء الفلاسفة جسم . . . وما دام يقبل التقسيم عندهم فهو جسم، وجوهر إن لم يقبل ذلك. والفلاسفة بضدّهم»: سقطت بكاملها من [Ṣ]، وحصل لغطٌ في ذكرها في [ف، ل] كما أشرنا من قبل إلى هذا في الحواشي أعلاه حول تصنيف محتويات هذين المخطوطين في هذا السياق من النصّ. < ويبرز في هذا المقطع التباين بين الفلسفة وعلم الكلام في كيفيّة فهم الطبيعة والموجودات والتكلّم حول الجواهر

واعلَمْ يا أخي[(١)] بأنّ كلّ خطٍّ مستقيمٍ مفروضٍ في الوهمِ، فلا بدّ له من نهايتين، وهما رأساه، ويسمّيان النُّقطتين الوهمِيَّتين. وإذا توهّمتَ أنّه تحرّكت إحدى النُّقطتين وسكنتِ الأُخرى حتّى رَجعَت إلى حيثُ ابتدأت بالحركة، حدث في فِكرِك من ذلك سطحٌ مُدوَّرٌ وهميٌّ، وتكون النُّقطةُ الساكنةُ مركزَ الدائرةِ، والنُّقطةُ المتحرّكةُ التي قد حدثت في فِكرِك بحركتها مُحيطَ الدائرة.

ثمّ اعلَمْ بأنَ أوّلَ سطحٍ يحدُثُ من حركتِها ثُلثُ الدائرةِ، ثمّ رُبعُ الدائرةِ، ثمّ نِصفُ الدائرةِ، ثم الدائرةُ نفسُها[(٢)]. وإذا توهّمتَ أنّ الخطّ المقوّسَ الذي هو نِصفُ مُحيطِ الدائرةِ سَكَن رأساه جميعاً، وتحرّكَ الخطُّ نفسُه حتّى يرجع إلى حيث ابتدأ بالحركة، حدث في فكرك من حركتها جِسمٌ كُرِيٌّ.

<فصل ٢٠>

فقد بان لك[(٣)] بما ذكرنا أنّ الهندسةَ العقليّةَ هي النظرُ في الأبعادِ الثلاثةِ، التي هي[(٤)] الطولُ والعرضُ والعُمقُ، خِلواً من الأجسامِ الطبيعية، وذلك أنّ الناظرين في الهندسة الحِسِّيَّةِ التي

والأعراض، حيث تأثّر الفلاسفة بأقوال أرسطوطاليس في هذا الصدد بينما نزع المتكلّمون إلى استلهام نظريّتهم من المّذهب الّذريّ (atomism) عند الإغريق في القول بالجزء الذي لا يقبل القسمة>.

(١) «يا أخي»: سقطت من [ع، ن، ف، ك]، وردت في [S̤].

(٢) «نفسها»: زيادة في [ع، ن، ك]، سقطت من [S̤].

(٣) «لك»: سقطت من [ن].

(٤) «التي هي»: سقطت من [ع]، وردت في [ن، S̤].

تقدّم ذِكرُها إذا ارتاضوا فيها وقَوِيَت أفكارُهم[١] بالنظر فيها[٢]، انتزعوا هذه الأبعادَ[٣] الثلاثةَ عن تلك المقاديرِ الثلاثةِ[٤]، التي هي الخطُّ والسطحُ والجِسمُ، وصُوَرِها في نفوسهم، ونظروا فيها خِلواً من الهَيُولى، فيكون عند ذلك جوهرُ نفوسِهم لتلك الأبعاد المصّورة فيها[٥] كالهَيُولى، وهي فيها كالصورة. ويسمّونها مقاديرَ في الأخرى[٦] مِساحيّةً، يعني الأبعادَ[٧]، ويستغنون عن النظر إلى المقادير الحِسِّيَّة. ثم يتكلّمون عليها [١٠٣] ويخبرون عن أجناسها وأنواعها وخواصِّها، وما يعرِض فيها من المعاني إذا أُضيف بعضُها إلى بعض. فيقولون:

الخطُّ هو مِقدارٌ ذو بُعدٍ واحدٍ، والسطحُ هو مِقدارٌ ذو بُعدَينِ، والجِسمُ هو مِقدارٌ ذو ثلاثةِ أبعادٍ، والخطُّ المستقيمُ هو أقصرُ خطٍّ وَصَلَ بين النُّقطتين، والنُّقطةُ رأسُ الخطِّ، والخطُّ المقوّسُ هو الخطُّ[٨] الذي لا يمكن أن يُفرضَ عليه ثلاثُ نُقط[٩] على سَمتٍ

(١) «أفكارهم»: أفكار نفوسهم [ط].
(٢) «فيها»: سقطت من [ع]، وردت في [Ṣ].
(٣) «الأبعاد»: الأفكار [ن].
(٤) «عن تلك المقادير الثلاثة»: عن الكائنات [ن]. زيادة في [ع]، سقطت من [ن، Ṣ].
(٥) «ونظروا فيها خلواً من الهيولى، فيكون عند ذلك جوهر نفوسهم لتلك الأبعاد المصّورة فيها»: لتلك الأبعاد المصوّرة [Ṣ].
(٦) «في الأخرى»: زيادة في [ع]، سقطت من [ن، ك، Ṣ].
(٧) «يعني الأبعاد»: زيادة في [ع، ن]، سقطت من [Ṣ].
(٨) «الخطّ»: سقطت من [ع، ن]، وردت في [Ṣ].
(٩) «نقط»: فقط [Ṣ].

واحِدٍ، والزوايا تماسّ[١] ما بين[٢] خطّين على |ع ٢١ ظ| غير[٣] استقامة. والشكلُ ما[٤] أحاط به خطٌّ واحِدٌ أو خُطوطٌ، كالمثلَّثِ والمربَّعِ وغيرِهما[٥]. والدائرةُ شكلٌ يحيط به خطٌّ واحِدٌ يقال له المحيطُ[٦]، [٧] وفي داخله نُقطةٌ كلُّ الخطوطِ المستقيمةِ[٨] المُخرَجَةِ منها إليه متساويةٌ. والمثلّثُ شكلٌ يحيط به ثلاثةُ خطوطٍ وله[٩] ثلاثُ زوايا. والمربّعُ شكلٌ يحيط به أربعةُ خطوطٍ وله أربعُ زوايا قائماتٍ[١٠]. وعلى هذا القياس[١١] والمِثال سائرُ ما يتكلمون به في أشكال الهندسة من غير إشارةٍ إلى جسمٍ من الأجسام الطبيعيّة.

---

(١) «تماسّ»: زيادة في [ع]، سقطت من [ṢX].
(٢) «ما بين»: سقطت من [ع، ن]، وردت في [ṢX].
(٣) «غير»: سقطت من [ن].
(٤) «ما»: سقطت من [ن].
(٥) «كالمثلّث والمربّع وغيرهما»: زيادة في [ع]، سقطت من [ṢX].
(٦) «يقال له المحيط»: سقطت من [ع]، وردت في [ṢX].
(٧) «أو خطوط، كالمثلّث والمربّع وغيرهما. والدائرة شكل يحيط به خطّ واحد يقال له المحيط»: سقطت من [ن].
(٨) «المستقيمة»: سقطت من [ع، ن]، وردت في [ṢX].
(٩) «له»: زيادة في [ع]، سقطت من [ṢX].
(١٠) «قائمات»: سقطت من [ع، ن]، وردت في [ṢX].
(١١) «القياس»: سقطت من [ن].

فصل < ٢١ >[١] في حقيقة الأبعاد في الهَندَسَةِ العَقليّة[٢]

إعلَمْ[٣] بأنّ كثيراً من المهندسين والناظرين في العلوم يظنّون أنّ لهذه الأبعادِ الثلاثةِ، أعني الطولَ والعرضَ والعُمقَ، وجوداً بذاتِها وقِوامِها، ولا يدرون أنّ ذلك الوجودَ إنّما هو في جوهرِ الجِسمِ أو في جوهرِ النفس، وهي لها[٤] كالهَيُولى وهي فيها كالصورةِ إذا انتزعتها القوّةُ المفكّرةُ من المحسوسات.

ولو علِموا أنّ الغرضَ الأقصى من النظر في العلوم الرياضيّة إنّما هو أن ترتاض أنفُسُ المتعلّمين بأن يأخذوا صورَ المحسوسات من طريق القوى الحسّاسة وتصوّرها في ذاتها بالقوّة المفكّرة، حتى إذا غابت المحسوساتُ عن مشاهدة الحواسّ لها، بقيت تلك الرسومُ التي أدّتْها القوى الحسّاسةُ إلى القوّةِ المتخيِّلةِ، والمُتخيِّلةُ[٥] إلى القوّة المفكّرة، والمفكّرةُ أدّتْ إلى القوّة

(١) < تغيّر هنا ترتيب محتويات المخطوط [ط] عن سائر المخطوطات الأخرى، حيث ذكرت فيه الجداول الهندسيّة ومحتوياتها العدديّة التي ظهرت في بقيّة المخطوطات في الموضع الذي يوافق الفصل ٢٦ في متن النصّ أدناه، وذلك ورد في سياق وصف ميّزات الأشكال المتّسعة والشكل ذي ستّة عشر بيتاً أو ذو الخمسة والعشرين وما شاكلها في صورها وأوصافها وخصائصها الحسابيّة والرمزيّة > .

(٢) «في حقيقة الأبعاد في الهندسة العقليّة»: سقطت من [ع، ن، ط، ك]، وردت في [S̱].

(٣) زيادة في [ع، ن، ف، ك، ل] :«أيّها الأخ البارّ الرحيم أيّدك الله وإيّانا بروح منه».

(٤) «لها»: فيها [ل]. سقطت من [ع]، وردت في [S̱].

(٥) «إلى القوّة المتخيّلة، والمتخيّلة»: سقطت من متن النص في [ع] ولكن الناسخ أضافها في الهامش، وقد وردت أيضاً في [ن، ك، ل، S̱].

الحافظةِ، مُصوَّرةً[١] في جوهر النفس، فاستغنت عند ذلك النفسُ عن استخدامها [١٠٤] القوى الحسّاسة في إدراك المعلومات عند نظرها إلى ذاتها، ووجدت[٢] صور المعلومات كلِّها في جوهرها، فعند ذلك استغنت عن الجسد، وزهِدت في الكون[٣] معه، وانتبهت من نوم الغفلة، واستيقظت[٤] من رقدة الجهالة، ونهضت بقوّتها واستقلّت بذاتها، وفارقت الأجسامَ وخرجت من بحر الهَيُولى ونجت من أسرِ الطبيعة، وأُعتِقت من عبوديّةِ الشهوات الجسمانيّة، وتخلّصت من حَرقةِ الاشتياقِ إلى اللذّات الجِرمانيّة[٥]، وشاهدت عالمَ الأرواحِ[٦]، وارتقت إلى هناك حيثُ قال: ﴿إِلَيۡهِ يَصۡعَدُ ٱلۡكَلِمُ ٱلطَّيِّبُ وَٱلۡعَمَلُ ٱلصَّٰلِحُ يَرۡفَعُهُۥۚ﴾[٧]؛ أراد به النّفسَ الزكيّة[٨]، وجوزيت بأحسنِ الجزاء، وهذا هو الغرضُ الأقصى من النظر في العلوم الرياضيّة التي كانوا يُخرِّجون بها أولاد الفلاسفة[٩] وتلامذة الحكماء[١٠]. هكذا مذهبُ إخوانِنا

---

(١) «مُصوَّرة»: مقصورة [S].

(٢) «ووجدت»: ووجدانها [ع].

(٣) «الكون»: السكون [S].

(٤) «استيقظت»: سقطت من [ع]، وردت في [S].

(٥) < «الجرمانية»: أي الجسمانية >.

(٦) زيادة في [ل]: «الروحانيّ».

(٧) < القرآن: فاطر ١٠ >.

(٨) «حيث قال: ﴿إليه يصعد الكلم الطيّب، والعمل الصالح يرفعه﴾؛ أراد به النفس الزكيّة»: سقطت من [ع، ن، ف، ك، ل]، وردت في [S].

(٩) «الفلاسفة»: الحكماء [S].

(١٠) «الحكماء»: القدماء [S].

الأبرارِ[١] الكرامِ الفُضلاء[٢]،[٣] وفّقك الله وإيّانا سبيلَ الرَّشادِ إنّه رؤوف بالعباد[٤]. وإليه يدعو إخوانُنا الأبرارُ الرُّحماءُ[٥]، أيّدهم الله وإيّانا بروح منه حيث كانوا في البلاد. إنّه كريمٌ جوّاد[٦] [٧].

* * * * * *

فصل < ٢٢ > في خواصّ الأشكال الهندسيّة[٨]

إعلَمْ يا أخي، أيّدك الله وإيّانا بروح منه، بأنّ للأشكال الهندسيّة خواصَّ، كما أنّ للأعدادِ خواصَّ[٩]، ولمجموعِها

(١) «الأبرار»: زيادة في [ع]، سقطت من [Ṣ].

(٢) «الفضلاء»: زيادة في [ع]، سقطت من [Ṣ].

(٣) «الكرام الفضلاء»: الرحماء [ن].

(٤) «وفّقك الله وإيّانا سبيل الرشاد إنه رؤوف بالعباد»: سقطت من [ع، ن، ف، ك]، وردت في [Ṣ].

(٥) «وإليه يدعوا إخواننا الأبرار الرحماء»: سقطت من [ن].

(٦) «إنه كريم جوّاد»: سقطت من [ن، ف، ك، ل].

(٧) «وإليه يدعو إخواننا الأبرار الرحماء، أيّدهم الله وإيّانا بروح منه حيث كانوا في البلاد. إنه كريم جوّاد»: زيادة في [ع]، سقطت من [Ṣ]. <أضاف ناسخ مخطوط [ع] هنا: «تمّت الرسالة»، وذلك من دون أن تكون قد تمّت فعلاً في المخطوط عينه، أو في المخطوطات الأخرى، ناهيك أيضاً عن أن هذه الرسالة لم تنجز عند ذلك الحدّ المشار إليه هنا تبَعاً لما ورد في طبعة دار صادر [Ṣ]. وقد ختم مخطوط [ن] هنا من دون أن يكون له بقيّة في النصّ، والختام ورد على هذا الشكل عنده: «ولله الحمد تمّت الرسالة الثانية بحمده». وليس هنالك أيّة إشارة في [ط، ل] أن هذه الرسالة قد تمّت. ولكن مجمل الفصل ٢١ أعلاه قد سقط من [ط]> .

(٨) «في خواص الأشكال الهندسيّة»: سقطت من [ع، ف، ك، ل]، وردت في [Ṣ].

(٩) «كما أن للأعداد خواص»: زيادة في [ع]، سقطت من [ك، Ṣ].

خواصَّ أيضاً، كما أنّ لمفرداتها خواصَّ أيضاً[1]. وقد بيّنا في رسالة الأرِثماطيقي طرفاً من خواصِّ العددِ، فنريد أن نذكرَ في هذا الفصل طرفاً من خواصِّ الأشكالِ الهندسيّةِ، ليكونَ تنبيهاً للناظرين في هٰذين العِلمَين على الغرض منهما[2]، |ع ٢٢ و| ويكونَ أيضاً إرشاداً لطالبي خواصِّ الأشياء وكيفيّةِ المسلكِ[3] فيها.

ونبدأ أوّلاً[4] بذكرِ المثلّثاتِ، إذ كانت هي أوّل الأشكالِ الهندسيّةِ[5]. فنقول:

[١٠٥] إنّ الشكلَ المثلّثَ هو الذي له ثلاثةُ أضلاع وثلاثُ زوايا؛ وهو سبعةُ أنواع: أوّلُها المتساوي الأضلاعِ الحادُّ الزوايا مِثلُ هذا[6]:

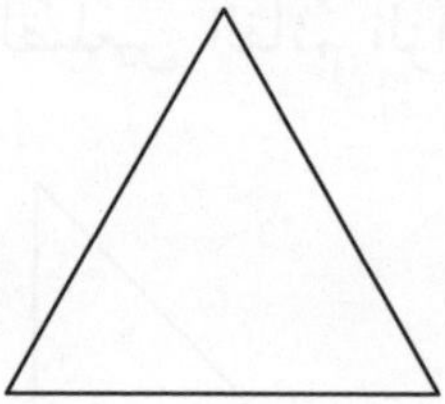

(١) «كما أن لمفرداتها خواص أيضاً»: زيادة في [ع، ل]، سقطت من [ك، ṣ].
(٢) «على الغرض منهما»: سقطت من [ع]، وردت في [ṣ].
(٣) «المسلك»: المسالك [ع].
(٤) «أولاً»: الآن [ع].
(٥) <أضاف هنا ناسخ مخطوطات [ع، ك] العبارة التالية: «كما بيّنا في رسالة جومَطْريا»، كما لو أن الرسالة قد تمّت فعلاً قبل الابتداء بالفصل ٢٢، وكما أشرنا إلى ذلك في الحاشية أعلاه عندما فرغنا من الفصل ٢١. وقد وردت الإشارة إلى نفس العبارة حول رسالة الجومَطْريا في طبعة دار صادر [ṣ]>.
(٦) «الحادّ الزوايا مثل هذا»: سقطت من [ع]، وردت في [ṣ].

والثاني الحادُّ الزوايا المتساوي الضِّلعَين مِثلُ هذا[١]:

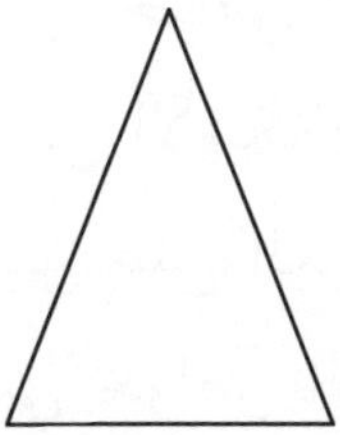

والثالثُ الحادُّ الزوايا[٢] المختلفُ الأضلاعِ كهذا[٣]:

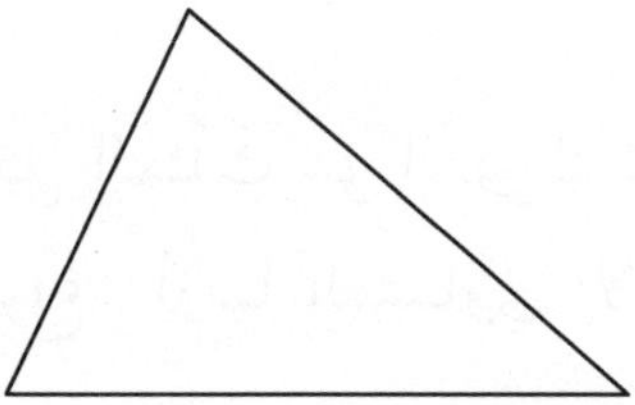

والرابعُ المتساوي الضِّلعَين القائمُ الزاويةِ مِثلُ هذا:

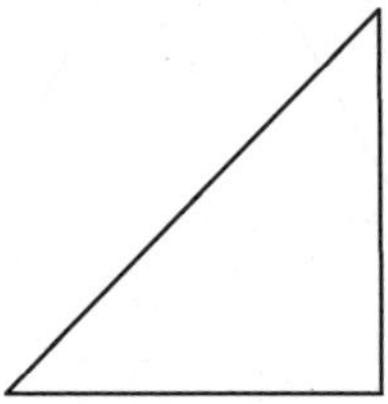

والخامسُ القائمُ الزاويةِ المختلفُ الأضلاعِ مِثلُ هذا:

(١) «مثل هذا»: سقطت من [ع]، وردت في [ل، S]. <وينطبق هذا على بقيّة الفصل>.

(٢) «الحادّ الزوايا»: سقطت من [ع]، وردت في [ل، S].

(٣) «كهذا»: سقطت من [ع]، وردت في [ل، S].

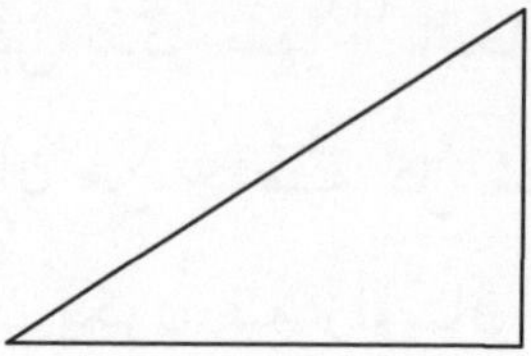

والسادسُ المنفرجُ الزاويةِ المتساوي الضِّلعَين هكذا[١] :

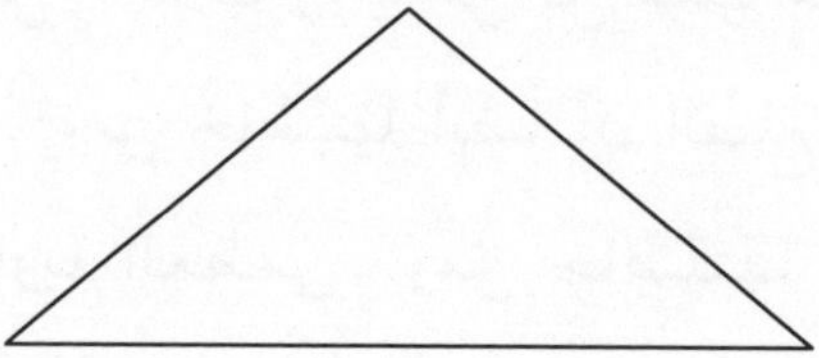

[١٠٦] والسابعُ المنفرجُ الزاويةِ المختلفُ الأضلاعِ مِثلُ هذا :

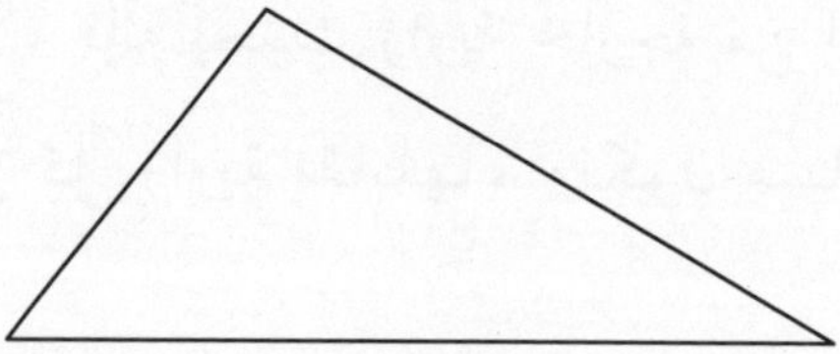

فصل < ٢٣ > في بيان تلك الخواصّ[٢]

واعلَمْ يا أخي[٣] بأنّ لِكلِّ واحدٍ من هذه المثلّثاتِ خاصِّيَّةً ليست للآخَرِ، فقد تبيّنَ ذلك في كتاب أوقليدِس[٤] في المقالة

---

(١) «هكذا»: مثل هذا [ك، ل]. سقطت من [ع]، وردت في [S].

(٢) «في بيان تلك الخواص»: سقطت من [ع، ف، ك، ل]، وردت في [S].

(٣) زيادة في [ف، ل]: «أيّدك الله وإيّانا بروح منه».

(٤) <أي كتاب أوقليدِس في الأصول (أو في الأركان، أو الأُسطُقسّات: Elements, Stoikheia)، وهو يشتمل على ثلاثة عشر مقالة في العلوم الهندسيّة الرياضيّة> .

الأولى ببراهينها، ولكن نذكر منها[1] الخاصِّيَّةَ التي تشتمل على سَبعتِها كلِّها. وذلك أنّ من خاصِّيَّةِ كلِّ شكلٍ مثلّثٍ، أيَّ مثلّثٍ كان، أنّه لا بدّ من أن يكونَ فيه زاويتان حادّتان، فأمّا الزاويةُ الثالثةُ فيمكن أن تكونَ حادّةً أو قائمةً أو منفرجةً. ومن خاصِّيَّتها أيضاً أن تكون كلُّ[2] ثلاثِ زوايا من كلِّ مثلّثٍ مجموعُها مساوٍ[3] لزاويتين قائمتين. ومن خاصِّيَّتها أيضاً أن الضِّلعَ الأطولَ من كلِّ مثلّثٍ بوتَرِ الزاوية العظمى. ومن خاصِّيَّتها أنّ كلَّ ضِلعَين مجموعَين من كلِّ مثلّثٍ[4] أطولُ من الضِّلعِ الثالث. ومن خاصِّيَّتها أيضاً أنّه إذا أُخرج ضِلعٌ من أضلاعِه، أيَّ ضِلعٍ كان، على استقامته[5]، فإنّه يُحدِثُ زاويةً خارجةً من المثلّث، وتكون هي[6] أكبرَ من كلِّ زاويةٍ تقابلها، ويكون مساوياً للداخِلَتين المقابلتَين لها.

ومن خاصِّيَّتها أيضاً[7] أنّ ضربَ مَسقَطِ الحَجَرِ من كلِّ مثلّثٍ في نِصفِ قاعدتِها هو مساحةُ ذلك المثلّث.

---

(١) «منها»: سقطت من [ع]، وردت في [Ṣ].
(٢) «تكون كلّ»: زيادة في [ع]، سقطت من [Ṣ].
(٣) «مجموعها مساوٍ»: مجموعات مساويات [ع].
(٤) «من كلّ مثلّث»: سقطت من [ع]، وردت في [Ṣ].
(٥) «أنه إذا أخرج ضلع من أضلاعه، أيّ ضلع كان، على استقامته»: أن كلّ ضلع من أضلاعه إذا أخرج على استقامة [ع].
(٦) «هي»: سقطت من [ع]، وردت في [Ṣ].
(٧) «أيضاً»: سقطت من [ع]، وردت في [Ṣ].

< فصل ٢٤ >

وأمّا خاصِّيَّةُ المثلّثِ القائمِ الزاويةِ فهي[١] أنّ مربّعَ وترِ الزاويةِ القائمةِ مساوٍ للمربّعَين الكائنَين من الضِّلعَين.

ومن خاصِّيَّةِ المثلّثِ الحادِّ الزاويةِ أنّ مربّعَ الوتر أقلُّ من مربّعِ[٢] الضِّلعَين [١٠٧] الباقيَين بمقدار مربّعِ الضِّلعِ الذي وقع عليه العُمودُ فيما بين مَسْقَط العُمودِ والزاويةِ مرّتين.

ومن خاصِّيَّةِ المثلّثِ المنفرجِ الزاويةِ أنّ مربّعَ الوترِ أكبرُ[٣] من مربّعِ[٤] الضِّلعَين بمقدار مربّعِ أحدِ الضِّلعَين فيما هو خارجٌ منه إلى مَسْقَطِ العُمودِ مرّتين مِثلُ هذا[٥]:

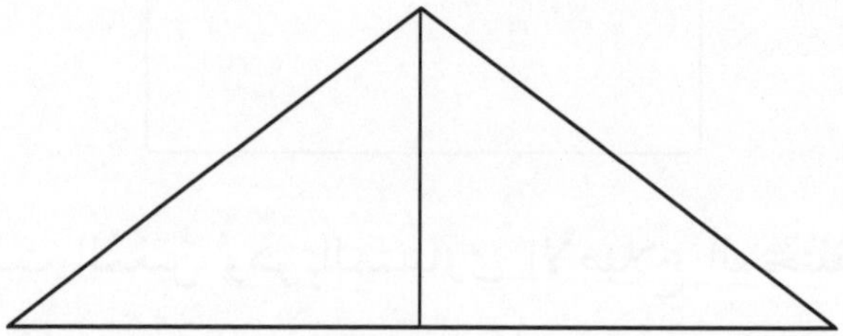

< فصل ٢٥ >[٦]

وأمّا الشكلُ المربّعُ فهو الذي له أربعةُ أضلاعٍ وأربعُ زوايا؛ وهو خمسةُ أنواع:

---

(١) «فهي»: سقطت من [ع]، وردت في [S].

(٢) «مربّع»: مربّعيّ [ع].

(٣) «أكبر»: أكثر [S].

(٤) «مربّع»: مربّعيّ [ع].

(٥) «مثل هذا»: سقطت من [ع]، وردت في [S].

(٦) <لم تذكر مواصفات الأشكال الواردة في هذا الفصل ضمن متن النصّ في مخطوط [ف]، وقد وردت بدلاً عن ذلك في الهامش>.

أوّلُها المتساوي الأضلاعِ القائمُ الزوايا مثل هذا[1]:

والثاني المستطيلُ القائمُ الزوايا، المتساوي |ع ٢٢ ظ| كلِّ ضِلعَين متقابلَين مِثلُ هذا:

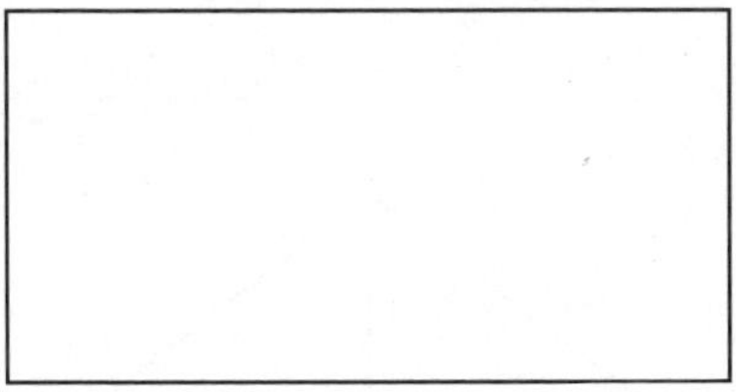

والثالث المُعَيَّنُ وهو المتساوي الأضلاعِ المختلفُ الزوايا مِثلُ هذا:

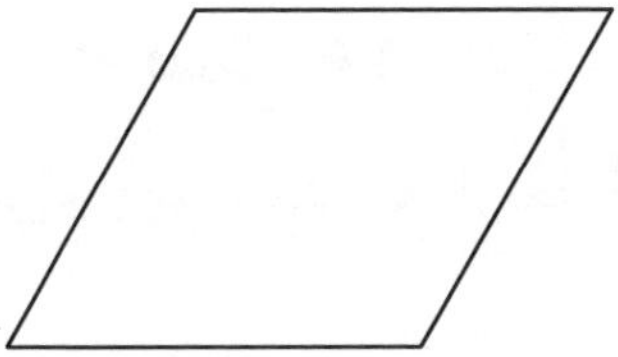

والرابع الشَبيهُ بالمُعَيَّن وهو المتساوي كلِّ ضِلعَين متقابلَين منه[2] مِثلُ هذا:

---

(١) «مثل هذا»: سقطت من [ع]، وردت في [S]. < وينطبق هذا على بقيّة الفصل >.

(٢) «منه»: زيادة في [ع]، سقطت من [S].

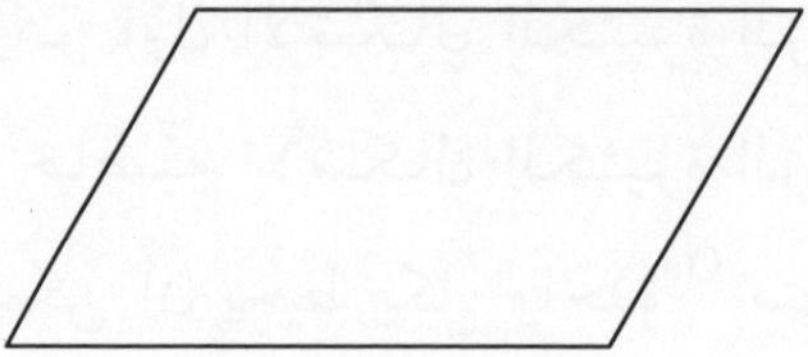

[١٠٨] والخامسُ المختلفُ الأضلاعِ والزوايا، ويُسمّى المُنجَرِفُ[١]، مِثلُ هذا:

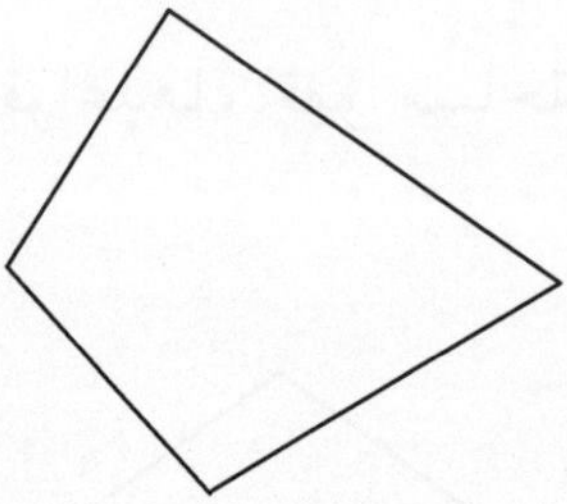

واعلَمْ يا أخي بأنّ لكلِّ واحدٍ من هذه الأشكال خواصٌّ يطول شرحُها، ولكن نذكر الخاصِّيَّةَ التي تشمَلُها كلَّها وهي أنّ كلَّ مربّعٍ، أيَّ مربّعٍ كان، فإنّ زواياه الأربعَ مجموعةً[٢] تكون مساويةً[٣] لأربعِ زوايا قائمةٍ. وأنّ كلَّ مربّعٍ يمكن أن ينقسم بمثلّثين، وإن زيد عليه مثلّثٌ آخَرُ صار منها شكلاً مخمّساً[٤].

وأمّا الشكلُ المخمّسُ فهو الذي يُحيط به خمسةُ أضلاعٍ، وله

(١) «ويسمّى المُنجَرِف»: زيادة في [ع]، سقطت من [S]. <لفظة «المُنجَرِف» تدلل على انجراف هذا الشكل الهندسيّ، وتكاد أن تكون مطموسة كما ظهرت في مخطوط [ع]، وهي تتقبل التشكيل الذي يمكن قراءته كذلك على أن الكلمة المخطوطة هنا هي «المُنحَرِف» للدلالة على انحراف هذا الشكل الهندسيّ. ولقد فضّلنا هنا التسمية «المُنجَرِف» كونها أكثر استخداماً في هذه الصناعة>.

(٢) «فإن زواياه الأربع مجموعة»: أربع زواياه مجموعاً [ل].

(٣) «مجموعة تكون مساوية»: مجموعات تكون مساويات [ع].

(٤) «شكلاً مخمّساً»: شكل مجسّم [ع]، شكلاً مثلّثاً [ك].

خمسُ زوايا، وهو أوّلُ الأشكالِ الكثيرةِ الزوايا المتساوي الأضلاع. ومن خاصِّيَّةِ الأشكالِ الكثيرةِ الزوايا المتساويةِ الأضلاعِ[١] أنّه يمكن أن يحيط بكلِّ واحدةٍ[٢] منها دائرةٌ، ويمكن أن تحيط هي[٣] أيضاً بدائرة. وإنّ كلَّ شكلٍ منها الذي هو أكثرُ زوايا، فهو أكثرُ وأوسعُ مِساحةً من الذي هو أقلُّ منه، إذا كان المحيطُ بها مِقداراً واحِداً، وإنّ ضربَ عمودٍ واحِدٍ من تلك المثلّثات في نِصفِ قواعدِها، فهو مساحةُ ذلك الشكلِ الكثيرِ الزوايا؛ وهذا مِثالُه[٤]:

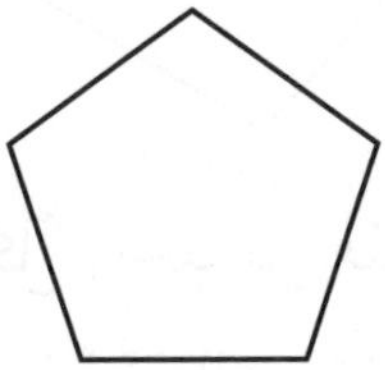

ومن خاصِّيَّةِ الشكلِ المسدّسِ المتساوي الأضلاعِ أنّ كلَّ ضِلعٍ من أضلاعِه مساوٍ لنِصفِ قُطرِ الدائرةِ التي تُحيط به. وبالجُملةِ ما من شكلٍ إلّا وله خاصِّيَّةٌ أو عِدّةُ خواصّ تركنا ذِكرَها مخافةَ التطويل.

فأمّا خَواصُّ الشكلِ المستديرِ فقد أفردها أوقليدِس في المقالة

---

(١) «ومن خاصّيّة الأشكال الكثيرة الزوايا المتساوية الأضلاع»: زيادة في [ع]، سقطت من [Ṣ].

(٢) < صححناها إلى «واحدة» بدلاً من «واحد» > .

(٣) < صححناها من «يحيط هو» إلى «تحيط هي» > .

(٤) «وهذا مثاله»: مثال ذلك [ل]، سقطت من [ف، ك].

الثالثة من كتابه في الأُسطُقسّات[١]، ولكن نذكر منها طرفاً، فنقول:

[٢]إنّ الشكلَ[٣] المستديرَ هو سطحٌ يُحيط به خطٌّ واحِدٌ، وأنّ مركزَه في وسطِه، وأنّ أقطارَه كلَّها متساويةٌ، وأنّه أوسعُ من[٤] كلِّ شكلٍ كثيرِ الزوايا إذا كان الذي يحيط به سطحاً[٥] واحداً.[٦] وهو يشارك الدائرةَ في خواصِّها، ونِسبتُه من سائرِ الأجسامِ كنسبةِ الدائرةِ من سائرِ السطوح. وقد تبيّنَ [١٠٩] خواصُّ هذا الشكل في المقالة الأخيرة من كتاب أوقليدِس بشرحٍ وبراهينَ.

فأقول[٧]: وبالجُملةِ إنّك لو تأمّلتَ يا أخي غرضَ ما في[٨] أوقليدِس من البيان وعُلِمَ ما في سائرِ كتُبِ الهندسةِ، لوجدتَ < فيها >[٩] كلِّها أيضاً[١٠] البحثَ عن خواصِّ المقاديرِ ومعرفةِ

---

(١) «أفردها أوقليدِس في المقالة الثالثة من كتابه في الأُسطُقسّات»: أفرِد لها في المقالة الثالثة من كتاب أوقليدِس [ل]، أفردها أوقليدِس في مقالة من كتابه [ك، وفي هامش ل]، أفرَد لها أوقليدِس مقالةً من كتابه [Ṣ].

(٢) زيادة في [ك]: «فنقول».

(٣) زيادة في [ل]: «المسدّس». < وهذا غير صحيح هندسياً >.

(٤) «أنه أوسع من»: سقطت من [ف].

(٥) «سطحاً»: مقدار [ع].

(٦) زيادة في [ف]: "إن كلّ الأشكال موجودة فيه بالقوة. < ثمّ أضاف ناسخ [ل] في هذا السياق: «وأما الشكل الكريّ فهو جسم يحيط به سطح واحد وهو يشارك الدايرة في خواصها ونسبته من ساير الأجسام كنسبة الدايرة من ساير السطوح» >.

(٧) «فأقول»: زيادة في [ع]، سقطت من [ل، Ṣ].

(٨) «ما في»: زيادة في [ع، ل]، سقطت من [Ṣ].

(٩) < «فيها»: إضافة هنا من المحقق >.

(١٠) «أيضاً»: زيادة في [ع]، سقطت من [Ṣ]. < ورد بعد ذلك في [Ṣ] العبارة التالية: «إنما هو». وقد أسقطناها من النصّ للحفاظ على سلاسة القراءة >.

حقائِقها، التي هي الخطوطُ والسطوحُ والأجسامُ، وما يعرِضُ فيها من الأبعاد والزوايا والمناسبات التي بين بعضِها وبعض.[١]

فصل <٢٦>

وإذ قد بيّنا طرفاً من خواصِّ الأشكالِ في هذه الرسالة، وقبلها طرفاً من خواصِّ العدد في رسالة الأرِثماطيقي، فنريد أن نذكرَ |ع ٢٣ و| طرفاً من خواصِّ مجموعهما، وذلك أنّه إذا جُمِعَ بين بعض الأعداد وبين بعض الأشكال الهندسيّة ظهر منها خواصُّ أُخَرُ لا يتبيَّن في كلٍّ واحِدٍ منهما بمجرّده. مِثالُ ذلك إذا كُتِب التسعةُ الأعداد[٢] في الشكلِ المُتَسَّعِ على هذه الصورة فإنّ خاصِّيَّتهِ في الشكلِ المتّسعِ أنّه كيفما عُدَّ كانت الجُملةُ خمسة عشرَ، مِثلَ هذا[٣]:

٢	٧	٦
٩	٥	١
٤	٣	٨

وهكذا الستّةَ عشرَ إذا كُتِب في الشكل ذي الستّةَ عَشَرَ بيتاً[٤]

(١) <المقاطع التي وردت في الفصل ٢٦ أدناه ظهرت في مخطوط [ط] بعد الفصل ٢١>.

(٢) «الأعداد»: الآحاد [ع، ف، ل].

(٣) «مثل هذا»: سقطت من [ع]، وردت في [ل، S]. <وينطبق هذا على بقيّة الفصل>.

(٤) «بيتاً»: سقطت من [ع]، وردت في [ل، S].

على هذه الصورة فإنّ من خاصِّيَّتِهِ أنّه كيفما عُدَّ كانت الجُملةُ أربعةً وثلاثين[١]، مِثلَ هذا:

١	١٥	١٤	٤
١٢	٦	٧	٩
٨	١٠	١١	٥
١٣	٣	٢	١٦

[١١٠] وهكذا الخمسةُ والعشرون إذا كُتِبَ في الشكل ذي الخمسةِ والعشرين بيتاً على هذه الصورة[٢] فإنّ من خاصِّيَّتِهِ أنّه كيفما عُدَّ كانت الجُملةُ خمسةً وستّين، مِثلَ هذا:[٣]

٢٥	١٢	٤	٣	٢١
٨	١٩	٦	١٧	١٥
١٦	٢	١٣	٢٤	١٠
١١	٩	٢٠	٧	١٨
٥	٢٣	٢٢	١٤	١

(١) < أضاف ناسخ مخطوط [ع] في متن النصّ: «يُعتمد في صورة الجدّول على ما في الورقة». ولعل ذلك خطّ من قبله عن سهوٍ أو عدم دراية، كما لو أنّ ذلك التنبيه كان المراد منه إرشاد الرسّام في تصويره للجدول المشار إليه في ورقة التعليمات >.

(٢) «إذا كتب في الشكل ذي الخمسة والعشرين بيتاً على هذه الصورة»: سقطت من [ع]، وردت في [S].

(٣) < كلّ ما ذكر في متن النصّ هنا وأدناه، من نهاية المقطع الذي يصف الشكل ذي الخمسة والعشرين بيتاً حتى نهاية ذكر الشكل ذي الأحد والثمانين بيتاً،

وهكذا الستّةُ والثلاثون إذا كُتِب في الشكل ذي الستّةِ والثلاثين بيتاً على هذه الصورة فإنّ من خاصِّيَّتِهِ أنّه كيفما عُدَّ كانت الجُملةُ مِئةً وأحدَ عشرَ، مِثلَ هذا:

١٨	٢٣	٥	٣٢	٢٢	١١
٢٠	١٣	٣٠	٧	١٦	٢٥
٣	٤	٣٦	٣٥	٦	٢٧
٣٤	٣٣	٢	١	٣١	١٠
١٥	٢٦	٢٩	٨	١٩	١٤
٢١	١٢	٩	٢٨	١٧	٢٤

وهكذا التسعةُ والأربعون إذا كُتِب في الشكل ذي التسعةِ والأربعين [١١١] بيتاً على هذه الصورة فإنّ من خاصِّيَّتِهِ أنّه كيفما عُدَّ كانت الجُملةُ مئةً وخمسةً وسبعين، مِثلَ هذا:

٤٩	٤٥	٦	٩	٨	١١	٤٧
٤٦	٣٥	١٦	١٧	٢٠	٣٧	٤
٤٨	٣٢	٢٨	٢١	٢٦	١٨	٢
٧	٣١	٢٣	٢٥	٢٧	١٩	٤٣
١٢	١٤	٢٤	٢٩	٢٢	٣٦	٣٨
١٠	١٣	٣٤	٣٣	٣٠	١٥	٤٠
٣	٥	٤٤	٤١	٤٢	٣٩	١

سقطت جميعها من مخطوطات [ط، ف] ومن متن المخطوط [ع]؛ ولكن هذه الأشكال وردت في هوامش [ع] بأحجام متباينة >.

وهكذا الأربعةُ والسِّتون إذا كُتِب في الشكلِ ذي الأربعةِ
والسِّتين بيتاً على هذه الصورةِ فإنَّ من خاصِّيَّتِهِ أنَّه كيفما عُدَّ كانت
الجُملةُ مئتين وستِّين، وهذه صورتُها:

٤٥	٣٦	٢٩	٢٠	١٣	٤	٦١	٥٢
١٩	٣٠	٣٥	٤٦	٥١	٦٢	٣	١٤
٤٤	٣٧	٢٨	٢١	١٢	٥	٦٠	٥٣
٢٢	٢٧	٣٨	٤٣	٥٤	٥٩	٦	١١
٤٢	٣٩	٢٦	٢٣	١٠	٧	٥٨	٥٥
٢٤	٢٥	٤٠	٤١	٥٦	٥٧	٨	٩
٤٧	٣٤	٣١	١٨	١٥	٢	٦٣	٥٠
١٧	٣٢	٣٣	٤٨	٤٩	٦٤	١	١٦

وهكذا الأَحَدُ والثمانون إذا كُتِب في الشكلِ ذي الأَحَدِ
والثمانين [١١٢] بيتاً على هذه الصورةِ فإنَّ من خاصِّيَّتِهِ أنَّه كيفما
عُدَّ كانت الجُملةُ ثلاثمئةً وتسعةً وستِّين، وهذه صورتُها[١]:[٢].

---

(١) >عرض هذا الشكل الأخير ومحتوياته العدديّة على بعض أساتذة الرياضيّات
ولم يستطيعوا تصويبه<.

(٢) >كما ذكرنا أعلاه فإن كلّ المقاطع التي وردت منذ وصف الشكل ذي
الخمسة والعشرين بيتاً حتى نهاية وصف الشكل ذي الأحد والثمانين بيتاً،
سقطت جميعها من مخطوطات [ط، ف] ومن متن المخطوط [ع]؛ ولكن هذه
الأشكال وردت في هوامش [ع] بأحجام متباينة<.

٧٨	٦٥	٦٤	٢٧	١	١٨	١٩	١٧	٨٠
٢٥	٥	٤٧	٤٩	٦٨	٣٩	٤٠	٧٤	٢٢
٤٦	٤٥	٦	٥٠	١٥	٤٤	٧٣	٣٣	٥٧
٣٤	٤٣	٤٨	٧	١٦	٧٢	٣٧	٥٢	٦٠
٦٩	٥٦	٧١	٧٢	٣١	٤١	١٤	١٢	٣
٢٩	٤٢	٣١	١١	٦٦	٧٩	٣٤	٥١	٢٦
٣٢	٣٠	٩	٣٦	٦٧	٢٤	٧٧	٣٥	٥٩
٥٤	٨	٢٣	٥٧	١٣	٢٨	٥٣	٧٥	٥٨
٢	٦١	٦٢	٦٣	٨١	٥٥	٢٠	٢١	٤

وعلى هذا المثالِ سائرُ الأعدادِ والأشكال، إذا جَمَعْتَ بينها ظهرت منها خواصٌّ أُخَرُ(١).

وأمّا منافِعُها والفائدةُ منها، فقد ذكرنا ذلك(٢) في رسالة الطِّلَّسْماتِ والعَزائم(٣)، وأوردنا(٤) طرفاً منها(٥)، ولكن نذكر منها في هذا الفصل مثالاً واحداً ليكون دلالةً على صِدْق ما قلنا.

(١) «وعلى هذا المثال سائر الأعداد والأشكال، إذا جَمَعْتَ بينها ظهرت منها خواص أُخَر»: زيادة في [ع]، سقطت من [Ṣ].

(٢) «ذلك»: زيادة في [ع]، سقطت من [ف، Ṣ].

(٣) < «رسالة الطِّلَّسْمات والعَزائم» هي الرسالة الأخيرة من رسائل إخوان الصفاء، وهي اثنتان وخمسون رسالة. وكما أفاد محقق [Ṣ] فإن لفظ «العَزائم» الذي وضع في العنوان قد عني بها هنا «الرُّقى»، أي قراءة آيات من القرآن الكريم على ذوي الآفات لرجاء الشفاء أو لحلول بعض من السكينة عليهم >.

(٤) «وأوردنا»: زيادة في [ع]، سقطت من [ف، Ṣ].

(٥) «منها»: سقطت من [ع]، وردت في [Ṣ].

فنقول: إنّ من خاصِّيَّةِ هذا الشكلِ المُتَسَّعِ ومنفعتِه تسهيلَ الوِلادةِ إذا كُتب على خزَفَتين لم يُصبْهما الماءُ وعلَّقَتهما على المرأة التي ضربها الطَّلق، وإن اتّفق أن يكونَ القمرُ في التاسِع متّصِلاً بربِّ التاسِعِ سهّلَ الوِلادةَ، أو بِربّ بيته من التاسع وما شاكل ذلك من المُتَسَّعات[1]:

ح	ج	د
ا	ه	ط
و	ز	ب

[١١٣] وعلى هذا الطريق سلك أصحابُ الطِّلَّسْمات في نَصْبِها[2]، وذلك[3] أنّه ما من شيء من الموجودات الرياضيّة والطبيعيّة والإلٰهيّة إلّا وله خاصِّيَّةٌ ليست لشيءٍ آخرَ من الموجودات[4]، ولمجموعاتها خواصٌّ ليست لمفرداتها من الأعدادِ والأشكالِ والصور والمكان والزمان والعقاقير والطُّعوم والألوان والروائح[5] والأصوات والكلمات والحروف والأفعال[6] والحركاتِ؛ فإذا جمعتَ بينها على النِّسَبِ[7]

---

(١) «فنقول: إن من خاصّيّة هذا الشكل المتّسع ومنفعته تسهيل الولادة إذا كتب على خزفتين لم يصبهما الماء وعلقتهما على المرأة التي ضربها الطَّلق... سهل الولادة، أو بربّ بيته من التاسع وما شاكل ذلك من المتسّعات»: سقطت من [ل].

(٢) «وعلى هذا الطريق سلك أصحاب الطِّلَّسْمات في نصبها»: سقطت من [ل].

(٣) «وذلك»: إعلم يا أخي [ل].

(٤) «من الموجودات»: زيادة في [ع]، سقطت من [S].

(٥) «الروائح»: الأرايح [ع].

(٦) «والحروف والأفعال»: والأفعال والحركات [ل]، والأفعال والحروف [S].

(٧) «النّسب»: النّسبة [ع].

التأليفيّةِ[١]، ظهرت خواصُّها وأفعالُها. والدليلُ على صحّةِ ما قلنا أفعالُ التِّرياقاتِ والمراهِمِ والشَّرِباتِ، وألحانُ الموسيقى[٢]، وتأثيراتُها في الأجسادِ والنفوسِ جميعاً ممّا لا خفاء به عن كلِّ ذي لُبٍّ حكيمٍ فيلسوفٍ؛ كما بيّنا طرفاً من ذلك في رسالة الموسيقى.[٣]

فصل < ٢٧ > في ثَمَرَةِ هذا الفن[٤]

واعلَمْ يا أخي[٥] بأنّ النظرَ في عِلم الهندسة الحِسّيةِ يُعينُ على الحِذقِ في الصَّنائع، والنظرُ في الهندسةِ العقليّةِ، ومعرفةُ خواصِّ العدد والأشكال، يُعينُ على فَهمِ كيفيّة[٦] تأثيرات |ع ٢٣ ظ| الأشخاص الفلكيّةِ في الأشخاصِ السِّفْلِيّة الطبيعيّة وعلى فَهمِ تأثيراتِ كيفيّةِ[٧] أصوات الموسيقى[٨] في نفوس المستمعين،

---

(١) «النّسب التأليفيّة»: نسبة واحد وثلاثة [ف]. < والمراد من القول نسبة «واحد وثلاثة» غير واضح >.

(٢) «الموسيقى»: الموسيقيّات [ع].

(٣) < كما ذكرنا أعلاه، فإن محتويات الفصل ٢٦ وردت كلّها في مخطوط [ط] بترتيب مخالف لما ذكر في المخطوطات الأخرى وفي طبعة دار صادر، وذلك من خلال ظهورها في موضع يتوافق مع بداية الفصل ٢١ في سياق النصّ أعلاه >.

(٤) «في ثَمَرَةِ هذا الفن»: سقطت من [ع، ف، ل]، وردت في [Ṣ].

(٥) «يا أخي»: زيادة في [ع]، سقطت من [ل، Ṣ].

(٦) «كيفيّة»: سقطت من [ع]، وردت في [ل، Ṣ].

(٧) «الأشخاص السّفليّة الطبيعيّة وعلى فهم تأثيرات كيفيّة»: زيادة في [ع]، سقطت من [ل، Ṣ].

(٨) «الموسيقى»: الموسيقار [ل].

والنظرُ في كيفيّاتِ تأثيرات هذين الجنسين[١] في مُنفَعِلاتها يُعينُ على فَهمِ كيفيّة تأثيرات النفوس المُفارِقة[٢] في النفوس المُتَجَسِّدَة في عالم الكون والفساد.[٣]

وفي عِلمِ[٤] الهندسةِ العقّليّةِ للناظرين طريقٌ إلى الوصول إلى معرفتها بعون الله وهدايته.[٥]

تمّت رسالة الجومَطْريا، الرسالة الثانية من القسم الأوّل؛ والحمد لله وحده، والصلوٰة على محمّد وآله بعده[٦]؛ ويتلوها رسالة في مدخل عِلم النجوم، وهي الثالثة من القسم الأوّل من الأربعة الأقسام[٧] [٨].

---

(١) «هذين الجنسين»: الحسّ [Ṣ].

(٢) <كما أفاد محقق [Ṣ] «النفوس المُفارِقة» هي بخلاف النفوس المُتَجَسِّدَة> .

(٣) < خُتِم هنا مخطوط [ف] بالقول: «تمّت الرسالة الثانية في الهندسة وتتلوها الرسالة الثالثة في النجوم المسمّاة أسطرُنوميا». ولكنه للغط ما أو سهواً نراه يعاود ذكر كيّفيّة تكوّن الأشكال من النّقط حسب ما يحصل عند حاسة البصر وتبعاً بذلك لما ورد أعلاه في الفصل ١٠ من نصّ هذه الرسالة> .

(٤) «علم»: سقطت من [ع]، وردت في [Ṣ].

(٥) <خُتِم هنا المخطوط [ط] بالقول: «تمّت الرسالة الثانية في الهندسة من رسايل إخوان السعادة صان الله أقدارهم والحمد لله ربّ العالمين وصلواته على رسوله محمّد وآله أجمعين وحسبنا الله ونعم الوكيل». ولكنه للغط ما نراه يعاود ذكر الأشكال الهندسيّة وخواصها في مقطع يطاول العشرة أسطر ومن ثمّ يختم الرسالة مجدداً بتكراره طرفًا من خاتمته الأولى> .

(٦) «الرسالة الثانية من القسم الأوّل، والحمد لله وحده والصلوٰة على محمّد وآله بعده»: زيادة في [ع]، سقطت من [Ṣ].

(٧) «ويتلوها رسالة في مدخل علم النجوم، وهي الثالثة من القسم الأوّل من الأربعة الأقسام»: سقطت من [ع]، وردت في [Ṣ]. <والأصوب هنا القول: «من الأقسام الأربعة»، بدلاً من «الأربعة الأقسام»> .

(٨) <خُتِم هنا مخطوط [ل] بالقول: «تمّت الرسالة والحمد لله ربّ العالمين». ثمّ قدّم رسالة الأسطرُنوميا على أنها المدخل إلى علم النجوم> .

مُلحَق رِسالَة الهَندَسَة

من المخطوط [ط]

خُطَّ مقطع مطوّل في الأوراق ٢٧-٢٨ من المخطوط [ط]، في موضع قارب وسط الفصل ١٤ من النصّ المحقّق لرسالة الهندسة؛ حسب ما أشرنا إلى ذلك أيضاً في معرض تحقيقنا لتلك الرسالة.

ما ورد في هذا المقطع له قيمة معرفيّة تتوافق مع موضوع رسالة الهندسة تبعاً لما ورد كذلك في بعض جوانبه من الدلالة على موضوع الفصل الأوّل من المقالة الأولى من كتاب أوقليدِس في الأصول. ولكننا آثرنا عدم وضع هذا المقطع في متن النصّ المحقّق أو في الحواشي مخافة التطويل، ولكونه كذلك قد خرج عن سياق السرد في الرسالة وابتعد عن الأصل الذي نُقل عنه، كما أشير إلى ذلك في آخر المقطع نفسه. كذلك، لم يظهر له مثيل آخر في بقيّة المخطوطات المعتمدة في التحقيق النقدي، أو في طبعة دار صادر. لهذا أوردناه هنا في هذا الملحق كما خُطَّ حرفياً في [ط]، مع بعض من التشكيل والتحريك، ومع القليل من التدقيق وضبط الشكل.

ترجمناه أيضاً إلى اللغة الإنكليزيّة في ملحقٍ (Appendix) وضع بعد نهاية ترجمتنا الإنكليزيّة لرسالة الهندسة الموسومة بالجومَطريا.

وقد نُسبت محتويات المقطع هذا في مخطوط [ط] إلى أقاويل من عُرِّفَ هناك بلقَب «الناسخ»؛ لعلّه شخص آخر مغاير في هويّته لمن خطَّ النصّ بيده فعليّاً، حسب ما ذُيِّلَ في خاتمة [ط] لرسالة الهندسة، بذكر اسم وكنية ولقب الخطّاط على الشكل التالي وفقاً لما تمكّنا من قراءتِهِ من خطّه شبه المطموس:

للعبد الضّعيف

صالِح المفاخِر الحُسن

بن سَعد

من أهل شَهرَباد

لُقِّب

«المَشهور بِعَصَه»

والمقطع الذي ورد في الأوراق ٢٧-٢٨ من المخطوط [ط] خُطَّ على الشكل التالي:

< البداية على وسط السطر ١٠ من القسم الأيسر من الورقة ٢٧ من [ط]> «قال الناسخ: 'من خواص المسدّس المتساوي الأضلاع، الذي يعمل من الدايرة أيضاً، أن ضلعه يساوي نصف قطر الدايرة. فبفتح البركار، الذي تدار به الدايرة، يقسم محيط الدايرة بستّة أقسام متساوية. ومن خواصّه أن مسدّسات الأضلاع والزوايا، المساوية بعضها البعض، إذا ضمّ بعضها إلى بعض تُرَكّب وتُكَثف وتَملأ السطح الذي تكون فيه، من غير أن يكون بينهما خلاء'. كما قال. 'وهذه الخاصّيّة ليست لشيء من الأشكال سوى المسدّس والمربّع والمثلّث. والسبب في ذلك أن

الخطوط المستقيمة التي تتقاطع على نقطة واحدة، يكون مجموع زواياها مساوياً لأربع زوايا قائمة. وقد تبيّن ذلك في الشكل من المقالة الأولى من كتاب أقليدِس، فيما كان مقدار زاوية الشكل مقداراً يتمّ من انضمام بعضها إلى بعض أربع زوايا قائمة. فإن زواياها إذا اجتمعت على نقطة واحدة مشتركة بينها، منها < بداية القسم الأيمن من الورقة ٢٨ من [ط] > وحدها يمكن أن يمتلىء السطح من غير زيادة ولا نقصان. فإن منها وحدها تتمّ أربع زوايا قائمة. أما المسدّس فإن مقدار زاويته قائمة وثلث، لأنه ينقسم إلى أربع مثلّثات وزوايا كلّ مثلّث قائمتان، فيكون ثمان قوايم، وكلّ واحد منها قايمة وثلث، فيتم من ثلاث زوايا منه إذا اجتمعت على نقطة واحدة أربع قوايم. وأما المربّع فلإن مقدار زاويته قايمة واحدة، فأربعة منها تكون أربع زوايا قايمة. وأما المثلّث فلأن مقدار زاويته ثلثا قايمة، فيتمّ من ستّة منها أربع قوايم. وأما ما سواها من الأشكال، فإن اجتماع زواياها إما أن ينقص من أربع قوايم أو يزد عليها، ولا يتمّ منها أربع قوايم قطّ. مثال ذلك المخمّس، فإن مقدار زاويته قايمة وخُمس لأنه ينقسم إلى ثلٰث مثلّثات، ويكون مجموع زواياها ستّ قوايم، كلّ واحدة منها قايمة وخُمس، فلا يتمّ منه أربع قوايم. فإنه إذا اجتمع ثلٰثة منها نقص منها الأربع قوايم. فإنها تكون ثلٰث قوايم وأربعة أخماس قايمة. وزاوية المسبّع بمثل ذلك البيان، فإنه ينقسم إلى خمس مثلّثات، ومقدارها يكون قايمة وثلثة أسباع قايمة. ومقدار زاوية المثَمَّن قايمة ونصف. ومقدار زاوية المتَّسِع قايمة وخمسة أتساع قايمة. ومقدار زاوية المعَشَّر قايمة وثلٰثة أخماس قايمة. فليس

يتمّ شيء من زوايا هذه الأشكال، ولا فيما بعدها، أربع قوايم. فإذاً هذه الخاصّيّة هي لهذه الثلٰثة، أعني المثلّث والمربّع والمسدّس فحسب دون غيرها. ولذلك تَجِد جميع أعمال الطبيعة والحيوانات إما دايرة، وهو الشكل التامّ، وإما مثلّثات أو مربّعات أو مسدّسات في دواير، أو في أقسام دواير‘. إنقضى كلام الناسخ المذكور ورجع إلى الأصل».[١]

(١) <أي أنه تابع هنا القول: «وهكذا ينسج العنكبوت شبكته في زوايا البيوت. . .»، وحسب ما ورد في وسط الفصل ١٤ من متن نصّ رسالة الهندسة>.

فهرس المصطلحات

فهرس الأعلام